NATO ASI Series

Advanced Science Institutes Series

A series presenting the results of activities sponsored by the NATO Science Committee, which aims at the dissemination of advanced scientific and technological knowledge, with a view to strengthening links between scientific communities.

The Series is published by an international board of publishers in conjunction with the NATO Scientific Affairs Division

A Life Sciences B Physics	Plenum Publishing Corporation London and New York
C Mathematical and Physical Sciences D Behavioural and Social Sciences E Applied Sciences	Kluwer Academic Publishers Dordrecht, Boston and London
F Computer and Systems Sciences G Ecological Sciences H Cell Biology	Springer-Verlag Berlin Heidelberg New York London Paris Tokyo Hong Kong

Series H: Cell Biology Vol. 33

The ASI Series Books Published as a Result of Activities of the Special Programme on CELL TO CELL SIGNALS IN PLANTS AND ANIMALS

This book contains the proceedings of a NATO Advanched Research Workshop held within the activities of the NATO Special Programme on Cell to Cell Signals in Plants and Animals, running from 1984 to 1989 under the auspices of the NATO Science Committee.

The books published as a result of the activities of the Special Programme are:

Vol. 1: **Biology and Molecular Biology of Plant-Pathogen Interactions.** Edited by J.A. Baily. 1986.

Vol. 2: **Glial-Neuronal Communication in Development and Regeneration.** Edited by H.H. Althaus and W. Seifert. 1987.

Vol. 3: **Nicotinic Acetylcholine Receptor: Structure and Function.** Edited by A. Maelicke. 1986.

Vol. 4: **Recognition in Microbe-Plant Symbiotic and Pathogenic Interactions.** Edited by B. Lugtenberg. 1986.

Vol. 5: **Mesenchymal-Epithelial Interactions in Neural Development.** Edited by J.R. Wolff, J. Sievers, and M. Berry. 1987.

Vol. 6: **Molecular Mechanisms of Desensitization to Signal Molecules.** Edited by T.M. Konjin, P.J.M. Van Haastert, H. Van der Starre, H. Van der Wel, and M.D. Houslay. 1987.

Vol. 7: **Gangliosides and Modulation of Neuronal Functions.** Edited by H. Rahmann. 1987.

Vol. 9: **Modification of Cell to Cell Signals During Normal and Pathological Aging.** Edited By S. Govoni and F. Battaini. 1987.

Vol. 10: **Plant Hormone Receptors.** Edited by D. Klämbt. 1987.

Vol. 11: **Host-Parasite Cellular and Molecular Interactions in Protozoal Infections.** Edited by K.-P. Chang and D. Snary. 1987.

Vol. 12: **The Cell Surface in Signal Transduction.** Edited by E. Wagner, H. Greppin, and B. Millet. 1987.

Vol. 19: **Modulation of Synaptic Transmission and Plasticity in Nervous Systems.** Edited by G. Hertting and H.-C. Spatz. 1988.

Vol. 20: **Amino Acid Availability and Brain Function in Health and Disease.** Edited by G. Huether. 1988.

Vol. 21: **Cellular and Molecular Basis of Synaptic Transmission.** Edited by H. Zimmermann. 1988.

Vol. 23: **The Semiotics of Cellular Communication in the Immune System.** Edited by E.E. Sercarz, F. Celada, N.A. Mitchison, and T. Tada. 1988.

Vol. 24: **Bacteria, Complement and the Phagocytic Cell.** Edited by F.C. Cabello and C. Pruzzo. 1988.

Vol. 25: **Nicotinic Acetylcholine Receptors in the Nervous System.** Edited by F. Celementi, C. Gotti, and E. Sher. 1988.

Vol. 26: **Cell to Cell Signals in Mammalian Development.** Edited by S.W. de Laat, J.G. Bluemink, and C.L. Mummery. 1989.

Vol. 27: **Phytotoxins and Plant Pathogenesis.** Edited by A. Graniti, R.D. Durbin, and A. Ballio. 1989.

Vol. 31: **Neurobiology of the Inner Retina.** Edited by R. Weiler and N.N. Osborne. 1989.

Vol. 32: **Molecular Biology of Neuroreceptors and Ion Channels.** Edited by A. Maelicke. 1989.

Vol. 33: **Regulatory Mechanisms of Neuron to Vessel Communication in the Brain.** Edited by F. Battaini, S. Govoni, M.S. Magnoni, and M. Trabucchi. 1989.

Regulatory Mechanisms of Neuron to Vessel Communication in the Brain

Edited by

Fiorenzo Battaini

Department of Experimental Medicine and Biochemical Sciences
II University of Rome, Via Orazio Raimondo, 00173 Rome, Italy

Stefano Govoni

Department of Pharmacobiology
University of Bari, Via Re David 4, 70100 Bari, Italy

Maria Sandra Magnoni

Institute of Pharmacological Sciences, Faculty of Pharmacology,
University of Milano, Via Balzaretti, 9, 20133 Milano, Italy

Marco Trabucchi

Department of Experimental Medicine and Biochemical Sciences,
II University of Rome, Via Orazio Raimondo, 00173 Rome, Italy

Springer-Verlag Berlin Heidelberg New York
London Paris Tokyo Hong Kong
Published in cooperation with NATO Scientific Affairs Division

Proceedings of the NATO Advanced Research Workshop on Regulatory Mechanisms of Neuron to Vessel Communication in the Brain held in Salò, Italy, September 3–8, 1988

ISBN-13:978-3-642-74154-8 e-ISBN-13:978-3-642-74152-4
DOI: 10.1007/978-3-642-74152-4

Library of Congress Cataloging-in-Publication Data. NATO Advanced Research Workshop on Regulatory Mechanisms of Neuron to Vessel Communication in the Brain (1988; Salò, Italy). Regulatory mechanisms of neuron to vessel communication in the brain / edited by Fiorenzo Battaini . . . [et al.]. p. cm.—(NATO ASI series, Series H. Cell biology: vol. 33) "Proceedings of the NATO Advanced Research Workshop on Regulatory Mechanisms of Neuron to Vessel Communication in the Brain, held in Salò, Italy, September 3–8, 1988"—T.p. verso. "Published in cooperation with NAATO Scientific Affairs Division." Includes index.
1. Cerebral circulation—Congresses. 2. Microcirculation—Congresses. 3. Brain—Blood-vessels—Innervation—Congresses. 4. Brain—Metabolism—Regulation—Congresses. I. Battaini, Fiorenzo. II. North Atlanctic Treaty Organization. Scientific Affairs Division. III. Title. IV. Series. [DNLM: 1. Blood-Brain Barrier—congresses. 2. Brain—physiology—congresses. WL200 N279r 1988] QP108.5, C4N368 1988 612.8'24—dc20 DNLM/DLC for Library of Congress 89-19666.

Softcover reprint of the hardcover 1st edititon 1989

2131/3140-543210 – Printed on acid-free-paper

PREFACE

Different molecular mechanisms cooperate to the regulation of brain microcirculation: hormones, local metabolic products, transmitters. Great interest has recently emerged in literature on the neuronal systems controlling the blood-brain barrier and on the reciprocal interaction between neuronal activity and vascular supply.

In particular, this volume on the NATO Advanced Research Workshop held in Salo' (BS), Italy, September 3-8, 1988, outlines the importance of neuron-vessel communication in the regulation of cerebral microenvironment. The problem is approached from a multidisciplinary point of viev, with contributions coming from basic and clinical areas. Stimulating results emerge from in vivo imaging techniques which allow the investigation of the hemodynamic and metabolic activities of the brain in various physiological and pathological conditions, such as aging or dementia. This methodology is of fundamental importance to acquire deeper knowledge on the relationship between cerebral blood flow, metabolism and higher brain functions.

Another stimulating aspect addressed in this volume is the role of the glia in maintaining the homeostasis of the neuronal microenvironment, by "buffering" the excess of ions in the extracellular space and regulating the vascular transport processes.

The complex interplay neuron-glia-vessel may indicate new perspectives in the therapy of cerebrovascular disorders, suggesting as a target for pharmacological intervention the neuronal (or glial) cells participating to the exchange of ions and substances responsible for the extension of the ischemic damage.

Finally, a chapter of the book presents new hypothesis on the function of cerebral endothelium in the genesis and development of multiple sclerosis lesions. The classical view of the brain as an immunological privileged organ is challenged by the observation that alteration of the blood brain barrier induced by elements of the immune system may be the starting factor for the pathological injuries.

In the book we have added part of the discussions which have greatly contributed to the success of the workshop.

In conclusion, we whould like to thank the National Research Council of Italy (CNR) who in addition to NATO has provided support to the meeting.

We whould like to express our sincere appreciation to Mrs B.R. Chilo', General Secretary of the Nutrition Foundation of Italy and all the members of our laboratory who have collaborated in the organization of this workshop.

The Editors

LIST OF CONTRIBUTORS

BATTAINI F., Dept. Experimental Medicine and Biochemical Sciences, Chair of Toxicology, Univesity of Rome "Tor Vergata", Via Orazio Raimondo, 00187 Rome, Italy

BETZ A.L., Dept. Pediatrics, Surgery and Neurology, University of Michigan, D3227 Medical Professional Bldg, Ann Arbor, MI 48109-0718, U.S.A.

BURNS E.M., Ohio State University, Columbus, Ohio, U.S.A.

CANGIANO C., Dept. Internal Medicine, University of Rome "La Sapienza", Viale dell'Universita' 37, 00185 Rome, Italy.

CARDELLI-CANGIANO P., Dept. Human Biopathology, University of Rome "La Sapienza" Viale Regina Elena 324, 00185 Rome, Italy.

CERVOS-NAVARRO J., Institute of Neuropathology, Free University of Berlin, Hindenburgdamm 30, 1000 Berlin 45, West Germany

CREMER J.E., MRC Cyclotron Unit, Hammersmith Hospital, Du Cane Road, London W 12 OHS, U.K.

CUNNINGHAM V.J., MRC Cyclotron Unit, Hammersmith Hospital, Du Cane Road, London W 12 OHS, U.K.

EDVINSSON L., Dept. Internal Medicine, University Hospitals Lund, 22185 Lund, Sweden.

FREEDMAN L., Pharmacology Dept., Pennwalt Corporation, 755 Jefferson Road, Rochester, New York 14623, U.S.A.

GOVONI S. Pharmacobiology Department, University of Bari, Via Re David 4, 70125, Bari, Italy.

HARGREAVES J. Merck, Sharp and Dohme, Neuroscience Research Centre,Terling Park, Harlow, Essex CH20 2QR. U.K.

HARTMANN B.K., Division of Neuroscience Research in Psychiatry, Dept. Psychiatry, Univrsity of Minnesota Medical

School, Minneapolis, MN 55455, U.S.A.

HERTZ L., Dept. Pharmacology, Universityof Saskatchewan, Saskatoon, Saskatchewan, S7N OWO, Canada

HORWITZ B., Laboratory of Neurosciences, National Institutes on Aging, N.I.H., Bldg. 10, Rm. 6C 603,. Bethesda, Md 20892, U.S.A.

JOO F., Laboratory of Neuropathologfy and Neuroanatomical Sciences, Institute of Neurological and Communicative Disorders and Stroke, N.I.H., Bldg. 36 Rm. 4D04, Bethesda, Md 20892, U.S.A.

KRUCKEBERG T.W., University of Iowa, Iowa City, U.S.A.

KURCEWICZ I., Unite'de Recherches sur l'Epilepsie, INSERM U 97 2ter rue d'Alesia, 75014 Paris, France.

LOU H., The John F. Kennedy Institute, Dpt. Neuropediatrics, Gl. Landevej 7, 2600 Glostrup, Denmark.

LOUVEL J., Unite'de Recherches sur l'Epilepsie, INSERM U 97 2ter rue d'Alesia, 75014 Paris, France.

MAGNONI M.S., University of Milan , Institute of Pharmacological Sciences, Via Balzaretti 9, 20133 Milan, Italy

MC CREEDY A. Pharmacology Dept., Pennwalt Corporation, 755 Jefferson Road, Rochester, New York 14623, U.S.A.

MC CULLOCH J. , Wellcome Neuroscience Group, University of Glasgow, Glasgow, G61 1QH, U.K.

ORKAND R.K., Institute of Neurobiology, University of Puerto Rico Medical Sciences Campus, 201 Blvd. del Valle, San Juan PR 00901, U.S.A.

PALMER G.C., Pharmacology Dept., Pennwalt Corporation, 755 Jefferson Road, Rochester, New York 14623, U.S.A.

PUMAIN R., Unite'de Recherches sur l'Epilepsie, INSERM U 97

2ter rue d'Alesia, 75014 Paris, France.

RAPOPORT S.I., Laboratory of Neurosciences, National Institutes on Aging, N.I.H., Bldg. 10, Rm. 6C 603,. Bethesda, Md 20892, U.S.A.

SALLY A., Pharmacology Dept., Pennwalt Corporation, 755 Jefferson Road, Rochester, New York 14623, U.S.A.

SAMPAOLO S., Institute of Neuropathology, Free University of Berlin, Hindenburgdamm 30, 1000 Berlin 45, West Germany

SIESJO B.K., University of Lund, Laboratory for Experimental Brain Research, Floor EA-5, Lund Hospital, 221 85 Lund, Sweden.

SMITH M-L., University of Lund, Laboratory for Experimental Brain Research, Floor EA-5, Lund Hospital, 221 85 Lund, Sweden.

SOKOLOFF L., Laboratory of Cerebral Metabolism, National Institute of Mental Health, 9000 Rockville Pike, Bldg. 36, Rm. 1A05, Bethesda, Md 20892, U.S.A.

TRABUCCHI M., Dept. Experimental Medicine and Biochemical Sciences, Chair of Toxicology, Univesity of Rome "Tor Vergata", Via Orazio Raimondo, 00187 Rome, Italy.

TRAUGOTT U., Dept. of Pathology (Neuropathology) and Neurology and the Rose F. Kennedy Center for Reserch in Mental retardation and Human development, Albert Einstein College of Medicine, The Bronx, New York, U.S.A.

CONTENTS

THE ULTRASTRUCTURAL BASIS OF CAPILLARY PERMEABILITY

Elizabeth M. Burns (1) Thomas W. Kruckeberg (2)

(1) Neurophysiologist, The Ohio State University, Columbus, Ohio
(2) Biologist, The University of Iowa, Iowa City, Iowa

ABSTRACT

The highest level of integrative brain function requires homeostasis and isolation of the internal milieu of the brain. Homeostasis and isolation are accomplished by a constellation of brain capillary wall morphological, biochemical, and physiological mechanisms that constitute the blood-brain barrier (BBB). The ultrastructural basis of the BBB consists of tight interendothelial junctions, the absence of endothelial pores, a paucity of pinocytotic vesicles, and an increased numerical density of endothelial mitochondria. The endothelial membrane is supported by pericytes, basal lamina, and astrocytic end-foot processes.

KEY-WORDS

Blood-brain barrier, tight interendothelial junctions, endothelial pores, pinocytotic vesicles, endothelial mitochondria, pericytes, basal lamina, glial end-foot processes.

NATO ASI Series, Vol. H 33
Regulatory Mechanisms of Neuron to Vessel Communication in the Brain. Edited by F. Battaini et al.

INTRODUCTION

According to classic Starling explanation, transcapillary exchange is determined essentially by the hydrostatic pressure on each side of the capillary wall, the osmotic pressure of protein in plasma and interstitial fluid, and the intrinsic properties of the capillary wall (Milnor, 1980). Of these determinants of transcapillary exchange, the intrinsic properties of the capillary wall have been the most difficult to explicate. Capillary walls are of three types: continuous, consisting of contiguous endothelial cells with numerous intercellular channels or pores (4 - 5 nm in diameter); fenestrated, endowed with intracellular endothelial fenestrations (0.1 um in diameter), as in renal glomeruli; and discontinuous, containing relatively large intercellular gaps (0.1 - 1 um in width) through which formed elements of the blood can pass, as in splenic or hepatic sinusoids (Milnor, 1980).

Central nervous system (CNS) and peripheral organ capillaries are supported by a well developed basement membrane, or a basal lamina. The capillaries in the CNS are also supported by pericytes and astrocytic end-foot processes. Channels or pores in capillary interendothelial junctions and pinocytotic vesicles or invaginations of the endothelial membrane as well as membrane fenestrations are associated with capillary permeability. For example, the transcapillary exchange of water soluble hydrophilic substances is limited to transmural channels or pores (Pappenheimer, 1953); transport of proteins may occur via pinocytotic vesicles (Rapoport, 1976); however, lipid soluble substances (oxygen, carbon dioxide, and some nonelectrolytes) diffuse freely through the capillary walls (Diamond & Wright, 1969). The ultrastructural basis of capillary permeability therefore includes: pores or channels in the interendothelial regions; fenestrae; and, vesicles.

THE CONCEPT OF A BLOOD-BRAIN BARRIER

Prior to the studies of Ehrlich and Goldmann, capillaries had been observed to be exclusionary relative to transcapillary passage of molecules of numerous substances, with the greatest amount of restriction being found in the CNS capillaries. These observations provided a basis for the barrier concept. Initially the barrier in brain vessels primarily was considered to be a passive protective mechanism and only later came to be regarded as an active regulatory mechanism. However, early in the ontogeny of the concept of a blood-brain barrier (BBB), it was noted that transcapillary exchange rates and concentrations of various substances at steady states differed markedly in brain as compared with other organs. The concept of a BBB was advanced by the work of Ehrlich who reported in 1885 that the intravenous injection of selected vital dyes stained peripheral organs and tissues but did not stain the brain (Lee, 1971). Goldmann, a student of Ehrlich, in a classic experiment performed in 1913, showed that direct injection of dye into the cerebrospinal fluid (CSF) readily stained the entire brain but did not stain any other organs or tissues of the body (Lee, 1971). Further, Goldmann hypothesized that brain capillaries subserved the barrier function that excluded intravenously administered vital dyes from entry into the brain or restricted intracerebrally administered vital dyes to the brain. These early studies revealed, however, that capillaries in the posterior pituitary, the median eminence, area postrema, tuber cinereum and in the pineal body were devoid of barrier mechanisms.

During the past three decades a vast amount of research has been directed toward elucidation of the role of the capillary wall with respect to brain blood-tissue exchange. It is well established that the direction and the rate of brain transcapillary exchange are dependent upon intrinsic properties of the brain capillary wall and that the relative impermeability of brain capillary walls to hydrophilic molecules and water is a key factor involved in interorgan (brain versus muscle) variation in transcapillary exchange (Cserr & Bundgaard, 1986). The substrate of the constellation of complex, precisely interacting morphological, biochemical, and physiological mechanisms that subserve brain transcapillary exchange is known as the BBB.

THE BLOOD-BRAIN BARRIER

The brain requires a stable internal environment different from that of the other organs of the body in order to function at its highest integrative level. Fluctuations in extracellular fluid (ECF) concentrations of hormones, amino acids, nutrients, vitamins, and ions occur in response to the ingestion of a meal, abstention from food or fluids, exposure to stress, and especially in response to strenuous exercise. Small fluctuations in the concentrations of these substances are tolerated by the peripheral organs. However, the CNS has a unique microenvironment, a disturbance of which would interfere with the processes of CNS integration. For example, the behavior of neurons is directly related to their electrophysiological activity which in turn is dependent upon the chemical composition of their environment. Relatively insignificant changes in potassium ion concentration would thus adversely affect neuronal activity. First, this unique microenvironment of the CNS demands, on both a short- and long-term basis, a dynamic homeostasis in order to insure that individual neurons, synapses, and neuronal systems may receive, process, store, transfer, integrate, retrieve, and utilize the billions of bits of information essential to normal neurological function. Secondly, and simultaneously, the entire, internal environment of the CNS, including its supportive tissues, must be maintained in isolation from that of the blood. The evolution in parallel of the highest integrative level of CNS function and the most efficient BBB is supportive of the hypothesis that the homeostasis and isolation of the ECF microenvironment of central neurons facilitates brain integrative functions (Cserr & Bundgaard, 1986). Thus far all species that have a BBB are able to achieve and maintain both isolation and homeostasis of the internal environment of the brain. Isolation of the microenvironment is based primarily on the tightness or impermeability of the BBB to hydrophilic molecules whereas numerous other barrier mechanisms (e.g., active transport, facilitated transport, CSF and ECF production, and biotransformation) are basic to homeostasis. BBB integrity is essential to normal brain function.

The site of the BBB was definitively established by the studies of Brightman, Karnovsky, and Reese in the late 1960's (Reese & Karnovsky, 1967; Bightman & Reese, 1969). They injected into experimental animals electron dense markers that could be identified in electron micrographs to show that the brain capillary endothelium is the site of barrier mechanisms. Neither horseradish peroxidase (HRP, MW 39,800 and particle diameter 5 - 6 nm) nor microperoxidase (MW 1,880 and particle diameter 2 nm) passed from the blood to brain tissue following intravenous injection, nor from the CSF into the capillary after injection into the CSF. Electron dense tracers of various MWs and particle diameters, including lanthanum (both colloidal and ionic separately) and ferritin, were used also in an effort to elucidate further BBB structure (Rapoport, 1976). More recently Richards (1978) demonstrated that 5-hydroxydopamine (MW 256 and particle diameter 0.5 to 0.7 nm) did not pass from CSF into the lumen of the choroid plexus capillary in the rat.

Brain capillaries are relatively impermeable as compared with capillaries of most other regions of the body. They are characterized by: the presence of tight interendothelial junctions (zonulae occludens, less than 2 nm in width) between contiguous endothelial cells; the absence of endothelial channels or pores and fenestrations; a paucity of pinocytotic vesicles or invaginations of the endothelial membranes (Reese & Karnovsky, 1967); and a several-fold increase in the numerical density of endothelial mitochondria as compared with capillaries in other regions of the body (Oldendorf, Cornford & Brown, 1977).

Tight interendothelial junctions between contiguous endothelial cells constitute perhaps the single most important feature of the BBB with respect to isolation of the internal environment of the brain from blood. The electron microscopic studies of Brightman, Karnovsky, and Reese (Reese & Karnovsky, 1967; Bightman & Reese, 1969) clearly showed overlapping lateral edges of endothelial cells sealed together by tight junctions that formed a continuous belt around the capillary at each interendothelial junction. On the luminal side of the tight junction complex (after intravenous injection) molecules of the electron dense marker could be visualized in the proximal portion of the cleft, however their passage had been stopped abruptly at the first tight junction. Also, following CSF administration, molecules of the marker could be visualized in the abluminal basal lamina and into the first abluminal portion of the interendothelial cleft; again, the electron dense marker molecules were abruptly blocked at the first tight junction. The width of the tight junctions along the interendothelial cleft, as measured in electron micrographs, is less than twice the width of the endothelial cell luminal or abluminal membrane, which suggests that fusion of apposing membranes was complete and permanent.

A second characteristic feature of brain endothelium revealed by the same electron microscopic studies is the very low frequency of the type of vesicle (pinocytotic) generally associated with transcapillary exchange in peripheral organs (cardiac or muscle cells). There was no evidence of discharge of HRP from the luminal to the abluminal side of brain endothelium (Reese & Karnovsky, 1967). However, it was shown previously that vascular endothelium in cardiac and skeletal muscle transports material by vesicular filling on the capillary lumenal side and vesicular discharging on the abluminal side of the capillary (Palade & Bruns, 1964). Bundgaard (1986) has shown electron microscopically, by three-dimensional reconstructions based on consecutive ultrathin sections, that vesicular structures observed in brain endothelium are parts of invaginations of the cell membrane. He proposed that the intercellular clefts, not vesicles, are the most likely diffusion pathways for hydrophilic substances during normal and pathological conditions. Although there is some evidence in favor of this possibility, further study is needed to clarify this aspect of transcapillary exchange.

The tight interendothelial junctions and the virtual absence of endothelial pores, channels, vesicles, and fenestrae render the capillary wall relatively impermeable and enable it to behave like a tight epithelium. Bundgaard (1982) determined the electrical resistance across the endothelium of the frog cerebral capillary. Although the frog brain capillary lacks a glial sheath, it has the fine structural features of cerebral capillaries. Membrane current, based on cable theory (Bradbury, 1984), is 1000 times as great across the brain capillary wall as across the mesenteric capillary wall and 100 times as great as across the muscle capillary wall. Specific resistance of brain endothelium proved to be of the same magnitude as that measured across amphibian tight epithelium and erythrocyte membranes (Crone & Christensen, (1981). This further proves the tightness of the brain capillary endothelium.

Numerous substances needed by the brain require active transport for transcapillary exchange. Energy for active transport is derived from oxidative phosphorylation. The numerical density or content of endothelial mitochondria is related to oxidative phosphorylation and may predict the work capability of the BBB (Oldendorf, Cornford, & Brown, 1977).

Schmidley and Wissig (1986), found that ruthenium red stained the basement membrane of fenestrated capillaries and the epithelium of choroid plexuses but not the basement membrane of the BBB. This suggests that the basement membrane or basal lamina of the BBB, like its endothelium, may be structurally and perhaps functionally unique. These authors point out that there is evidence, from an embryological perspective, that the basal lamina of the BBB is a composite structure, one part belonging to the endothelium and the other belonging to astrocytic end-foot processes. Fusion of these two components effectively obliterates interstitial space. This fact has important implications relative to the prevention of edema formation in brain tissue.

The BBB effectively maintains the electrolyte composition of brain extracellular fluid significantly different from that of the blood or of the ECF of non-nervous tissues. It controls the rate of transcapillary exchange of fixed acids and bases thereby regulating the pH of the microenvironment of central neurons (Lambertson, 1980). Also, the BBB prevents the passage of protein from blood into brain ECF and CSF except in brain regions devoid of barrier structures (e.g., the posterior pituitary, median eminence, area postrema, tuber cinereum, pineal body). Active transport, carrier-mediated transport, and various other biological systems enable the BBB to maintain CNS homeostasis (Davson, 1976) while the tightness of the capillary wall to hydrophilic molecules primarily enables CNS isolation (Cserr & Bundgaard, 1986). Homeostasis and isolation thus support normal brain function.

Cerebral capillaries comprise 99% of the BBB (Pardridge, 1983). Brain endothelium is a blood-brain interface that regulates transcapillary exchange between the circulating blood and brain ECF. The choroid plexus epithelium is an interface between circulating blood and CSF that regulates exchange between the blood and CSF, thus constituting a blood-CSF-barrier (B-CSF-B). Structurally, the BBB and B-CSF-B are very similar. A barrier function is attributed to the rings of tight junctions surrounding the cells in the middle layer of the arachnoid membrane. These junctions have been shown to be impervious to HRP and other electron dense tracers excluded by the BBB. The B-CSF-B and the arachnoid barrier comprise less than 1% of the BBB, however also subserve the maintenance of homeostasis and isolation of the CNS internal environment. Barrier type capillaries are located at the endoneurium and perineurium of peripheral nerve fibers and nerve bundles respectively. The eye contains two barrier systems, a blood-aqueous and a blood-vitreous barrier. The ophthalmic barrier systems regulate both the production and composition of aqueous and vitreous humors respectively, and thus protect the ionic milieu of the retina (Rapoport, 1976).

The adult human brain constitutes only approximately 2% of total body weight. Nevertheless, under essentially normal resting conditions, the brain receives roughly 15% of the cardiac output, takes up 20 - 35% of the oxygen absorbed by the lungs, and utilizes 25% of the glucose consumed. These facts emphasize the richness of brain microvascularity in comparison with that of most peripheral organs. Only the vascularity primarily of the thyroid and kidneys surpasses that of the brain. These facts emphasize the importance of the morphological, biochemical, and physiological mechanisms that constitute the BBB.

THE ONTOGENY OF THE BLOOD-BRAIN BARRIER

The ontogeny of the structural correlates of the BBB was described by Donahue and Pappas (1961), Donahue (1964), Caley and Maxwell (1970), Hannah and Nathaniel (1974), Singh and Nathaniel (1975), Dyson, Jones, and Kendrick (1976), and Stewart and Hayakawa (1987). Their results suggest that the BBB is either absent or poorly developed at birth in most animals and that its maturation varies from species to species. More evidence is needed relative to the degree of maturity of the BBB perinatally.

Saunders, Mollgard and co-workers disagree with the commonly held belief that the BBB is immature in fetuses or the newborn but that at least one of the BBB sites at this stage of development is different from that of the adult (Saunders & Mollgard, 1984; Mollgard & Saunders, 1986; Mollgard, Balslev, Lauritzen, & Saunders, 1987). They studied the BBB in the fetal chick, rat, monkey, and human brain. They found that the main sites of the BBB in the developing sheep brain include: cerebral capillary endothelium; choroid plexus epithelium; the pia-arachnoid membrane; and the CSF-brain barrier across the neuroependyma (not present in the adult). From day 20 through day 150 postconceptual in sheep (term birth at 150 days) they found no evidence of morphological changes in tight interendothelial junctions. Thus they concluded that if the BBB is defined as the presence of tight junctions between brain endothelial cells and choroid epithelial cells then clearly the BBB is present very early in brain development in fetal sheep. Their definition of the BBB deals with only one of several morphological features that may contribute to the tightness of the brain capillary wall and fails to consider either the biochemical or physiological features of the BBB. Adinolphi (1985) pointed out the need for awareness that during ontogeny the brain is particularly sensitive to environmental toxic agents which may gain access into the fetal bloodstream and possibly may cross the BBB. Further BBB ontogenetic studies are indicated.

Light microscopic studies (Rowan & Maxwell, 1981a) indicate that vascular proliferation takes place during postnatal cortical development in response to metabolic demands imposed by brain maturational processes. The mechanism of induction of vascular proliferation within the CNS, however, has not been completely elucidated. Electron microscopic studies by the same researchers (Rowan & Maxwell, 1981b) showed that alkaline phosphatase cytochemistry is capable of demonstrating the presence of proliferating vasculature within the CNS, since the alkaline phosphatase reaction product is visible in early vascular sprouts.

Craigie, while studying changes in vascularity in the brainstem and cerebellum (1924) and in the cerebral cortex (1925) of the albino rat between birth and maturity, observed marked regional variation in brain vascularity. He concluded that the richness of the capillary supply in various brain regions is directly related to functional activity during ontogeny. He proposed that the increased metabolic rate associated with heightened neuronal activity would require a greater blood supply than the metabolic rate associated with either cellular differentiation or cellular growth. Craigie's findings have been corroborated by the recent studies of Rowan and Maxwell (1981a, 1981b).

AGE DIFFERENCES ACROSS THE LIFESPAN IN THE BLOOD-BRAIN BARRIER

A number of studies have shown differences in the BBB at different ages across the lifespan (Schwink & Wetzstein, 1966; Bar, 1978; Bar & Wolff, 1973; Burns, Kruckeberg, Comerford, & Buschmann, 1979; Burns, Kruckeberg, and Gaetano, 1981; Burns, Kruckeberg, Gaetano, & Schulman, 1983). Schwink and Wetzstein (1966) observed thinning of capillary walls in subcommissural organs in the rat between birth and 18 months postnatally. Bar and Wolff (1976) found that capillary branching in the rat peaks at postnatal day 20 and then declines until maturity. They observed that capillary branching density paralleled changes in aerobic metabolism and established, in the rat, that capillary branching ceased at one month postnatally, with growth thereafter being due largely to elongation of endothelial cells rather than to cell division. Therefore, it appears that brain capillary proliferation during ontogeny is directly related to aerobic metabolism (Bar & Wolff, 1973), probably being induced by the maturational demands of the rapidly growing brain. Thinning of cerebral capillary walls in rats was observed between birth and 18 months of age (Schwink & Wetzstein, 1966) and in the nonhuman primate between four and twenty years of age (Burns et al., 1979), being due almost entirely to thinning of the endothelial component. Loss of cerebral capillary endothelial cells with age occurs in the rat (Bar, 1978). This cellular loss is thought to be compensated for, at least in part, by cellular elongation. Capillary mean diameter decreases significantly in rats between 23 and 30 months of age whereas capillary mean length per unit volume of brain increases significantly with age (Bar, 1978). The latter, however, may reflect a concentration of existing vessels within a decreased brain volume.

Age differences in cerebral capillary morphology were investigated in 4-, 10-, and 20-year old Macaque monkeys and in 1-, 14-, 35-, 180-, and 800-day old Sprague-Dawley rats (Burns, et al., 1981; 1983). The following age differences were revealed: a significant decrease in cerebral capillary wall thickness in the frontal cortex in the Macaque monkey but not in the rat; a significant increase in the thickness of basal lamina (BL) of brain capillaries in rats but not in Macaque monkeys, however there was a marked increase in this parameter in the monkey between 4 and 10 years of age; a significant decline in cerebral capillary mitochondrial content in monkeys whereas a significant decline in this parameter in rats was found only when the peak content at 35 days of age was contrasted with that at 800 days of age; and, aberrant tight interendothelial junctions and thickened BL in one of five 20-year-old monkeys. These findings suggest impairment of BBB mechanisms in both older rats and monkeys. Because of species differences in the timing of ontogenetic events, it is not possible to compare early postnatal events in the rat and monkey based on these studies. The three early ages in rats (days 1, 14, and 35 postnatally) are not comparable with any of the ages of the monkeys. The 180-day-old rat is somewhat comparable with the 4-year-old monkey, however the 800-day-old rat is not very closely comparable with the 20-year-old monkey. Older rats definitely should be studied.

Thinning of capillary walls in the Macaque monkey is attributable almost entirely to a decrease in the cross-sectional area of its endothelial component. This is consistent with the endothelial cell loss observed with increasing age in rats coupled with compensatory elongation of remaining endothelial cells (Bar, 1978). Thus, the changes occurring with age in the endothelium of the monkey cerebral capillary probably reflect compensatory thinning after cell loss. Studies in the rat showed a significant increase in the thickness of basal lamina (BL), a slight increase in capillary lumen diameter, and no change in overall thickness of the capillary wall, which taken together suggest a trend toward thinning of the endothelial component of the capillary wall (Burns et al., 1981).

The decline with age in brain capillary endothelial mitochondrial content coupled with thinning of the endothelial component of the cerebral capillary wall suggests a decline in work capability of the brain capillary per unit length as well as an increased vulnerability of brain capillaries to mechanical damage during aging. These changes may interfere with various regulatory mechanisms, e.g., the ability of the BBB to maintain normal ionic differentials between the blood and brain ECF.

The significant increase in rat cerebral capillary BL thickness with increasing age (Burns et al., 1981) is of interest relative to changes observed in BL of peripheral capillaries in diabetes mellitus. Thickened BL, containing pericyte debris, was observed in skeletal muscle capillaries in human diabetes mellitus (Tilton, Hoffmann, Kilo, & Williamson, 1981), a disease with exaggerated age-dependent accumulations in the BL of skeletal muscle capillaries (Vracko, Pecoraro, & Carter, 1980). The thickening of peripheral capillary BL in the human has been thought to be a by-product of endothelial cell renewal, the mechanism of which is not clearly understood (Vracko, Thorning, & Huang, 1979). Whether thickening of the BL in rat cerebral capillaries is related to endothelial cell renewal is not known.

Brain endothelial cells may prove to be postmitotic. Generally, the BL of brain capillaries is not visible prior to arrival of astrocytic end-foot processes at the site of the capillary wall. As mentioned above, one portion of the BL is thought to be of endothelial origin and the other portion of epithelial origin. Thus, thickening of the BL may be influenced not only by the endothelial cell but also by the presence or absence of astrocytic end-foot processes. In the Macaque monkey, at all ages studied (Burns et al., 1979), the BL of capillaries from occipital cortex was significantly thicker than that of capillaries from the frontal cortex. Thickened BL in cortical capillaries has been reported in patients with Alzheimer's disease and was thought to be associated with alteration of BBB permeability (Mancardi, Perdelli, Rivano, Leonarde, & Bugiani, 1980). The BL is an important site of enzyme activity. For example, ATPase activity has been localized in the BL of brain capillaries and in astrocytic end-foot processes (Joo, 1968). Thus, the thickening of the BL, as observed in brain capillaries, may affect both the permeability properties and metabolic function of the BBB, either of which could lead to a leaky capillary wall. Undoubtedly even slow leakage of molecules or ions normally prevented from entering brain ECF by an intact BBB would be deleterious to brain function.

The decline with increasing age in numerical density or content of endothelial mitochondria per cerebral capillary profile in Macaque monkeys and Sprague-Dawley rats (Burns et al., 1981, 1983) is of particular interest relative to the free radical theory of aging. It has been hypothesized that intrinsic mitochondrial senescence, a manifestation of oxygen toxicity, is related to changes occurring in the aging brain (Miquel, Economos, Fleming, & Johnson, 1980). Briefly, this hypothesis proposes that free radicals and lipid peroxides, by-products of the reduction of oxygen, are injurious to mitochondria in fixed postmitotic cells, thus rendering those organelles incapable of dividing because of damaged lipids in mitochondrial membranes and damaged mitochondrial DNA. Therefore, the observed decrease in the mitochondrial content of postmitotic cells may be of importance in aging.

Decline in cerebral capillary endothelial mitochondrial content coupled with thinning of the endothelial component of brain capillary walls suggests a decline in work capability of the cerebral capillary per unit length as well as an increased vulnerability of the cerebral capillary to mechanical damage during aging. These changes may interfere with various regulatory mechanisms, e.g., they may compromise the ability of the BBB to maintain normal ionic differentials between the blood and the ECF of brain tissue. Since neuronal function may be affected by slight alterations in concentrations of various ionic species in the brain ECF, particularly the potassium ion, serious neurological consequences might result.

These findings relative to age differences in cerebral capillaries in nonhuman primates (Burns et al., 1979, 1981, 1983) and in rats (Bar, 1978; Bar & Wolff, 1976; Burns et al., 1983; Schwink & Wetzstein, 1966; Wolff & Bar, 1976) accentuate the need for additional research concerning the effects of aging on brain capillaries, particularly on interendothelial tight junctions, endothelium, endothelial mitochondria, and on the cerebral capillary basal lamina.

Differences in numerical density of cerebral cortical capillaries in the rat across the lifespan probably reflect the rapid increase in volume of CNS tissue that occurs during the brain growth spurt and the decreasing volume of CNS tissue observed during aging as well as the changes occurring in the temporal pattern of brain vascularization.

Capillary changes reported in aging human cerebral cortex include: increased mean capillary diameters and volumes, increased capillary length per unit volume of cortex (Hunziker, Frey, & Schulz, 1974), a decreased specific surface area and shortened mean inter-capillary distance (Hunziker & Schweizer, 1977; Hunziker, Abdel'Al, Frey, Veteau, & Meier-Ruge, 1978; Hunziker, Abdel'Al, Shulz, & Schweizer, 1978; Hunziker, Abdel'Al, & Schulz, 1979). In the putamen (Meier-Ruge, Hunziker, Schulz, Tobler, & Schweizer, 1980; Shulz, Abdel'Al, Hunziker, & Meier-Ruge, 1980) capillary diameter does not change with increasing age although length per unit volume increases and inter-capillary distance decreases progressively with increasing age.

Stewart and coworkers (Stewart, Magliocco, Hayakawa, Farrell, Del Maestro, Girvin, Kaufmann, Vinters, & Gilbert, 1987) found in the human that capillary walls in gray matter normally were thinner than those in white matter, but that during aging white matter capillary walls decreased in thickness until they were of similar thickness to those in gray matter. They attributed the thinning of capillary walls during aging to loss of pericytes in addition to thinning of the endothelial cytoplasm. They found no change in endothelial mitochondria with increasing age.

The increase in lumenal diameter of cerebral capillaries during human aging (Hunziker et al., 1979) may be due to loss of endothelial cells with subsequent cellular elongation and thinning of remaining cells, leading to a larger capillary lumen. Attenuation of capillary walls during human aging might increase the vulnerability of cerebral capillaries to mechanical damage and thus contribute to neurological pathology associated with aging.

The evolution of the highest level of integrative brain function occurred in parallel with that of an efficient BBB. The constellation of complex, precisely interacting morphological, biochemical and physiological mechanisms that constitute the BBB isolates the internal environment of the brain from that of the blood and maintains a dynamic homeostasis of the entire microenvironment of central neurons, neuronal systems, synapses, and their supporting tissues, thereby enabling the human central nervous system to function normally.

REFERENCES

Adinolfi M (1985) The development of the human blood-CSF-brain barrier. Dev Med and Child Neurol 27:532-537.
Bar T (1978) Morphometric evaluation of capillaries in different laminae of rat cerebral cortex by automatic image analysis: Changes during development and aging. In: J Cervos-Navarro, E Betz, G Ebhardt, R Ferszt, B Wullenweber (eds). Advances in Neurology, vol 20, Pathology of Cerebrospinal Microcirculation, Plenum Press, New York, pp 1-9.
Bar T, Wolff J R (1973) Quantitative Beziehungen zwischen der Verzweig ungsdichte und Lange von Capillaren im Neocortex der Ratte wahrend der postnatalen Entwicklung. Z Anat EntwGesch 141:207-228.
Bar T, Wolff J R (1976) Development and adult variations of the wall of brain capillaries in the neocortex of rat and cat. In: J Cervos-Navarro, E Betz, F Matakas, R Wullenweber (eds) The Cerebral Vessel Wall, Raven Press, New York, pp 1-6.
Bradbury M B W (1984) The structure and function of the blood-brain barrier. Fed Proc 43:186-190.
Brightman M W & Reese T S (1969) Junctions between intimately apposed cell membranes in the vertebrate brain. J Cell Biol 40:648-677.
Bundgaard M (1982) Ultrastructure of frog cerebral and pial microvessels and their impermeability to lanthanum ions. Brain Res 241:57-65.
Bundgaard M (1986) Pathways across the vertebrate blood-brain barrier: Morphological viewpoints. In H F Cserr (ed) The Neuronal Microenvironment. Annals of the New York Academy of Sciences. 481:7-19.
Burns E M, Kruckeberg T W, Comerford L E, Buschmann MB T (1979) Thinning of capillary walls and declining numbers of endothelial mitochondria in the cerebral cortex of the aging primate, Macaca nemestrina. J Geront 34:642-650.
Burns E M, Kruckeberg T W, Gaetano P K (1981) Changes with age in cerebral capillary morphology. Neurobiol of Aging, 2:285-291.
Burns E M, Kruckeberg T W, Gaetano P K, Shulman L M (1983) Morphological changes in cerebral capillaries with age. In: J Cervos-Navarro, H I Sarkander (eds) Brain Aging: Neuropathology and Neuropharmacology, New York: Raven Press, pp 115-132.
Caley D W, Maxwell D W (1970) Development of the blood vessels and extracellular spaces during postnatal maturation of rat cerebral cortex. J Comp Neurol 138:31-48.
Craigie E H (1924) Changes in vascularity in the brain stem and cerebellum of the albino rat between birth and maturity. J Comp Neurol 38:27-48.
Craigie E H (1925) Postnatal changes in vascularity in the cerebral cortex of the male albino rat. J Comp Neurol 39:301-324.
Crone C, Christensen O (1981) Electrical resistance of a capillary endothelium. J Gen physiol 77:1349-1371.
Cserr H F, Bundgaard M (1986) The neuronal microenvironment: A comparative view. In H F Cserr (ed) The Neuronal Microenvironment. Annals of the New York Academy of Sciences. 481:1-7.
Davson H (1976) The blood-brain barrier. J Physiol 255:1-28.
Diamond J M, Wright E M (1969) Biological membranes: The physical basis of ion and nonelectrolyte selectivity. Annu Rev Physiol 31:581-646.
Donahue S (1964) A relationship between fine structure and function of blood vessels in the central nervous system of rabbit fetuses. Am J Anat 115:17-26.
Donahue S, Pappas G D (1961) The fine structure of capillaries in the cerebral cortex of the rat at various stages of development. Am J Anat 108:331-347.

Dyson S E, Jones D G, Kendrick W L (1976) Some observations on the ultrastructures of developing rat cerebral capillaries. Cell Tissue Res 173:529-542.
Hannah R S, Nathaniel E J H (1974) The postnatal development of blood vessels in the substantia gelatinosa of rat cervical cord - an ultrastructural study. Anat Rec 178:691-710.
Hunziker O, Frey H, Schulz U (1974) Morphometric investigations of capillaries in the brain cortex of the cat. Brain Res 65: 1-11.
Hunziker O, Schweizer A (1977) Postmortem changes in stereological parameters of cerebral capillaries. Beitr Path 161: 244-255.
Hunziker O, Abdel'Al S, Frey H, Veteau M-J, Meier-Ruge W (1978) Quantitative studies in the cerebral cortex of aging humans. J Geront 24:27-31.
Hunziker O, Abdel'Al S, Schulz U, Schweizer A (1978) The architecture of cerebral capillaries in aged human subjects with hypertension. In: J Cervos-Navarro, E Betz, R Wullenweber (eds). Advances in Neurology, Vol. 20: Cerebrospinal Microcirculation, Raven Press, New York pp. 471-478.
Hunziker O, Abdel'Al S, Schulz U (1979) The aging human cerebral cortex: A stereological characterization of changes in the capillary net. J Geront 34:345-350.
Joo F (1968) Effect of inhibition of adenosine triphosphatase activity on the fine structural organization of the brain capillaries. Nature 219:1378-1379.
Lambertson C J (1980) Chemical control of respiration at rest. In: V B Mountcastle (ed) Medical Physiology, C V Mosby, St. Louis, 1774-1827.
Lee J C (1971) Evolution in the concept of the blood-brain barrier phenomenon. Progress in Neuropathol 1:84-145.
Mancardi G L, Perdelli F, Rivano C, Leonarde A, Bugiani O (1980) Thickening of the basement membrane of cortical capillaries in Alzheimer's disease. Acta Neuropath 49: 79-85.
Meier-Ruge W, Hunziker O, Schulz U, Tobler H-J, Schweizer A 1980) Stereological changes in the capillary network and nerve cells of the aging human brain. Mechs Aging Dev 14:233-243.
Milnor W R (1980) Capillaries and lymphatic vessels. In: V B Mountcastle (ed) Medical Physiology, C V Mosby, St. Louis, pp 1085-1093.
Miquel J, Economos A C, Fleming J, Johnson J E, Jr. (1980) Mitochondrial role in cell aging. Expl Geront 15: 575-591.
Molgard K, Balslev Y, Lauritzen B, Norman R (1987) Cell junctions and membrane specializations in the ventricular zone (germinal matrix) of the developing sheep brain: a CSF-brain barrier. J Neurocytol 16:433-444.
Mollgard K, Saunders N R (1986) The development of the human blood-brain and blood-CSF barriers. Neuropath Applied Neurobiol 12:337358.
Oldendorf W H, Cornford M E, Brown W J (1977) The large apparent work capability of the blood-brain barrier: A study of the mitochondrial content of capillary endothelial cells in brain and other tissues of the rat. Ann Neurol 1:409-417.
Palade G E, Bruns R R (1964) Structure and function in normal muscle capillaries. In: M D Sipperstein, A R Colwell, K Meyers (eds) Small blood vessel involvement in diabetes mellitus. Baltimore, Garamound/Pridemark. pp 45-55.
Pappenheimer J R (1953) Passage of molecules through capillary walls. Physiol Rev 33:387-423.
Pardridge W M (1983) Brain metabolism: A perspective from the blood-brain barrier. Physiol Rev 63:1481-1535.
Rapoport S I (1976) Blood-Brain Barrier in Physiology and Medicine, Raven Press, New York.
Reese T S, Karnovsky M J (1967) Fine structural localization of a blood-brain barrier to exogenous peroxidase. J Cell Biol 34: 207-217.

Richards J G (1978) Permeability of intercellular junctions in brain epithelia and endothelia to exogenous amine: Cytochemical localization of extracellular 5-hydroxydopamine. J Neurocytol 7:61-70.
Rowan R A, Maxwell S (1981a) Patterns of vascular sprouting in the postnatal development of the cerebral cortex of the rat. Am J Anat 160: 247-255.
Rowan R A, Maxwell S (1981b) An ultrastructural study of vascular proliferation and vascular alkaline phosphatase activity in the developing cerebral cortex of the rat. Am J Anat 160:257-265.
Saunders N R, Mollgard K (1984) Development of the blood-brain barrier. J Dev Physiol 6:45-57
Schmidley J W, Wissig S L (1986) Basement membrane of central nervous system capillaries lacks ruthenium red-staining sites. Microvasc Res 32:300-314.
Schulz U, Abdel'Al S, Hunziker O, Meier-Ruge W (1980) Quantitative morphological changes in capillaries and neurons in the aging human putamen. Microsc Acta 4: Suppl 4, 135-139.
Schwink A, Wetzstein R (1966) Die Kapillaren in Subcommissuralorgan der Ratte. Z Zellforsch 73:56-88.
Singh D N P, Nathaniel E J H (1975) Postnatal development of blood vessels (capillaries) in the rat olfactory bulb: A light and ultrastructural study. Neurosci Lett 1:203-208.
Stewart P A, Hayakawa (1987) Interendothelial junctional changes underlie the developmental 'tightening' of the blood-brain barrier. Devel Brain Res 32:271-281.
Stewart P A, Magliocco M, Hayakawa K, Farrell C L, Del Maestro C F, Girvin J, Kaufmann J C E, Vinters H V, Gilbert J (1987) A quantitative analysis of blood-brain barrier ultrastructure in the aging human. Microvasc Res 33:270-282.
Tilton R G, Hoffmann P L, Kilo C, Williamson J R (1981) Pericyte degeneration and basement membrane thickening in skeletal muscle capillaries of human diabetics. Diabetes 30:326-334.
Vracko R, Thorning, Huang T W (1979) Basal lamina of alveolar epithelium and capillaries: Quantitative changes with aging and in diabetes mellitus. Am Rev Resp Dis 120:973-983.
Vracko R, Pecoraro R E, Carter W B (1980) Overview article: Basal lamina of epidermis muscle fibers, muscle capillaries, and renal tubules: Changes with aging and in diabetes mellitus. Ultrastruct. Path. 1:559-574.
Wolff J R, Bar T (1976) Development and adult variations of the pericapillary glial sheath in the cortex of rat. In: J Cervos-Navarro, E Betz, F Matakas, R WullenWeber (eds) The Cerebral Vessel Wall. Raven Press, New York, pp 7-13.

DISCUSSION

Chaired by L. Edvinsson, M. Trabucchi.

S. Rapoport:

I would like to make some comments on two particular points

1) the role of vesicular transport at the level of BBB as well as other capillary regions.

2) BBB permeability during aging, both in humans and in animal models

1) At the capillary level two possible pathways have been described: one is represented by the tight junctions of the endothelial wall and the second by pinocytotic vesicles within the capillary endothelium.
From electromicroscopic studies, intracellular vesicles apparently move from the intravascular to the extravascular space over a short period of time. However, Kobayashi in 1970, used fixed capillaries to show that there is no movement of vesicles which could account for a transport across endothelial cells. This was the first attack on the concept of vesicular transport.
Bundgaard et al. (1979), using sections of mesenteric capillaries, demonstrated that the apparent vesicles are part of a vacuose system but do not form transcellular channels. Another series of observations presents further evidence against the existence of vesicular transport (Rapoport and Robinson, 1986; Shinowara et al., 1982; Michel et al., 1984). It has been demonstrated in frog peripheral nerves, where capillaries and perineural cells are rich in apparent vesicles, that vascularly injected peroxidase or ionic lanthanum remain within the vasculature. This finding suggests that the presence of vesicular elements in serial sections has little to do with transport processes. In addition, the same authors confirmed by electro-microscopy that each apparent vesicle was an invagination of the membrane.
On the basis of these observations I do think that we should investigate the role of such microenvironmental regions, which are caveoli and not isolated cytoplasmic vesicles.

- Kobayashi S, 1970. Arch Histol Jpn, 32:81-86.
- Bundgaard M, et al, 1979. Proc Natl Acad Sci USA, 76:6439-6442.
- Rapoport SI, Robinson PJ, 1986. Ann NY Acad Sci, 481:250-267.
- Shinowara NL, et al, 1982. Cell Tissue Res, 227:11-22.
- Michel ME, et al, 1984. Brain Res, 299:25-30.

However, under pathological conditions (hypertension, trauma, etc) we cannot preclude the possibility of involvement of vesicular formation. We still do not know in detail how the endothelium responds to injuries.
Lastly, tracer studies on the rate of penetration of compounds as a function of molecular weight, following the osmotic opening of the blood-brain barrier, show that the vesicular pathway is not a probable means of transport (Robinson, and Rapoport; 1987). A pore mechanism, consistent with opening of tight junctions, aided by water flow, would account for the differential changes in permeability of the substances in relation to their molecular weight.

2) The second comment is that no physiological studies, as far as I know, have ever demonstrated that BBB permeability changes with age, even though morphology has revealed slight modifications of the capillary endothelium and basement membrane. There is no evidence of abnormal cerebrospinal fluid proteins in healthy human subjects or rats (proteins are excluded to a large extent from the brain by the capillaries and enter very slowly through the choroid plexus by a process of ultrafiltration, so that the concentration is less then 0.190 ml plasma), (Kay et al., 1987; Rapoport et al., 1979).

L. Edvinsson:

One problem which is still controversial is where the BBB is located: also cerebral arteries and arterioles may have barrier properties, besides the capillaries?
Returning to the question of vesicular transport, there are data showing that the possibility exists of enhancing the passage of tracers by, for example, increasing blood pressure through injection of hyperosmotic solution. An interesting observation is that Vincristine, which blocks microtubular transport, may blunt this passage: this represents one type of evidence in support of the existence of vesicular transport across the capillaries.

E. Burns:

I think we should really look at what happens in diseased conditions as opposed to the normal. I still hold that in the normal situation there are very few vesicles at the capillary level.

- Robinson PJ, Rapoport SI, 1987. Am J Physiol, 253:R459-R466.
- Kay AD, et al, 1987. Neurobiol Aging, 8:21-25.
- Rapoport SI, et al, 1979. J Gerontol, 34:162-169.

J. Cervos Navarro:

The point, in fact, is to differentiate the transport in normal and pathological states. It has been observed that the injection of peroxidase after death in animals leads to an increase in pinocytotic vesicles, suggesting that this phenomenon is not an "intravital" event.
It seems therefore that peroxidase itself either penetrates the channels or increases the possibility of the membrane to form vesicles. However, since vesicle formation is observed post-mortem, we can not establish whether this event is intravitae or a simple artifact.
The problem now is to know whether the so called pinocytotic channels are caveoli deriving from the membrane, since in this case our view on the BBB needs reevaluation. At present, as morphologists, we are not yet able to differentiate between vesicles and invaginations; however, this does not exclude the possibility of the existence of vesicular transport.

L. Edvinsson:

What about the development of the BBB?

E. Burns:

Considering the electronmicroscopic aspect of the BBB and in particular the formation of tight junctions, there are recent studies reporting that at the time of birth the BBB is mature. However, I think that we should examine the degree of maturity also from a physiological and biochemical point of view.

J. Cervos Navarro:

One of the indicators of BBB immaturity or impairment in newborns is the passage of bilirubin, which in adults occurs in case of BBB disruption, for example in conditions of cerebral ischemia.
In elderly people above 90, plenty of collagen has been found in the capillaries of the spinal cord, in the space between the astrocytes and in the endothelial membrane. Interestingly, the same finding characterizes the spinal cord capillaries in children of 7 months of age.
From clinical observation and preliminary experimental data we have the impression that elderly people or aged animals are less prone to develop edema after cerebral injury, suggesting that the vascular permeability capacity is modified in senescence.

L. Edvinsson:

This would suggest that all the subtle morphological changes occurring with age have a functional meaning.

E. Burns:

I think so. For example, we have observed a decline in the number of capillary mitochondria with senescence, and this may indeed be related to the work capacity of the BBB. It has been observed that elderly people show a similar sensitivity as the children to dehydration and also develop symptoms of mental confusion afer 6-8 hours of fasting. This suggests the possibility that BBB in aging is not quite effective in certain circumstances.

S.I. Rapoport:

In 1927 Behnsen showed an increased distribution of the dye trypan blue in the developing brain despite of the fact that the fraction of nonbarrier regions (the regions that did not stop the entry of the tracer) is higher than in the mature brain.
Other authors (Ferguson and Woodburg, 1969), reported a high rate of penetration of sucrose and inulin into the cerebrospinal fluid of immature rats, for up to 2 weeks of age. It was not clear which pathway these substances took: through the cerebral capillaries or through the choroid plexus.
Therefore, I think that one of the ways to approach the study of the multimembrane system which is BBB is to define it in terms of permeability, structure, transport, etc. Each one of these parameters follows a different time course until we have the integrated effect in the mature animal.

- Behnsen G, 1927. Munch Med Wochenschr, 73:1143-1147.
- Ferguson RK, Woodburg DM, 1969. Exp Brain Res, 7:181-194.

STRUCTURAL AND FUNCTIONAL POLARITY OF THE BLOOD-BRAIN BARRIER

A. Lorris Betz
Departments of Pediatrics, Surgery and Neurology
University of Michigan
D3227 Medical Professional Building
Ann Arbor, MI 48109-0718
U.S.A.

The blood-brain barrier has often been considered a simple membrane that forms an impermeable wall to polar solutes while permitting lipid-soluble compounds to pass. Over the past two decades, our understanding of blood-brain barrier function has been advanced by recognition of the importance of carrier-mediated transport processes which facilitate the permeation of selected polar compounds across this lipid barrier. In most cases, however, kinetic models of these transport systems have considered only a single transport step in the movement of a compound from blood to brain. These single-membrane models are obviously oversimplifications since the structure which forms the blood-brain barrier, the brain capillary endothelium, is a monolayer of interconnected cells with separate membrane surfaces facing the blood and the brain. The two opposing plasma membranes of the endothelial cell need not have identical permeability properties. Thus, a comprehensive representation of the blood-brain barrier must take into account the two separate membranes that solutes encounter as they move between blood and brain as well as the cytoplasm of the endothelial cell which may contain enzymes capable of reacting with the solutes while they are in transit through the cell.

The purpose of this paper is to review the data which suggest that the endothelial cell plasma membrane facing the blood (luminal membrane) is different than the membrane facing the brain (antiluminal or abluminal membrane). Functional studies of blood-brain barrier and brain capillary endothelial cell transport properties provide indirect evidence for these differences, while structural investigations of membrane proteins provide direct evidence.

NATO ASI Series, Vol. H 33
Regulatory Mechanisms of Neuron to Vessel Communication in the Brain. Edited by F. Battaini et al.

DEFINITION AND SIGNIFICANCE OF CELLULAR POLARITY

When the two opposing plasma membranes of a cell have different properties, that cell is said to be polar. This type of cellular polarity is typically seen in epithelial cells where different ion transporters are found on the serosal and mucosal sides of the cell. Often epithelial cells also have a prominent structural polarity with microvilli forming a brush

A good example of a polar epithelial cell which is familiar to neuroscientists is found in the choroid plexus. Here, the brush border surface of the choroid plexus epithelial cell faces the cerebrospinal fluid and the basal surface faces the blood vessels. Structural (Ernst, et al., 1986) and functional (Wright, 1972; Saito and Wright, 1983) studies have demonstrated that the important ion transporter, Na,K-ATPase, is found only in the brush border surface of the cell while Na/H and Cl/HCO_3 exchangers are located in the basal membrane. This arrangement of transporters underlies the ability of the choroid plexus epithelial cell to pump sodium and chloride in the blood to brain direction and thereby secrete spinal fluid. Thus, the active, energy requiring ion pump, Na,K-ATPase, transports sodium out of the cell and into the spinal fluid in exchange for potassium. This lowers the concentration of sodium within the epithelial cell by a small amount which is replaced by sodium entering the cell from the blood side in exchange for hydrogen ions. The net result is a unidirectional flux of sodium from blood to spinal fluid. Driven by the electrical gradient resulting from the sodium flux, chloride ions also cross the cell by separate transport systems and together the unidirectional flux of sodium and chloride produces an osmotic driving force for water. This type of active, transcellular transport could not occur if the Na,K-ATPase and the Na/H exchanger were found on both sides of the cell.

Cellular polarity, in fact, provides the basis for active transcellular transport of many types of solutes across epithelial cells. For example, the highly efficient absorption of sugars and amino acids by the intestine and kidney depends upon cellular polarity. As a result, large concentration gradients are produced for these solutes across the epithelial cell layers. Cellular polarity also permits an epithelial cell to take up and metabolize a substrate from one side while still maintaining a barrier which prevents the substrate from crossing to the other side. This could be an important mechanism for local elimination of chemicals that might otherwise produce toxicity if released systemically. Finally, polarity of hormone receptors could allow a cell to respond to agonists that are present on one side but not the other side of the cell. This may provide additional specificity in hormone action.

FUNCTIONAL POLARITY OF THE BLOOD-BRAIN BARRIER

Since cellular polarity is required for active transport across a cell layer, polarity of the blood-brain barrier can be demonstrated functionally by the presence of active transport against a gradient between blood and brain. The presence of active transcellular transport can be implied if significant concentration gradients exist between the blood and the brain's interstitial fluid for substances that are not rapidly produced or consumed by the brain. This appears to be the case for certain amino acids (Lerma, et al., 1986). Similarly, the concentration of potassium in the brain's extracellular fluid is maintained at a level that is slightly lower than, and independent of, changes in its plasma concentration (Katzman, 1976). These studies must be interpreted with caution, however, because of active uptake by neurons and glia and transport by the choroid plexus.

The presence of active transport and a concentration gradient between blood and brain have also been suggested by studies which show that specific transport inhibitors increase rather than decrease the content of a tracer within the brain tissue. This phenomena has been demonstrated using perchlorate to inhibit iodide transport (Davson and Hollingsworth, 1973) and probenecid to inhibit 5-hydroxyindoleacetic acid (Atack, et al., 1974) or para-aminohippuric acid transport (Ennis and Betz, unpublished). The results of such studies are best explained by the presence of an active efflux system operating from brain to blood which maintains a low concentration of the solute in the brain's interstitial fluid. When this transport system is inhibited, the rate of clearance of solute from brain is reduced and the solute accumulates in brain to a higher steady state level.

A greater rate of solute flux in one direction compared to the opposite direction across the blood-brain barrier is another functional indicator of blood-brain barrier polarity. The best described example of this situation is potassium. Several studies, including the elegant investigations of Hansen, et al. (1977), have demonstrated the extremely low permeability of the blood-brain barrier for potassium in the blood to brain direction. However, Bradbury and Stulcová (1970) described a concentration dependent efflux of potassium from the brain that is reduced by the Na,K-ATPase inhibitor, ouabain. This asymmetric transport of potassium between blood and brain suggests the presence of polar transport systems and it may well contribute to the stability of the potassium concentration in the brain's extracellular space. Operating in conjunction with asymmetrically distributed sodium transporters (Betz, 1983; Crone, 1985), the capillary Na,K-ATPase might also provide the driving force for secretion of fluid from blood to brain. Certain neutral amino acids (Lajtha and Toth, 1961; Lorenzo and Snodgrass, 1972; Murray and Cutler, 1970) and prostaglandins (Bito, et al., 1976) also appear to cross the barrier more readily in the brain to blood than in the blood to brain direction.

Additional evidence for transport asymmetry is found in experiments with isolated brain capillaries. Several compounds whose blood to brain flux is extremely low in vivo like potassium, small neutral amino acids, and catecholamines are readily accumulated within isolated brain capillaries (Betz and Goldstein, 1978; Goldstein, 1979; Hardebo and Owman, 1980; Spatz, et al., 1981). A reasonable explanation for these observations is that the compounds cannot cross the luminal membrane of the capillary to which they are exposed in vivo but they are transported across the antiluminal membrane which is exposed in vitro. However, the presence of cell types other than the endothelial cells in isolated microvessel preparations must be taken into account in interpreting such studies.

STRUCTURAL POLARITY OF THE BLOOD-BRAIN BARRIER

Studies of the specific components of the membranes of brain capillary endothelial cells provide more direct and convincing evidence for polarity of the blood-brain barrier. One approach is to purify membranes from isolated brain capillaries and then fractionate them into different populations trying to separate the membranes from the luminal and antiluminal sides of the endothelial cell. The content of various membrane markers is then examined in each fraction. Using this method, we were able to identify two membrane fractions that were equally enriched in alkaline phosphatase, but one fraction contained a much greater activity of Na,K-ATPase and a somewhat higher activity of 5'-nucleotidase (Betz, et al., 1980). Assuming that membranes from the two sides of the endothelial cell had segregated between the two different fractions, we concluded that alkaline phosphatase might be equally distributed between the opposing membranes while Na,K-ATPase and possibly also 5'-nucleotidase would be present primarily on only one side of the cell. Based upon the functional studies of potassium transport in vivo and in vitro, we suggested that the membrane fraction containing the Na,K-ATPase activity must be derived from the antiluminal side of the cell. Once again, the presence of other cell types within the isolated capillaries from which the membranes were made requires that caution be used in formulating conclusions.

Fortunately, a more direct method for demonstrating the location of enzymes within cells exists. With cytochemical techniques, a specific substrate is presented to an intact tissue either by vascular perfusion or in vitro incubation of a tissue sample. The reaction product is then trapped by a capture reagent and made electron-dense so that it can be observed using electron microscopy. The presence of electron-dense granules along the surface of the cell membrane, therefore, indicates the presence of enzyme activity within that membrane. The enzyme activity is further identified using specific substrates and/or inhibitors.

Using this approach in collaboration with Anthony Firth, we were able to demonstrate that Na,K-ATPase was indeed located primarily in one membrane of the endothelial cell, the antiluminal membrane, while alkaline phosphatase was present in both membranes (Firth, 1977; Betz, et al., 1980). Since then, others have confirmed these distributions (Inomata, et al., 1984; Kato and Nakamura, 1987; Kreutzberg and Toth, 1983; Latker, et al., 1987; Mori and Nagano, 1985;Vorbrodt, et al., 1981) and a number of other membrane-associated proteins have also been studied cytochemically (Table 1). While there are some conflicting results concerning the distributions of specific proteins, the general picture of polarity of the endothelial cell appears to be firmly established by these studies.

Lidinsky and Drewes (1983) used another approach to directly demonstrate polarity of the brain capillary. They labeled membrane proteins with radioactive iodine by either perfusing the isolated dog brain in situ or by incubating isolated brain capillaries in vitro and hypothesized

TABLE 1. CYTOCHEMISTRY OF BRAIN CAPILLARY ENDOTHELIAL CELLS

Activity	Species	Localization	Reference
Na,K-ATPase			
	rat	antiluminal	Firth, 1977
	rat	antiluminal	Betz, et al., 1980
	rat	antiluminal or both	Vorbrodt, et al., 1982
Alkaline Phosphatase			
	rat	both	Betz, et al., 1980
	mouse	luminal	Vorbrodt, et al., 1981
	rat	antiluminal or neither	Masuzawa, et al., 1983
	rat	luminal or both	Kreutzberg and Toth, 1983
	rat	both	Inomata, et al., 1984
	rat	both	Mori and Nagano, 1985
	guinea pig	both	Kato and Nakamura, 1987
	rat	both	Latker, et al., 1987
5'-Nucleotidase			
	rat	both	Inomata, et al., 1984
	mouse	antiluminal	Vorbrodt, et al., 1984a
Ca-ATPase			
	rat	antiluminal	Joó, 1979
	rat	antiluminal > luminal	Inomata, et al., 1984
	rat	both	Nag, 1987
Angiotensin Converting Enzyme			
	rat	luminal	Pickel, et al., 1986
γ-Glutamyl Transpeptidase			
	rat	luminal	Budi Santos and Bär , 1986
Adenylate Cyclase			
	rat	both	Joó, et al., 1975
	rat	both	Szumanska, et al., 1984
	rat	luminal or both	Vorbrodt, et al., 1984b
Guanylate Cyclase			
	rat	both	Karnushina, et al., 1980
Con-A Receptor			
	mouse	luminal	Vorbrodt, et al., 1984a
	rat	luminal	Nag, 1985
	mouse	antiluminal > luminal	Vorbrodt, et al., 1986
Podocalyxin			
	rat	luminal	Horvat, et al., 1986
HT7 Antigen			
	Chicken	luminal	Risau, et al., 1986

that, in the former experiment, the capillaries that were subsequently isolated from the perfused brain would have only their luminal membranes labeled while the incubated capillaries would have their antiluminal membrane labeled. Cell proteins were then extracted, separated by gel electrophoresis, and labeled proteins identified by autoradiography. The patterns of protein labeling between the two experiments were distinctly different indicating that the opposing sides of the endothelial cell contained different membrane proteins.

INDUCTION OF POLARITY IN BRAIN CAPILLARY ENDOTHELIUM

Brain capillaries are composed of endothelial cells which are unique in comparison to other microvascular endothelia. They are sealed together by continuous tight junctions, lack transcellular channels and cytoplasmic vesicles, and they have a polar distribution of membrane bound enzymes and transporters. Why is it that brain capillaries develop these special properties? Since endothelial cells are all derived from the same embryonic tissue, it seems likely that another cell type within the brain exerts an inductive influence on the capillary endothelial cell. Support for this hypothesis of tissue-specific induction of the blood-brain barrier was provided by Stewart and Wiley (1981) who found that abdominal blood vessels invading a small piece of brain tissue transplanted to the abdominal cavity of a chick embryo took on the properties of brain microvessels. A likely source for the inductive signal is the astrocyte (Goldstein,1988), a cell with processes that completely encircle the capillary in most areas of the brain. Indeed, astrocytes purified from immature brain and implanted on the avascular cornea stimulate the growth of new microvessels which have blood-brain barrier properties (Janzer and Raff, 1987). Furthermore, astroglioma cells in tissue culture induce the reappearance of γ-glutamyl transpeptidase in cultured brain capillary endothelial cells (DeBault and Cancilla, 1980). Although this enzyme is present in brain capillaries and absent in other vascular beds, it is lost when brain capillary endothelial cells are grown in tissue culture. Thus glial cells appear to elaborate a signal which causes endothelial cells to develop unique properties related to blood-brain barrier function.

There is also evidence that glial cells are responsible for inducing polarity in brain endothelial cells. Beck, et al. (1984) grew endothelial cells derived from mouse brain capillaries on one side of a filter membrane and C_6 glioma cells on the other side. The membranes were then placed in chambers and the rate of flux of a small neutral amino acid analogue, α-methylaminoisobutyric acid, was measured in each direction across the membrane. This tracer was chosen because in vivo it appears to be transported in the brain to blood direction but not the blood to brain direction (Betz and Goldstein, 1978). In the absence of glial cells, the tracer moved at equal rates in both directions. However, when glial cells were co-cultured on the opposite side of the membrane, the tracer moved more rapidly from the glial to the endothelial side than it did from the endothelial to the glial side. This effect was not observed when smooth muscle cells were co-cultured with endothelial cells. Thus, the presence of glial cells resulted in the asymmetric transport of a small neutral amino acid across this artificial blood-brain barrier.

CONCLUSIONS

A variety of experimental techniques have been used to demonstrate that the brain capillary endothelial cell is polar, both structurally and functionally. This cellular polarity has important implications for blood-brain barrier function. As in epithelial cells, polarity underlies the active transport of solutes across the barrier against a gradient. This may be important for the maintenance of a constant potassium concentration in brain, the inactivation of certain neurotransmitters, and the removal of metabolic wastes. In addition, transport of ions could result in fluid secretion by brain capillaries and provide an additional source of cerebrospinal fluid. Finally, polarity of membrane receptors would permit the cell to respond specifically and differentially to neurotransmitters, hormones, and other factors present in the brain as compared to those in the blood.

REFERENCES

Atack C, Bass NH, Lundborg P (1974) Mechanism for the elimination of 5-hydroxyindoleacetic acid from brain and cerebrospinal fluid of the rat during perinatal development. Brain Res 77:111-120

Beck DW, Vinters HV, Hart MN, Cancilla PA (1984) Glial cells influence polarity of the blood-brain barrier. J Neuropathol Exper Neurol 43:219-224

Betz AL, Firth JA, Goldstein GW (1980) Polarity of the blood-brain barrier: distribution of enzymes between the luminal and antiluminal membranes of brain capillary endothelial cells. Brain Res 192:17-28

Betz AL, Goldstein GW (1978) Polarity of the blood-brain barrier: Neutral amino acid transport into isolated brain capillaries. Science 202:225-227

Betz AL (1983) Sodium transport from blood to brain: inhibition by furosemide and amiloride. J Neurochem 41:1158-1164

Bito LZ, Davson H, Hollingsworth J (1976) Facilitated transport of prostaglandins across the blood-cerebrospinal fluid and blood-brain barriers. J Physiol 256:273-285

Bradbury MWB, Stulcová (1970) Efflux mechanism contributing to the stability of the potassium concentration in cerebrospinal fluid. J Physiol 208:415-430

Budi Santos AW, Bar TH (1986) Postnatal development of γ-GT activity in rat brain microvessels corresponds to capillary growth and differentiation. Int J Devl Neurosci 4:503-511

Crone C (1985) The blood-brain barrier: a modified tight epithelium. In: Bradbury MWB, Rumsby MG, Suckling AJ (eds) The Blood-Brain Barrier in Health and Disease. Ellis Norwood Ltd, Chichester, pp 17-40

Davson H, Hollingsworth JR (1973) Active transport of ^{131}I across the blood-brain barrier. J Physiol 233:327-347

DeBault LE, Cancilla PA (1980) γ-Glutamyl transpeptidase in isolated brain endothelial cells: Induction by glial cells in vitro. Science 207:653-655

Ernst SA, Palacios JR II, Siegel GJ (1986) Immunocytochemical localization of Na^+, K^+ -ATPase catalytic polypeptide in mouse choroid plexus. J Histochem Cytochem 34:189-195

Firth JA (1977) Cytochemical localization of the K^+ regulation interface between blood and brain. Experientia 33:1093-1094

Goldstein GW (1979) Relation of potassium transport to oxidative metabolism in isolated brain capillaries. J Physiol 286:185-195

Goldstein GW (1988) Endothelial cell-astrocyte interactions. A cellular model of the blood-brain barrier. Ann NY Acad Sci 529:31-39

Hansen AJ, Lund-Andersen H, Crone C (1977) K^+-permeability of the blood-brain barrier, investigated by aid of a K^+-sensitive microelectrode. Acta Physiol Scand 101:438-445

Hardebo JE, Owman CH (1980) Characterization of the in vitro uptake of monoamines into brain microvessels. Acta Physiol Scand 108:223-229

Horvat R, Hovorka A, Dekan G, Poczewski H, Kerjaschki D (1986) Endothelial cell membranes contain podocalyxin-the major sialoprotein of visceral glomerular epithelial cells. J Cell Biol 102:484-491

Inomata K, Yoshioka T, Nasu F, Mayahara H (1984) Ultracytochemical studies of capillary endothelial cells in the rat central nervous system. Acta Anat 118:243-248

Janzer RC, Raff MC (1987) Astrocytes induce blood-brain barrier properties in endothelial cells. Nature 325:253-257

Joó F (1979) The role of adenosine triphosphatase in the maintenance of molecular organization of the basal lamina in the brain capillaries. Front Matrix Biol 7:166-182

Joó F, Tóth I, Jancsó G (1975) Brain adenylate cyclase: its common occurrence in the capillaries and astrocytes. Naturwissenschaften 8:397

Karnushina I, Tóth I, Dux E, Joó F (1980) Presence of the guanylate cyclase in brain capillaries: histochemical and biochemical evidence. Brain Res 189:588-592

Kato S, Nakamura H (1987) Ultracytochemical localization of alkaline phosphatase activity in endothelial cells in chronic relapsing experimental allergic encephalomyelitis. Acta Neuropathol (Berl) 73:220-226

Katzman R (1976) Maintenance of a constant brain extracellular potassium. Federation Proc 35:1244-1247

Kreutzberg GW, Toth I (1983) Enzyme cytochemistry of the cerebral microvessel wall. Acta Neuropathol (Berl) Suppl VIII:35-41

Lajtha A, Toth J (1962) The brain barrier system-III The efflux of intracerebrally administered amino acids from the brain. J Neurochem 9:199-212

Latker CH, Shinowara NL, Miller JC, Rapoport SI (1987) Differential localization of alkaline phosphatase in barrier tissues of the frog and rat nervous systems: a cytochemical and biochemical study. J Comp Neurol 264:291-302

Lerma J, Herranz AS, Herreras O, Abraira V, Del Rio RM (1986) In vivo determination of extracellular concentration of amino acids in the rat hippocampus. A method based on brain dialysis and computerized analysis. Brain Res 384:145-155

Lidinsky WA, Drewes LR (1983) Characterization of the blood-brain barrier: protein composition of the capillary endothelial cell membrane. J Neurochem 41:1341-1348

Lorenzo AV, Snodgrass SR (1972) Leucine transport from the ventricles and the cranial subarachnoid space in the cat. J Neurochem 19:1287-1298

Masuzawa T, Shimabukuro H, Sato F (1983) Cytochemical study of alkaline phosphatase activity in microvessels of rat cerebral cortex. Neurol Med Chir (Tokyo) 23:511-514

Mori S, Nagano M (1985) Electron-microscopic cytochemistry of alkaline-phosphatase activity in endothelium, pericytes and oligodendrocytes in the rat brain. Histochem 82:225-231

Murray JE, Cutler RWP (1970) Transport of glycine from the cerebrospinal fluid. Arch Neurol 23:23-31

Nag S (1985) Ultrastructural localization of monosaccharide residues on cerebral endothelium. Lab Invest 52:553-558

Nag S (1987) Ultrastructural localization of calcium-activated adenosine triphosphatase (Ca^{2+}-ATPase) in cerebral endothelium. Lab Invest 57:52-56

Pickel VM, Chan J, Ganten D (1986) Dual peroxidase and colloidal gold-labeling study of angiotensin converting enzyme and angiotensin-like immunoreactivity in the rat subfornical organ. J Neurosci 6:2457-2469

Risau W, Hallmann R, Albrecht U, Henke-Fahle S (1986) Brain induces the expression of an early cell surface marker for blood-brain barrier-specific endothelium. EMBO J 5:3179-3183

Saito Y, Wright EM (1983) Bicarbonate transport across the frog choroid plexus and its control by cyclic nucleotides. J Physiol 336:635-648

Spatz M, Maruki C, Abe T, Rausch D, Abe K, Merkel N (1981) The uptake and fate of the radiolabeled 5-hydroxytryptamine in isolated cerebral microvessels. Brain Res 220:214-219

Stewart PA, Wiley MJ (1981) Developing nervous tissue induces formation of blood-brain barrier characteristics in invading endothelial cells: a study using quail-chick transplantation chimeras. Dev Biol 84:183-192

Szumanska G, Palkama A, Lehtosalo JI, Uusitalo H (1984) Adenylate cyclase in the microvessels of the rat brain. Acta Neuropathol (Berl) 62:219-224

Vorbrodt AW, Dobrogowska DH, Lossinsky AS, Wisniewski HM (1986) Ultrastructural localization of lectin receptors on the luminal and abluminal aspects of brain micro-blood vessels. J Histochem Cytochem 34:251-261

Vorbrodt AW, Lossinsky AS, Wisniewski HM, Moretz RC, Iwanowski L (1981) Ultrastructural cytochemical studies of cerebral microvasculature in scrapie infected mice. Acta Neuropathol (Berl) 53:203-211

Vorbrodt AW, Lossinsky AS, Wisniewski HM (1982) Cytochemical localization of ouabain-sensitive, K^+-dependent p-nitrophenylphosphatase (transport ATPase) in the mouse central and peripheral nervous systems. Brain Res 243:225-234

Vorbrodt AW, Lossinsky AS, Wisniewski HM (1984a) Ultrastructural studies of concanavalin a receptors and 5'-nucleotidase localization in normal and injured mouse cerebral microvasculature. Acta Neuropathol (Berl) 63:210-217

Vorbrodt AW, Szumanska G, Dobrogowska DH (1984b) Cytochemical studies of adenylate cyclase in the choroid plexus and brain vessels of rat and mouse. J Histochem Cytochem 32:275-284

Wright EM (1972) Mechanisms of ion transport across the choroid plexus. J Physiol 226:545-571

DISCUSSION

Chaired by Battistin L. and Rapoport S. I.

L. Edvinsson:

How long is the interval of time between the microvessel preparation and the measure of enzyme activites? Is the function of the various enzymes preserved?

L.Betz:

The procedure for microvessel preparation is rather fast, about 3 hours. The technique to further separate the luminal from the abluminal membrane is a typical procedure used for membrane isolation in other types of cells; in relatively sterile conditions many of the enzymatic activities are well preserved.

L. Hertz:

I have two questions: Do you have comments on the fact that K+ uptake into the astrocytes has a rather high Km (~ 4 mM) comparable to that of endothelial cells and contrasting with the values observed in neurons?
In the co-culture of endothelial and glial cells did you remove the glial cells before performing any measurement or did you evaluate also their contribution (by the presence of endfeet in the microvascular preparation) to the expression of endothelial properties?

L. Betz:

The fact that glial cells have the same Km for K^+ transport as the endothelium suggests that both cell types work in concert to remove the excess of K+ from the extracellular fluid in response to neuronal activity. We do not know, however, if the transport capacity changes in relation to the K+ concentration. It is also possible that K+ uptake from glial cells may represent a short-term temporary storage mechanism which, in case of cell overload, could be accompained by a long-term regulation at the endothelial level.
We did not remove glial cells before the transport experiments. This is a very important point to complete our studies. Another control experiment that we have to perform is to investigate the influence of smooth muscle cells versus glial cells in the expression and polarity of the various markers at the endothelial level.

E.MacKenzie:

If you co-culture astrocytes with endothelial cells derived from peripheral vessels such as aorta, would this lead to the expression of the attributes of cerebral endothelial cells?

L. Betz:

This is an important point which has not yet been investigated, as far as I'm aware. I think that one may anticipate that the source of endothelial cells should not be important, but this needs to be verified.

R. Orkand:

The interaction between glial and endothelial cells is also supported by studies showing that K+ channels seem to be aggregated near regions in contact with endothelial cells, suggesting a sort of polarity of the glial cells and that the two cell types cooperate in K+ clearance.
My questions is: what is the origin of the potential across the BBB, is it a transport potential or is it due to ion gradients?

L. Betz:

Early studies show that the potential across the BBB is very sensitive to changes in pH and pCO2. Whether the difference in millivolts (3-5 mV) found between the interstitial fluid and the blood represents totally capillary transport or also results in part from the choroid plexus is not known.

J. Mc Culloch:

What do you think is the significance of the asymmetrical transport of dopamine?

L. Betz:

I think this is a mechanism by which the brain can isolate itself from circulating catecholamines and at the same time participate with the other cells in the removal from the interstitial fluid of the catecholamines released as a results of neuronal activity.
The asymmetric transport in general can also provide to the capillaries a way to utilize a substance and still form a barrier if the brain does not require it.

G. Lucignani:

For some substances which normally do not penetrate the BBB it is possible to favor the transport into the cerebral paranchyma by attaching them to specific carriers. The studies on CBF are based on the assumption that, once the substance or tracer has penetrated the brain, it does not return back. Your presentation reports a different perspective, suggesting to verify the inverse transport from the brain to the blood for all the tracers. Do you have further comments on this point?

L. Betz:

The important point is to distinguish transport from metabolism. So far as Deoxyglucose, this compound is transported across the BBB to the brain where it is metabolized and trapped in the cells. Thus, there is a large concentration gradient between the two interfaces of the microvasculature which, if we ignore the metabolism, can lead us to conclude that there is an active trasport of this substance across the endothelium.
Another example is represented by opiates: codeine is relatively impermeable to the BBB, in contrast to heroin which is more lipophylic. Heroin penetrates the brain and there is metabolized to the less permeable compounds morphine and codeine by a non-specific cholinesterase present also at capillary level. Thus, heroin seems to be trapped in the cells and accumulate in the CNS as in the case of deoxyglucose.

S.I. Rapoport:

Can you further comment on the experiments involving transplantation of endothelial cells in the brain?

L. Betz:

This work was performed by Drs. Stewart and Wiley at the University of Toronto.
Embryonic quail was used as a source of brain tissue, and chick embryo as a recipient. The nuclei of these two species are different, so that it is possible to distinguish the origin of the cells. The piece of brain was transplanted into the abdominal cavity of the chick, allowed to grow up and later studied at the microscopic level using tracers (trypan blue) to evaluate the microvascular development. The capillaries revealed the presence of tight junctions; the nuclear characteristics were the same as host animals. In other words, they had the properties of chick blood vessels grown into the transplanted brain, and not endothelial cells present in the brain and developed to complete capillaries. The marker alkaline phosphatase, which in the quail is not present in the abdominal vessels, was found in the vessels which had invaded the brain.
In another set of experiments, parts of muscle from embryonic chick were transplanted into the brain: in this case the vessels growing into the transplant lost the barrier properties.

ACTIVE AND FACILITATED AMINO ACID TRANSPORT MECHANISMS ACROSS THE BRAIN CAPILLARY ENDOTHELIAL CELLS IN PHYSIOLOGICAL AND PATHOLOGICAL CONDITIONS

C. Cangiano (1), P. Cardelli-Cangiano (2)

(1) Department of Internal Medicine III, University of Rome "La Sapienza", Viale dell'Università 37, 00185 Roma, Italy

(2) Department of Human Biopathology, University of Rome "La Sapienza", Viale Regina Elena 324, 00185 Rome, Italy

ABSTRACT

The purpose of this paper is to clarify how isolated brain microvessels have been utilized to study the transport systems for different categories of amino acids through the blood-brain barrier. The isolation procedures available as well as the motphological aspects and the enzyme activities of isolated brain microvessels are also described. This *in vitro* equivalent of the blood-brain barrier has been used to study the kinetic characteristics and the possible cooperative role of the amino acid transport systems located with a polar distribution in the endothelial cell membranes of brain microvessels. Moreover, using this *in vitro* model it has been possible to observe some hypothesized modifications of the neutral amino acid transport across the blood-brain barrier possibly involved in the pathogenesis of the neurological disturbances observed in chronic diseases like liver cirrhosis and uremia.

KEY-WORDS

Blood-brain barrier, amino acid transport systems, isolated brain microvessels, glutamine, ammonia, hepatic encephalopathy, uremia.

NATO ASI Series, Vol. H 33
Regulatory Mechanisms of Neuron to Vessel Communication in the Brain. Edited by F. Battaini et al.

INTRODUCTION

The existence of a barrier between the blood and the brain was hypothesized in 1913 by Goldmann, but the exact location and function of the blood-brain barrier has been identified rather recently. More than 99% of the barrier between the blood and the brain is constituted by brain capillaries whereas a small number of capillaries perfusing the choroid plexus and other circumventricular organs represent the blood-CSF barrier. Although the blood-CSF plays an important role in the transport of some nutrients brain capillaries forming the blood-brain barrier are widely thought to comprise the major diffusion barrier between blood and brain. Defined by its impermeability to large molecular weight solutes (i.e. protein bound dyes), the blood-brain barrier plays an essential role in determining the rate at which different compounds can reach the CNS. This in turn depends for any particular molecule either on its lipid solubility, or on its specific affinity for a carrier-mediated transport system. The brain capillary endothelial cell membranes, which represent the anatomical site of this barrier, are sealed together by tight junctions; they do not contain fenestrations or transendothelial channels and they have little pycnotic activity, resulting therefore in a structure which resembles many characteristics of the epithelial tissue. The blood-brain barrier can therefore be described as a complete cellular layer with two distinct surfaces, the luminal and the antiluminal wall, seperated by the cytoplasmatic compartment. Furthermore, the cells forming the blood-brain barrier are more than a simple barrier between the blood and the brain; they contain a relatively high number of mithochondria,they are metabolically active and responsible for passive and active transport of nutrients between blood and brain.

The blood-brain transport of nutrients has been studied using two different methodological approaches. The _in vivo_ techniques which are based on arterial or venous injection of radioactive tracers followed by cerebral tissue sampling to measure the total radioactivity by scintillation, (Oldendorf, 1970; 1971) or, as reported more recently, by quantitative antoradiography (Hawkins, 1986). The in vitro studies are instead based upon isolation procedures to obtain purified suspensions of metabolically active brain microvessels. By these procedures the different metabolic activities of isolated brain microvessels can be studied under controlled conditions.

Isolation of brain microvessels

So far several different methods on isolation and purification of brain microvessels have been published (Goldstein, 1975; Mrsulja, 1976; Hjelle, 1978). Actually it is possible to obtain metabolically active suspensions of brain capillaries by choosing between a technique based on the homogenization of the gray matter (Hjelle, 1978; Cangiano, 1983; Cardelli-Cangiano, 1981), or the one using an enzymatic treatment (Williams, 1980; Cardelli-Cangiano, 1987) followed by a centrifugation step either on percoll gradient or in physiological buffer. In the following section some of the isolation procedures are briefly described.

Microvessels can be isolated from the grey matter of fresh bovine brains by three different procedures: Procedure A, by mechanical homogenization, according to the technique of Hjelle et al. (1978); procedure B, by a double collegenase digestion, according to the technique of Williams et al. (1980); procedures C, by combining procedure A with a single collagenase digestion, using either 0.75 (C1) or 7.5 mg (C2) of collagenase/ml, respectively. The following Buffers are used: Buffer I comprised 122 mM NaCl, 25 mM $NaHCO_3$, 10 mM glucose, 3 mM KCl, 1.4 mM $CaCl_2$, 1.2 mM $MgSO_4$, 0.4 mM K_2HPO_4, and 10 mM HEPES, adjusted to pH 7.4 and equilibrated with an atmosphere of 95%O2 + 5%CO2; Buffer II comprised 140 mM NaCl, 10 mM KCl, 0.1 mM CaCl2, 0.2 mM MgCl2, and 10 mM HEPES, pH 7.1. Protein concentration can be determined according to the procedure of Lowry et al. (1951).

In procedure A, the gray matter is homogenized by hand in Buffer I (1:1 wt/vol), using a loose Teflon pestle homogenizer. The homogenate is passed through a nylon sieve with 86-um pores, and the retained material is washed with a spray of ice-cold Buffer I, subjected to a second homogenization step, filtered again through the nylon sieve, and repeatedly washed with a spray of cold Buffer I for removal of the residual nonvascular tissue. The isolated brain microvessels are then collected in a plastic tube and resuspended in Buffer I, under a 95%O2 + 5%CO2 atmosphere until use.

In procedure B, the gray matter (10-11 g) is placed in a Petri dish containing 40 ml of Buffer II and 10 mg of serum albumin. Using a stainless steel blade, the gray matter is minced into fragments (circa 2mm in size), which are then washed by centrifugation at 50 g for 5 min and then resuspended in 10 ml of Buffer II containing 50 mg of bovine serum albumin and 75 mg of crude collagenase. The fragments obtained are incubated for 30 min at 37^{0} C. The resultant slurry is repeatedly aspirated and expelled through a Pasteur pipette so that large chunks of tissue are dispersed. The suspension is centrifuged twice at 1,000 g for 5 min, and the pellet is resuspended in 20 ml of Buffer II containing 5 mg of bovine serum albumin and 75 mg of collagenase/ml. After 20 min at 37^{0} C, the suspension is washed twice by centrifugation at 100 g for 5 min with Buffer I supplemented with 10 mg/ml of albumin, and the final pellet, resuspended in Buffer I, is passed through a 3-cm (inner diameter) column filled with 0.45-mm-diameter glass beads. The microvessels retained on such a column, even after washing with 500 ml of Buffer I, are then mechanically eluted, by gentle agitation with a glass rod, from the glass beads resuspended in 200 ml of Buffer I. The microvessels are finally collected by centrifugation at 100 g for 10min.

In procedures C1 and C2, the brain microvessel suspension obtained by procedure A is centrifuged at 100 g for 5 min and subsequently resuspended, for a final protein concentration of 10 mg/ml, in 10 ml of Buffer II supplemented with bovine serum albumin (5 mg/ml) to which either 7.5 mg (procedure C1) or 75 mg (procedure C2) of crude collagenase has been added. Each suspension is incubated for 20 min at 37^0 C, poured on a nylon sieve with 86-um pores, and rapidly washed with ice-cold Buffer I. The material retained on the sieve is collected in a plastic tube and placed at 0^0 C under a 95%O2 + 5%CO2 atmosphere until use.

Isolated brain microvessel morphology

The morphological aspect of the isolated microvessels obtained by any of the procedure used yelds short, often branched, tubular segments showing an intact luminal and abluminal plasma membrane surrounded by their own basement membrane, exhibiting minimal numbers of pericytes or other cellular contaminants. In our experience the mechanically prepared microvessels (procedure A) appear to consist of branching capillary segments enveloped in their own basement membrane, generally free from nerve cells or from contamination with astrocytic foot processes (Fig.1 a and b). Digestion with collagenase (Fig. 1 c and d) yield microvessels uncontaminated by nerve or glial cells, as in procedure A, and, in addition, show the presence of fenestrations in the external fibrous envelope. This is more evident and ubiquitous when the higher concentration of collagenase is used. Nevertheless, the microvessels maintain their overall tubular shape in all cases; moreover, independent of the isolation procedure used, the microvessels are virtually impermeable to trypan blue (Tennant, 1964). A more accurate measurement of passive

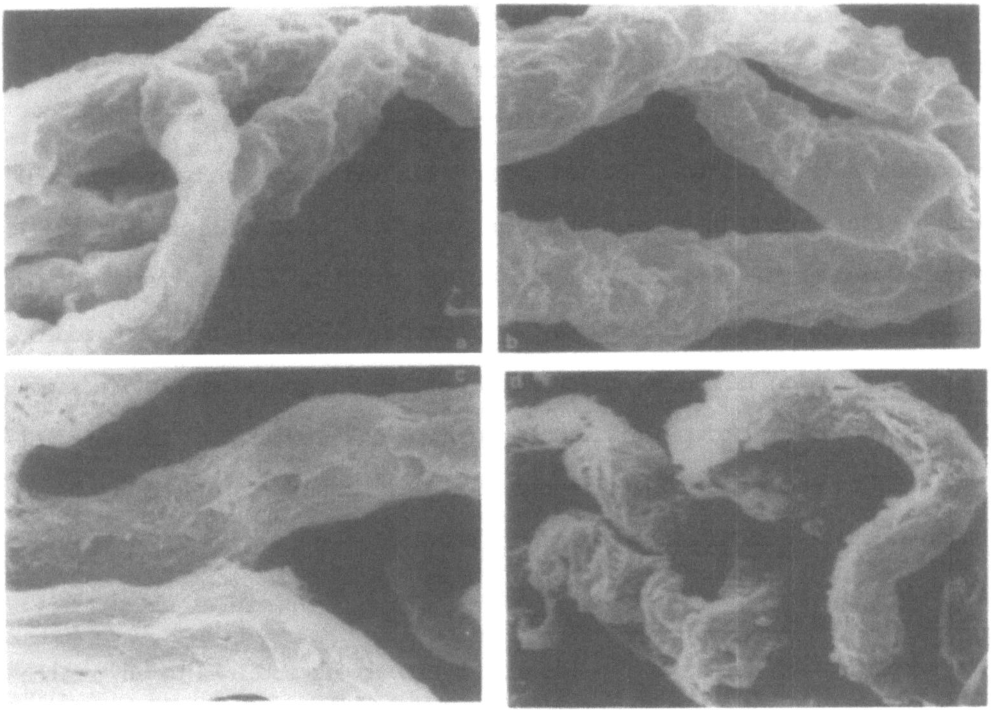

Fig. 1 Morphology of microvessels isolated from bovine brain. Scanning electron micrographs of brain microvessels isolated by procedure A (a and b), B (c), and C2 (d). The small blood vessels appear as short tubular segments free from glial or nerve cell contaminations. The basement membrane is surrounded by a collagenous envelope which is modified by collagenase digestion. Nevertheless the fenestrations seen after the enzymatic treatment on the external envelope do not compromise the capacity of the microvessels to exclude trypan-blue. Taken from Cardelli-Cangiano et al. (1987).

permeability of cell membranes, can also be expressed by the rate of carboxyfluorescein efflux from carboxyfluorescein diacetate preloaded microvessels (Strom, 1973). In the method using collagenase digestion a slight increase in the diffusion rate of the dye from the cells within the first 30 min was observed. Nevertheless, even after extensive collagenase digestion, the efflux of carboxyfluorescein during the first 30 min did not exceed 15% of the total intracellular dye content (Table I). The

TABLE I

CARBOXYFLUORESCEIN EFFLUX FROM ISOLATED BRAIN MICROVESSELS

Collagenase treatment	Carboxyfluorescin efflux (%)	
	5 min	30 min
None	6.08 ± 0.02	13.01 ± 0.10
0.75 mg/ml	5.92 ± 0.01	12.80 ± 0.11
7.50 mg/ml	6.12 ± 0.03	14.01 ± 0.20

After treatment with various concentrations of crude collagenase, the isolated microvessels, suspended in buffer at pH 7.4 at 150 ug of protein/ml, were loaded with carboxyfluorescin diacetate (50 ug/ml). The carboxyfluorescin efflux was followed at 37^0 C for the first 30 min, and values are expressed as percentage exit with respect to the total intracellular content. Data are mean ± SD values from three different experiments.

integrity and viability of the endothelial cells can be tested by measuring their ATP content: a mean value of 470 ± 20 pmol/mg of protein might be considered as the reference range of ATP content of the isolated brain capillaries under physiological conditions.

Isolated brain microvessel enzymatic activities

The isolated microvessels result to be enriched, with respect to the gray matter, in gamma-glutamyl transpeptidase and in alkaline phosphatase. Digestion with collagenase causes a severe decrease of some enzymatic activities such as alkaline phosphatase and gamma-glutamyltranspeptidase, while the activities of glutamine synthetase and 5'- nucleotidase are intimately associated with the microvessels independently of the isolation procedure used (Cardelli-Cangiano, 1987). These purified microvessels are metabolically active, they have the capacity to metabolize nutrients in their endocellular compartment and to transport many substances such as amino acids, carbohydrates, and fatty acids from the blood to the brain compartment and vice versa. Isolated brain microvessels are suitable for the study of the active and facilitated transport system activity of the blood-brain barrier at the luminal and abluminal side giving complementary information on in vivo investigations in which only the luminal membrane can be studied.

So far it can be pointed out that isolated microvessels, with similar morphological aspects, can show differences in their capacity to transport nutrients in dependence of the isolation procedure used, as for example, collagenase treated isolated vessels, are unable to transport MeAIB even though all the other metabolic functions are preserved.

Amino acids transport

Circulating amino acids are transported across the blood-brain barrier by one of the three individual transport systems that have an affinity for either neutral, basic or acidic amino acids (Pardridge, 1979). The study of blood-brain amino acid transport, which includes all processes involved in the movement of amino acid transport into as well as out of the brain, have been confirmed to characterize the movement of the amino acids from the blood to the brain. These studies brought evidence for an apparently homogeneous transport system charactererized by substrate specificity and saturability. However, the net movement of amino acids either into or out of the brain is the sum of influx and efflux. Since less is known about efflux than about influx, a complete account of the role of transport in the regulation of brain amino acid concentrations is not yet possible. It is now generally agreed that the location of the blood-brain amino acid transport system(s) is the plasma membrane of the microvessel endothelial cells. However, the close anatomical association of gial astrocites and microvessels suggests that these two cell types may cooperate in the transport phenomena, of which amino acid transport may be an example (Fig. 2).

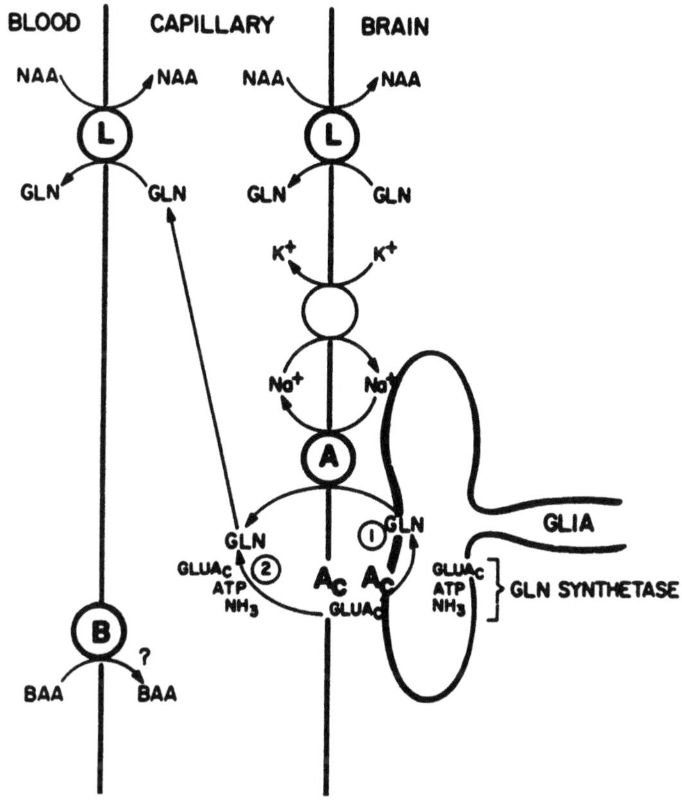

Fig. 2 Proposed distribution of the amino acid transport systems in a hypothetic schematic section of brain capillary endothelial cell.

Neutral amino acids

According to Christensen (1973), the transport of neutral amino acids into cells is mediated by a small number of transport systems. Two of these systems designated L (for leucine preferring) and A (for alanine preferring) are primarely responsible for the transport of neutral amino acid into cells. Two other systems ASC (for alanine, serine, cysteine) and Gly (for glycine) do not mediate aromatic amino acid transport. While there is no data to confirm or disprove the involvement of the ACS system in the blood-brain amino acid transport, the presence of the Gly system is unlikely since glycine uptake by brain could not be demonstrated by Oldendorf (1976). The characteristics of kinetic parameters for amino acid transport have been used by Christensen (1969) as criterion for segregating neutral amino acids into different transport classes. The two already mentioned transport systems, namely the L and the A systems have different transport characteristics. The L-system has preference for the large neutral (essential) amino acids and has a variable Vmax (maximum initial velocity) for the different amino acids. Transport via the L-system is sodium independent and equilibrative: i.e. the L-system does not concentrate amino acids within the cell. The A-system binds the small (nonessential) neutral amino acids and has a relatively constant Vmax among competing amino acids. The A-system is sodium-dependent, concentrative, and insulin-inducible (Pardridge, 1977; Cangiano, 1983b).

The properties of neutral amino acid transport into the brain are consistent with the existence of an active L-system and an absent or negligible A-system (Pardridge, 1975b; Wade, 1975; Toth, 1977). In addition to saturable transport, a nonsaturable or nearly nonsaturable passage of amino acids into brain from blood has been repeatedly demonstrated *in vivo* (Daniel, 1977). Similar results have also been obtained *in vitro* using brain microvessels isolated from the brain of different species of animals (Table II, fig. 3). The presence on the blood side of capillaries of a sole neutral amino acid transport system which mediates only amino acid exchange, but not net uptake, would set restrictions on the mechanism of brain amino acid uptake (Christensen, 1979). Any model of transport would, in fact, have to include as a condition that net transport of neutral amino acids as a group could not occur. While this condition would be consistent with maintaining the

TABLE II

KINETIC PARAMETERS OF AMINO ACID TRANSPORT BY ISOLATED BRAIN MICROVESSELS

	RAT			BOVINE		
	Kd	Km	Vmax	Kd	Km	Vmax
Phenylalanine	0.98	166±6	434±6	0.58	140±9	214±14
Tyrosine	-	-	-	0.68	171±8	278±10
Leucine	0.60	157±10	146±5	0.76	145±10	300±15
Lysine	0.50	101±8	111±6	1.50	103±8	242±13
Me AIB	-	-	-	0.90	280±10	418±13
Glutamic Acid	0.20	280±6	376±10	-	-	-

The initial rates of uptake were subjected to statistical analysis which allowed to evaluate the diffusional component Kd (min -1), the maximal velocity of transport Vmax (pmol.min. -1.mg protein -1), and the carrier affinity Km (uM). The values are expressed as the mean (±S.D.) of at least three different determinations.

the brain in a constant environment in the face of wide fluctuations in the blood content of neutral amino acids, it seems inconsistent with what is known about the brain's capacity to metabolize to a considerable extent the branched-chain amino acids (Chaplin, 1976) and to a lesser extent the aromatic amino acids. To satisfy the conditions of equal exchange, then, the brain must produce some nonessential system L substrates to exchange for those amino acids of which it is a net consumer.

Furthermore, recent studies in which the presence of system A activity was shown in isolated brain microvessel preparations seem to suggest that even though system A-transport cannot occur in the luminal side of capillary cells, such transport may occur instead in the abluminal, or brain side of capillaries (Betz, 1978). The system A activity could act to concentrate in

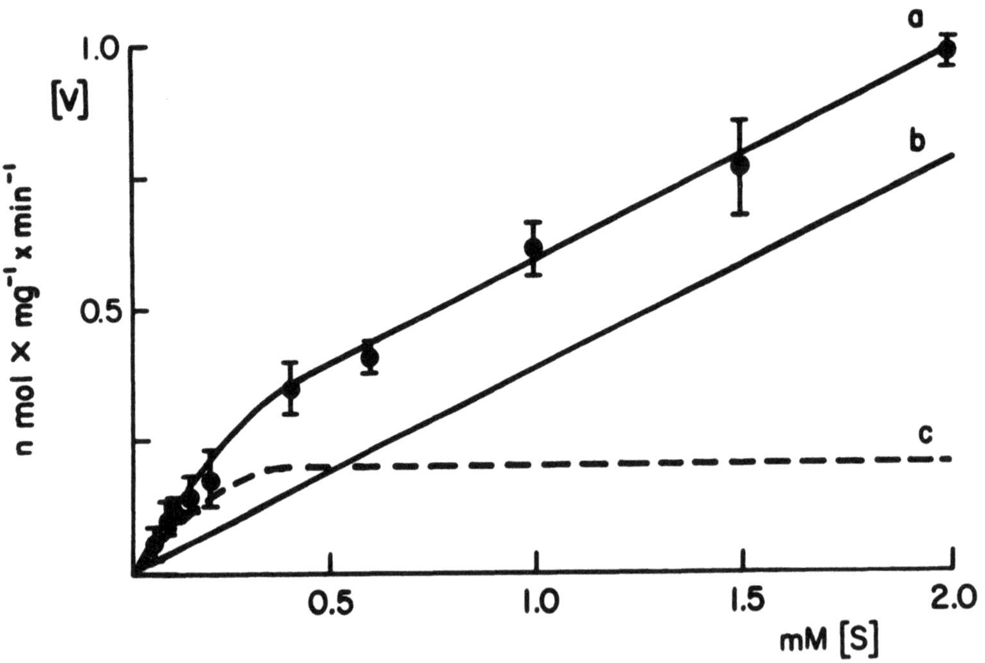

Fig. 3 Initial 2-min uptake of different concentrations of leucine from 0.025 to 2 mM by microvessels isolated from bovine brain. a, represents the experimental curve showing a saturable component superimposed to a non saturable one. b, is the mathematically calculated diffusional component. c, represents the saturable component obtained after the substraction of the diffusional component b.

the capillary cell nonessential amino acids such as alanine or glutamine which have a low affinity for the L-system (Cangiano, 1983a). These amino acids have little competitive effect on the uptake of essential neutral amino acids from the blood since their concentrations there are not high enough to overcome their low affinities for the transport system. If, however, they could be concentrated inside the capillary cells to high levels, they could act to 'drive' uptake of other neutral amino acids from blood in exchange from their own efflux as suggested by Christensen

(1979).

The hypothesis of a cooperative role between the abluminal A-system and the luminal L-system in the regulation of the movement of the neutral amino acids between the blood and the brain seems to be confirmed by a series of experiments in vitro.

If the intracellular concentration of a non essential amino acid such as glutamine is artificially increased by exposing endothelial cells to high external concentrations of such amino acid, the subsequent uptake of the large essential neutral amino acids shows an increase of the maximum initial velocity (Vmax) (Cangiano, 1983a). This phenomenon is characterized by the presence of an overshoot effect in the time course of large neutral amino acid uptake through the transport L-system. (Fig. 4).

When the pre-loading step with glutamine is performed either in the absence of sodium ions or in the presence of a high extracellular concentration of MeAIB, then the subsequent stimulation of the large neutral amino acid uptake is abolished. (Fig. 5). Similar results can be observed if the microvessels have been previously isolated using collagenase treatment which specifically produces a removal of the A-system from endothelial cell membranes (Fig. 6). These data strongly support the idea that at the blood-brain barrier exists a close cooperation between the A-system (at the brain side) and the L-system (at the blood side) in the regulation of blood to brain transport of neutral amino acids.

These observations which may be of importance in the understanding of the fine regulation of the brain omeostasis deserve further studies. Nontheless observations made in different pathological conditions involving the brain concentration of neurotransmitter precursor amino acids, phenylalanine, tyrosine and tryptophan, such as the encephalopathy occurring in the presence of severe hepatic or renal failure, seem to give importance to this hypothesized mechanism.

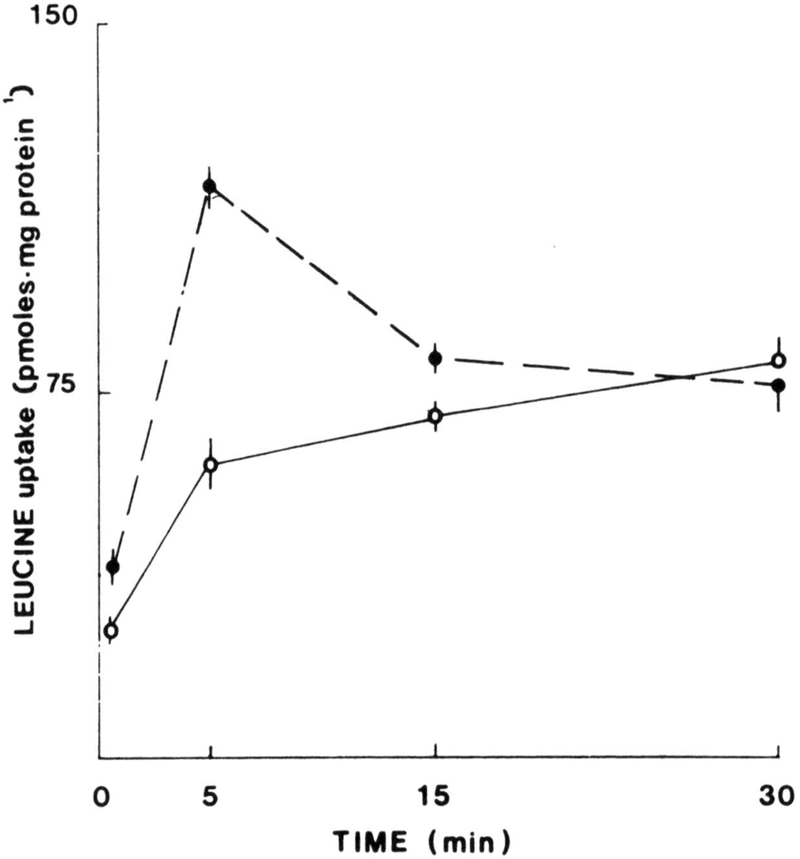

Fig. 4 Leucine uptake by glutamine preloaded microvessels. After preincubation (20 min at 37° C) with 20 mM glutamine the isolated microvessels are washed and resuspended in warmed (37° C) glutamine-free buffer and the subsequent 14C-leucine uptake followed for 30 min. Only in the glutamine preloaded microvessels (●) an overshoot effect is evident. Control microvessels (O) were subjected to the preloading step with a glutamine-free buffer. Data shown are the mean of three different experiments + S.D. Taken from Cangiano et al. (1983).

The high blood-to-brain transport of large neutral amino acids observed in portocaval anastomized rats has been causally involved in the patogenesis of the neurological syndrome which characterizes severe hepatic failure (i.e. hepatic encephalopathy). In 1979 James and co-workers proposed a unified theory for the pathogenesis of hepatic encephalopathy in which hyperammonemia, plasma amino acid imbalance and blood-brain amino acid

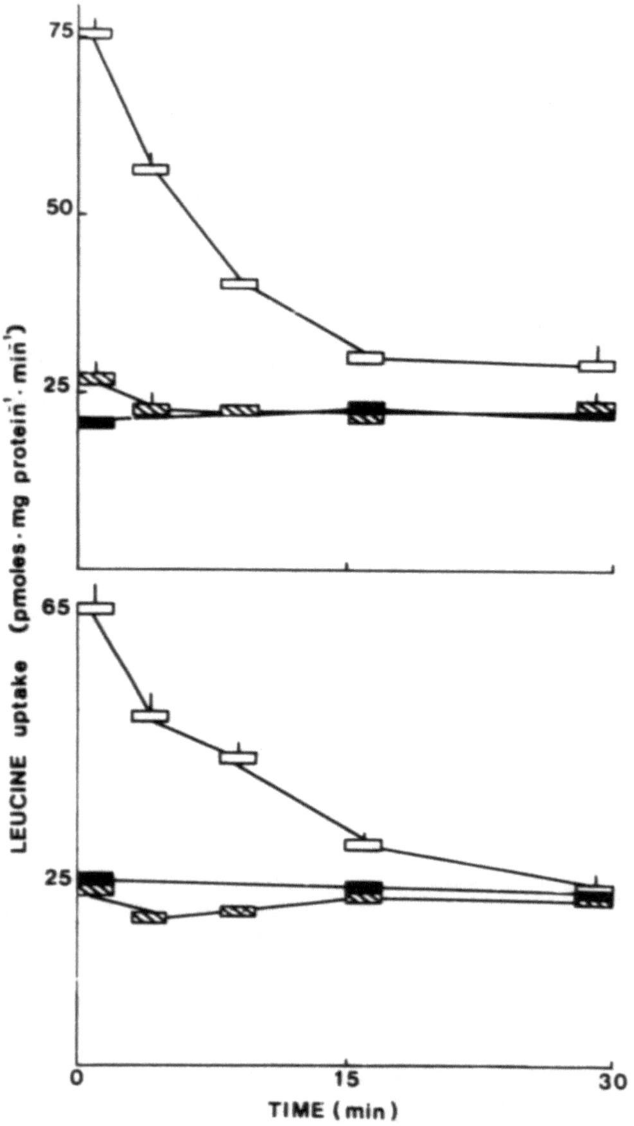

Fig. 5 Leucine transport by isolated microvessels measured at fixed time intervals after preloading with glutamine in different experimental conditions. The upper panel represents the leucine uptake after glutamine preloading either in the presence (open bars) or in the absence (slashed bars) of Na+ ions. The lower panel shows the leucine uptake after glutamine preloading in the presence of BCH (open bars) or MeAIB (slashed bars). Solid bars indicate, in both parts of the figure, the uptake by microvessels preincubated in glutamine-free buffer (control). The time scale on the abscissae indicates the minutes elapsed after resuspension in glutamine buffer and before the two minutes exposure to the 14C-labeled leucine. Data shown are the means +S.D. Taken from Cangiano et al. (1983).

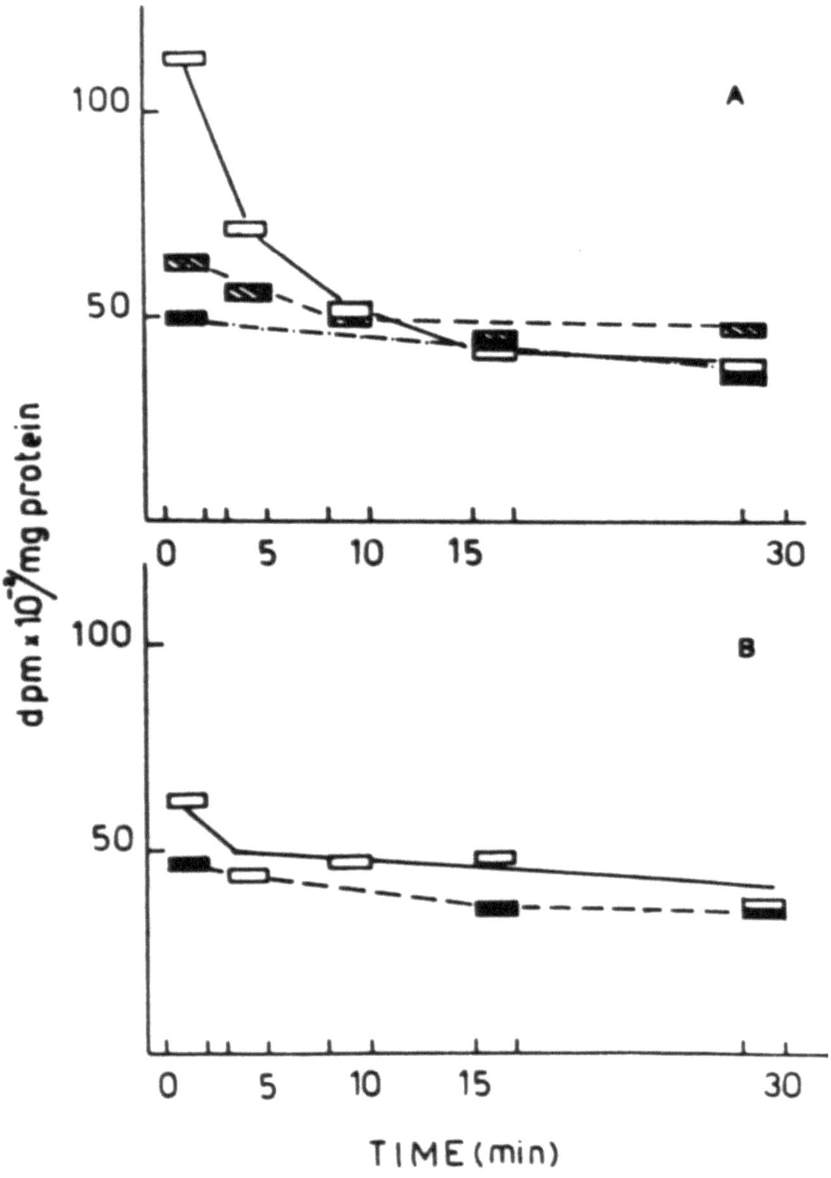

<u>Fig. 6</u> Stimulation by preloading with glutamine (20 mM, 20 min, 37^0 C) of pulsed (2 min) uptake of 14C leucine, measured at fixed intervals after the preloading step. A: microvessels preloaded with glutamine in the presence (open bars) or in the absence (slashed bars) of Na+ ions. B: collagenase treated microvessels (procedure B) after glutamine preloading in Na+ containing medium (open bars). Solid bars indicate uptake of the labeled amino acid by microvessels preincubated in glutamine-free Na+ containing buffer (control). Data are the mean values. Taken from Cardelli-Cangiano et al. (1987).

transport were involved. Acccording to this hypothesis, hyperammonemia, which usually occurs in chronic liver failure and often parallels the presence and the degree of hepatic encephalopathy, represents the "primus movens" of a series of events leading to the onset of hepatic encephalopathy (James, 1979). According to this hypothesis, hyperammonemia increases the influx of ammonia into brain where the excess of ammonia is "detoxified" by its reaction with glutamic acid to yield glutamine, a large neutral amino acid whose efflux from the brain is mediated by the large neutral amino acid carrier system. Glutamine has a low affinity for the neutral amino acid transport system of the blood-brain barrier (Oldendorf, 1976). Under normal, steady-state conditions, the rate of glutamine synthesis in brain is presumably matched by the rate of glutamine efflux from the brain. Increased synthesis of glutamine in the brain raises brain glutamine to a new steady-state concentration at which efflux again matches synthesis. The new efflux rate will result from the Km of glutamine for the neutral amino acid transport system at the concentrations of glutamine and competing neutral amino acids on the anti-luminal side of the endothelial cell capillary membranes. The central feature of this hypothesis is that the carrier L-system mediating the efflux of glutamine from brain also mediates influx into the brain of the other neutral amino acids according to their relative luminal (i.e. blood) concentrations and affinities for the transport system. Increased brain glutamine concentration probably also impairs the efflux of other neutral amino acids from the brain due to competition for available transport sites. Thus deranged plasma amino acid patterns and high brain glutamine act in concert to raise brain neutral amino acid concentrations and to disturb neurotransmitter metabolism in hepatic encephalopathy. This hypothesis, by linking together different metabolic features of hepatic encephalopathy, represents a corner-stone in the evolution of knowledge in this field. In the following years several experimental as well as clinical studies have substantially supported the hypothesis.

In vitro experiences, using isolated brain microvessel preparations, have shown that brain microvessels isolated from rats after portacaval anastomosis take up from the medium significantly larger amounts of neutral amino acids than microvessels from sham-operated controls (Fig. 7). Measurements of the kinetics showed a higher Vmax for phenylalanine and leucine uptake in shunted rats compared with control animals, whereas no differences were found in the Km values. In another series of experiments, glutamine preloaded microvessels, isolated from bovine brain, showed a stimulation of the subsequent uptake of other neutral amino acids by way of the sodium-independent L-system (Table III), while no effect on the uptake of either basic or acidic amino acids was observed (Cangiano, 1983a). Moreover in these experiments it was also demonstrated that the microvessels concentrate glutamine via the A-system which seems to act only at the anti-luminal side of endothelial cells, thus suggesting that the concentrative sodium-dependent A-system and the exchanging sodium-independent L-system can cooperate in the brain uptake of the large neutral amino acids. Using the same in vitro model of the blood-brain barrier, further studies showed that ammonia ions, at a concentration similar to that found in the plasma of patients with hepatic encephalopathy, cause a selective stimulation of the uptake by brain microvessels of large neutral amino acids (Cardelli-Cangiano, 1984). Such a stimulation was essentially due to an increase of the Vmax of the saturable component of the L-system-mediated transport (Table IV). Finally, it was apparently mediated by intracellular formation of glutamine which was then exchanged, through the transport L-system, for large neutral amino acids.

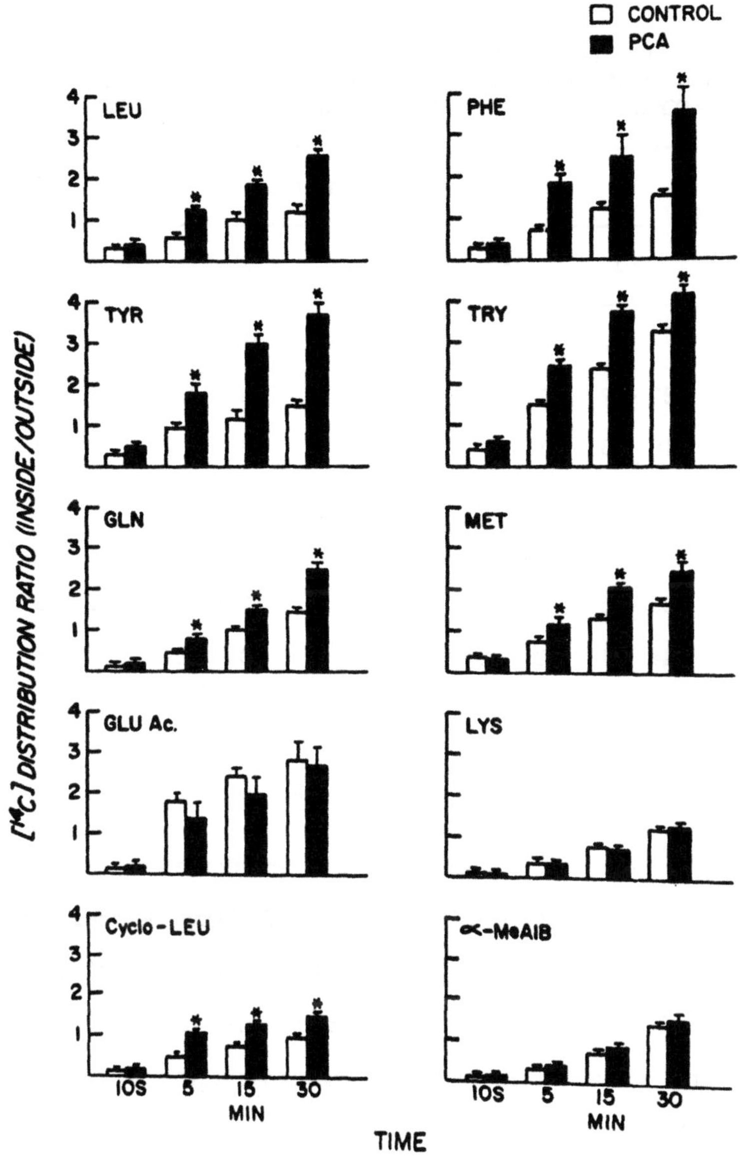

<u>Fig. 7</u> Amino acids transport by isolated brain microvessels obtained from portacaval anastomosis and sham-operated rats. The labeled amino acid uptake was followed at 37^0 C within the first 30 min either in microvessels obtained from portacaval shunted or by sham operated rats. A statistically signifcant increased uptake for the large neutral amino acids only, including cyclo-leucine which are transported by the L-system, is evident. No differences are demonstrable for the small neutral amino acid transport A-system, or for the transport systems specific for basic or acidic amino acid. Data shown are the mean of at least three different experiments + S.D. Taken from Cardelli-Cangiano et al. (1981).

TABLE III

[^{14}C] LEU 2-MIN UPTAKE MEASURED AT FIXED INTERVALS FOLLOWING 20-MIN PRELOADING OF THE MICROVESSELS SUSPENSIONS WITH DIFFERENT SOLUTIONS CONTAINING 20mM EACH OF THE AMINO ACIDS LISTED.

Amino acid	Time (min)				
	0-2	3-5	8-10	15-17	28-30
	pmol.mg protein^{-1}.min^{-1}				
MeAIB	21 ± 2	18 ± 3	19 ± 4	18 ± 4	16 ± 2
BCH	16 ± 3	18 ± 3	12 ± 3	13 ± 1.2	10 ± 2.5
Ala	35 ± 5	27 ± 3.4	24 ± 6	26 ± 3	24 ± 5
Ser	25 ± 2	22 ± 4	23 ± 7	18 ± 5	21 ± 3.4
Cys	25 ± 5	23 ± 2	24 ± 2	17 ± 4.5	22 ± 5
Met	144 ± 10[a]	106 ± 10[a]	66 ± 8[a]	30 ± 10[a]	24 ± 5
None	20 ± 2		23 ± 4		21 ± 3

[a]Significantly different from controls ($p < 0.001$)

Moreover, infusions of ammonia salt in normal rats resulted in a significant increase of the brain levels of large neutral amino acid and glutamine, whereas rats pretreated with MSO, a strong inhibitor of glutamine synthesis, and subsequently infused with ammonia had low brain levels of glutamine and normal brain levels of the other large neutral amino acids (Rigotti, 1985).

This hypothesis is further supported by clinical studies in which plasma and CSF amino acid levels were measured in cirrhotic patients during the evolution of hepatic encephalopathy. By using a multiple regression model, the possible relationship between CSF glutamine and plasma and CSF neutral amino acid levels was investigated (Cangiano, 1984). In the presence of

TABLE IV

KINETIC PARAMETERS OF AMINO ACID TRANSPORT BY ISOLATED BRAIN MICROVESSELS IN THE PRESENCE OR ABSENCE OF TWO AMMONIUM BICARBONATE CONCENTRATIONS.

	NH_4HCO_3	Kd	Vmax	Km
	mM	min.	$pmol.min^{-1}.mg^{-1}$ protein	uM
Leucine		0.76	300 ± 15	145 ± 10
	0.25	0.82	418 ± 10	146 ± 8
	1.00	0.74	599 ± 16	190 ± 12
Phenylalanine		0.58	214 ± 14	140 ± 9
	0.25	0.66	314 ± 11	135 ± 15
Lysine		1.50	242 ± 13	103 ± 8
	0.25	1.40	187 ± 16	95 ± 6

Initial rates of uptake, measured at various amino acid concentrations, were subjected to statistical analysis to obtain the values of the nonsaturable free diffusional component (Kd), of the maximal velocity (Vmax); and of the carrier affinity (Km). The values are expressed as the mean (±S.D.) of the three different determinations.

severe hepatic encephalopathy, CSF phenylalanine levels were closely related either to CSF glutamine or to the ratio between plasma phenylalanine and the sum of competing neutral amino acids (Fig. 8). More recently the amino acid uptake has been studied in brain microvessels isolated from rats with chromic uremia. (Cangiano, 1988). The rational to do so resiedes in a series of clinical and experimental observations in which hepatic and uremic encephalopathies have shown several similarities suggesting the existence of an at least partially common underlying pathogenic mechanism.

Brain microvessels isolated from uremic rats showed indeed a relatively increased uptake of large neutral amino acids with respect to sham-operated animals. This was mainly due to increased exchange transport activity

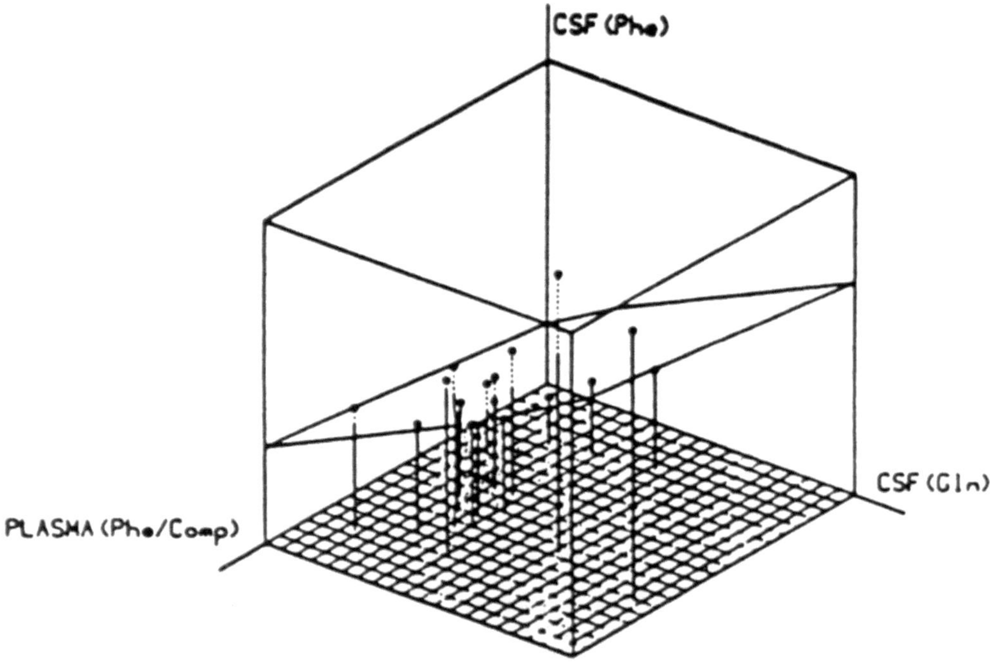

Fig. 8 Multiple regression analysis using phenylalanine as test neutral amino acid in 19 patients in grade 3-4 HE. ($r^2 = 0.7771$, $P < 0.001$). Taken from Cangiano et al. (1984).

(Vmax) of the L-system, rather than to changes in the affinity (Km) of the carrier system. (Fig. 9).

These data together with the significant rise of brain glutamine levels and an increased brain-to-plasma ratio of the sum of large neutral amino acids, are similar to those previously observed in rats with portocaval anastomosis and support the hypothesis that the neurological disorders seen in chronic renal failure may result from functional modifications of the L-system transport activity at the blood-brain barrier, similar to those observed in chronic liver failure.

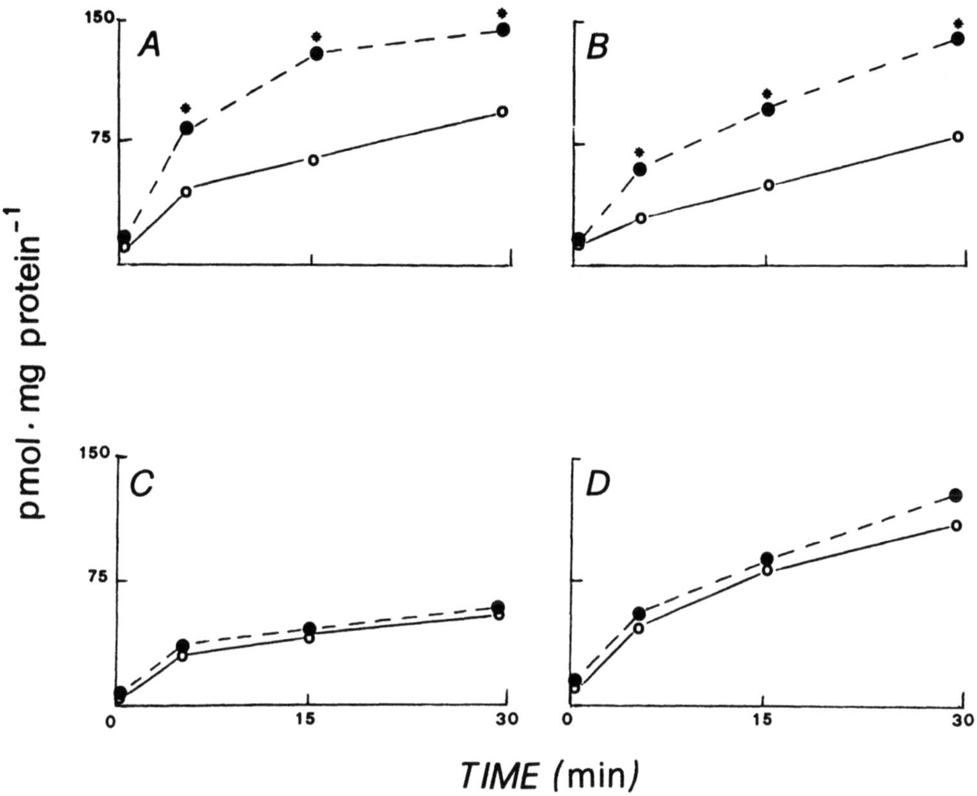

Fig. 9 Labeled amino acids uptake by isolated brain microvessels isolated from rats with experimental chronic renal failure (●) or sham operated rats (O). (A) phenylalanine, (B) leucine, (C) lysine, (D) MeAIB. The asterisks indicate that the difference from the control is statistically significant at a minimum of $p<0.05$. Taken from Cangiano et al. (1988).

Acidic amino acids.

Glutamate and aspartate which are two putative excitatory neurotransmitters, are transported from the luminal side of the brain endothelial cells to the brain compartment by a very low-capacity transport system (Oldendorf, 1976). Because the brain synthesizes the acidic amino acids, the presence of this transport system on the luminal side of the blood-brain barrier is not clear, but it might assume relevance mediating the efflux of acidic amino acids from the brain interstitial space to the bloodstream.

Furthermore, the rate of net efflux of glutamate from brain is nearly 20-fold greater than the rate of its unidirectional influx in vivo (Pardridge, 1979). Using isolated brain microvessels Hutchinson and co-workers (1985) identified a high-affinity acidic amino acid transport system operating on the antiluminal side of the endothelial cell membranes. This system, concentrative and energy dependent, is ouabain-sensitive, suggesting that the energy used to drive glutamate from brain to blood derives from coupling with the sodium gradient across the membrane of brain capillary endothelial cells.

Glutamate has been suspected as one of the toxic agents producing brain edema. (Baethmann, 1980; Chan, 1979; Van Harreveld, 1978). Thus the outward movement of glutamate from brain to blood may be involved in the control of the brain homeostasis.

Human isolated brain microvessels.

The transport of amino acids through the brain capillary endothelial cells has mainly been studied using microvessels isolated from experimental animals. However, rather recently it has been possible to characterize some of the transport systems of the blood-brain barrier also in human beings. Different procedures can be used for isolating human brain microvessels from byoptic and autoptic speciemen (Pardridge, 1988). The human brain capillaries are enriched in those enzymes specific of the mammalian blood-brain barrier such as gamma-glutamyl-transpeptidase, factor VIII antigen and a 46-K protein which might be associated with the blood-brain barrier tight junctions. The human isolated brain microvessels have the capacity to transport the large neutral amino acids via a passive facilitated transport system which has the same characteristics of the L-system of the other vertebrata. Another carrier system working only on the antiluminal surface as an energy dependent transport system specific for the small neutral amino acids has been instead identified with the A-system.

Conclusions

The opportunity of studying *in vitro* the blood-brain barrier represents *per se* a worthwile tool in the understanding of the mechanisms involved in the maintainance of the cerebral homeostasis. Several informations have been obtained and many hypotheses have been postulated or verified by studying the blood to brain transfer of molecules under different experimental conditions in animals. The near present and the next future in this field are represented by the new developing *in vitro* procedures to study the blood-brain barrier transport mechanisms in human beings. This will certainly help clarifying the role played by this sophisticated structure particularly in those pathological states in which an altered equilibrium between the blood and the brain represents a crucial step in the pathogenesis of a deranged central neurotransmission.

Acknowledgments.

The skilled assistance of Barbara Rodermond in preparing the manuscript is gratefully acknowledged.

REFERENCES

Baethmann A, Oettinger W, Rothenfusser W, Kempski O, Unterberg A, Geiger R (1980) Brain edema factors: current states with particular reference to plasma costituents and glutamate. In: Brain edema. Cerros-Navarro J, Ferszt R (eds) Raven press New York:171-195

Betz AL and Goldstein GW (1978) Polarity of the blood-brain barrier: neutral amino acid transport into isolated brain capillaries. Science 202:225-227

Cangiano C, Cardelli-Cangiano P, James JH, Rossi-Fanelli F, Patrini MA, Brackett KA, Strom R, Fischer JE (1983a) Brain microvessels take up large neutral amino acids in exchange for glutamine: cooperative role of Na +-dependent and Na+-independent systems. J Biol Chem 258:8949-8954

Cangiano C, Cardelli-Cangiano P, Cascino A, Patrini MA, Barberini F, Rossi-Fanelli F, Capocaccia L, Strom R (1983b) On the stimulation by insulin of tryptophan transport across the blood-brain barrier. Biochem Int 7: 617-627

Cangiano C, Cascino A, Del Signore S, Fiaccadori F, Ghinelli F, Muscaritoli M, Pelosi G, Sacchini D, Rossi-Fanelli F (1984) Cerebro-spinal fluid amino acid pattern in hepatic encephalopathy. In: Hepatic encephalopathy in chronic liver failure. Capocaccia L, Fischer JE, Rossi-Fanelli F (Eds) Plenum Press New York London:87-94

Cangiano C, Cardelli-Cangiano P, Cascino A, Ceci F, Fiori A, Mulieri M, Muscaritoli M, Strom R, Rossi-Fanelli F (1988) Uptake of amino acids by brain microvessels isolated from rats with experimental chronic renal failure. J Neurochem 51:1675-1681

Cardelli-Cangiano P, Cangiano C, James JH, Jeppson B, Brenner W and Fischer JE (1981) Uptake of amino acids by brain microvessels isolated from rats after portocaval anastomosis. J Neurochem 36:627-632

Cardelli-Cangiano P, Cangiano C, James JH, Ceci F, Fischer JE, Strom R (1984) Effect of ammonia on amino acid uptake by brain microvessels. J Biol Chem 259:5295-5300

Cardelli-Cangiano P, Fiori A, Cangiano C, Barberini F, Allegra P, Peresempio V, Strom R (1987) Isolated brain microvessels as in vitro equivalents of the blood-brain barrier: selective removal by collagenase of the A-system of neutral amino acid transport. J Neurochem 49: 1667-1675

Chan P, Fishman RA, Lee JL, Candelise L (1979) Effects of excitatory neurotransmitter amino acids on swelling of rat brain cortical slices. J neurochem 33: 1309-1315

Chaplin ER, Goldberg AL and Diamond I (1976) Leucine oxidation in brain slices and nerve endings. J Neurochem 26: 701-707

Christensen HN (1969) Some special kinetic problems of transport. Adv Enzymol 321:1-20

Christensen HN (1973) On the development of amino acid transport systems. Fedn Proc 32:19-28

Christensen HN (1979) Developments in amino acid transport, illustrated for the blood brain barrier. Biochem Pharmac 28:1989-1992

Daniel PM, Pratt OE, Spargo E (1977) The mechanism by which glucagon in release of amino acids from muscle and its relevance to fasting. Proc R Soc London B 196: 347-355

Goldmann EE (1913) Vitalfarbung am Zentralnervensystem. Berlin Eimer

Goldstein GW, Wolinsky JS, Csejtey J, Diamond I (1975) Isolation of metabolically active capillaries from rat brain. J Neurochem 25: 715-717

Hawkins RA (1986) Transport of essential nutrients across the blood-brain barrier of individual structures. Federation Proc 45: 2055-2059

Hjelle JT, Baird-Lambert J, Cardinale G, Spector S, Udenfriend S (1978) Isolated microvessels: the blood-brain barrier in vitro. Proc Natl Acad Sci USA 75:4544-4548

Hutchinson HT, Eisenberg HM, Haber B (1985) High-Affinity transport of glutamate in rat brain microvessels. Exp Neurol 85:260-269

James HJ, Ziparo V, Jeppsson B, Fischer JE (1979) Hyperammonemia, plasma amino acid imbalance, and blood-brain amino acid transport: unified theory of portal-systemic encephalopathy. Lancet ii :772-775

Lowry OH, Rosenbrough NJ, Farr AL, Randall RJ (1951) Protein measurement with the Folin phenal reagent. J Biol Chem 193:265-275

Mrsulja BB, Mursulja BJ, Fujimoto T, Klatzo L, Spatz M (1976) Isolation of brain capillaries: a simplified technique. Brain Res 110: 361-365

Oldendorf WH (1970) Measurement of brain uptake of radio-labeled substances using a tritiated water internal standard. Brain Res 24:372-376

Oldendorf WH (1971) Brain uptake of radiolabeled amino acids, amines, and hexose after arterial injection. Am J Physiol 221: 1629-1639

Oldendorf WH, Szabo J (1976) Amino acid assignment to one of the three blood-brain barrier amino acid carriers. Am J Physiol 230:94-98

Pardridge WM and Oldendorf WH (1975) Kinetic analysis of blood-brain barrier transport of amino acids. Biochem Biophys Acta 401:128-132

Pardridge WM (1977) The regulation of amino acid availability in the brain. In: Nutrition in the brain Raven Press New York:141204

Pardridge WM (1979) Regulation of amino acid availability to brain: selective control mechanisms for glutamate In: glutamic acid: Advanced in Biochemistry and Physiology. Filer L J Jr, Garattini S, Kare MR, Reynolds WA, Wurtman RJ (Eds) Raven Press New York :125-137

Pardridge WM (1988) New direction in blood-brain barrier research. In: Fourth colloquium in biological sciences: blood-brain transfer". Strand FL (eds) Annals of the New York Acc Scien: 50-60

Rigotti P, Jonung T, Peters JC, James JH, Fischer JE (1985) Methionine sulphoximine prevents the accumulation of large neutral amino acids in brain of portocaval-shunted rats J Neurochem 44:929-933

Strom R, Scioscia-Santoro A, Crifo C, Bozzi A, Mondovi B, and Rossi-Fanelli A (1963) The biochemical mechanism of selective heat sensitivity of cancer cells-IV. Inhibition of RNA synthesis. Eur J Cancer 9:103-112

Tennant JR (1964) Evaluation of the trypan blue technique for determination of cell viability. Transplantation 2:685-694

Toth J, Lajtha A (1977) Rate of exchange of free amino acids between plasma and brain in mice. Neurochem Res 2:149-160

Van Harreveld A (1978) Two mechansms for spreading depression in the chicken retina. J Neurobiol 9:419-431

Wade LA,Katzman R (1975) Synthetic amino acids and the nature of L-dopa transport at the blood-brain barrier. J Neurochem 25:837-842

Williams SK, Gillis JF, Matthews MA, Wagner RC, Bitensky MW (1980) Isolation and characterization of main endothelial cells: morphology and enzyme activity. J Neurochem 35:374-381

CENTRAL NORADRENERGIC INNERVATION OF CEREBRAL MICROVESSELS: IMPLICATIONS FOR AN AUTONOMIC NERVOUS SYSTEM WITHIN THE BRAIN.

Boyd K. Hartman

Division of Neuroscience Research in Psychiatry, Department of Psychiatry, University of Minnesota Medical School, Minneapolis, MN 55455, USA.

INTRODUCTION

There is now substantial evidence that neuronal circuits within the central nervous system directly influence various aspects of cerebral microvascular function. As research focuses on determining the exact anatomical site of this regulation (muscle layer, endothelial cells or even glial cells) and the precise mechanism of action (opening of tight junctions, pinocytosis, membrane permeability, etc.), it is important not to loose sight of the overall significance of this finding to general brain organization and function. That is, if central neurons are shown to functionally innervate non-neuronal elements then, by definition, this would constitute an autonomic nervous system within the brain.

Classically the brain is thought to be the one organ not regulated by an autonomic nervous system. Functions mediated by such a system in the periphery have traditionally been thought, in brain, to be controlled by autoregulatory mechanisms. Such mechanisms, it has been argued, make an autonomic nervous system unnecessary for the brain. This argument is not compelling since other organs which are

NATO ASI Series, Vol. H 33
Regulatory Mechanisms of Neuron to Vessel Communication in the Brain. Edited by F. Battaini et al.

served by the autonomic system also have autoregulatory mechanisms. Further it ignores the advantages of neurogenic mechanisms in terms of speed, discreteness, and most importantly their ability to anticipate the need for change based on other nervous system information. Thus, neurogenic mechanisms are able to maintain control within tighter tolerances than autoregulatory mechanisms.

If the brain has a neuronal system which is analogous to the peripheral autonomics, then one must also keep in mind that in the periphery microvascular control is only one part of the function of this system. In addition, each organ has special functions regulated by the same neurons. For example, in the heart there is rate of contracion, in the gut there is motility, in the islets there is insulin release, and so on. Such functions can best be described as affecting organ tone, or level of activity. Thus, an autonomic system has at least the following two functions: (1) to maintain organ tone as directed by the need of the organism as a whole and (2) to ensure the arrival of appropriate quantities of energy or other substrates and removal of biproducts of metabolism to maintain homeostatis (usually though not exclusively by control of the vascular system).

The purpose of this presentation is to review evidence supporting the concept that the central neoradrenergic system forms part of an autonomic nervous system within the brain.

ANATOMICAL EVIDENCE FOR A CENTRAL AUTONOMIC SYSTEM

The basic hypothesis is that the central noradrenergic system is an intrabrain autonomic system analogous to the sympathetic nervous system, except that it serves the brain. This hypothesis was originally based on the anatomical and organizational similarities between the two systems: (a) both originate from relatively small groups of neurons (e.g. the sympathetic ganglia vs the locus coeruleus), (b) both have relatively long, highly branched axon systems which innervate widely divergent areas, (c) both systems innervate by way of varicosed fiber plexes, (d) both systems contact both neuronal and non-neuronal structures, and (e) both systems have noradrenaline as their major neurotransmitter. It was our early observation with dopamine-beta-hydroxylase immunohistochemistry that noradrenergic terminal fibers of central origin were often closely associated with cerebral microvessels that most clearly suggested an autonomic like function for these neurons. The existence of a close association between the central noradrenergic system and the microvasculature has been confirmed at the light and electron microscopic level in several laboratories since that time. Furthermore, adrenergic receptors have now been demonstrated on microvessels as well as other non-neuronal elements in brain.

Early objections that specialized synaptic contacts with microvessels were absent (sympathetic neurons do not make synaptic contacts with the vessels they innervate) and that the distances between central noradrenergic terminals and the microvessels were too great to influence microvascular function (the distances are much less than those that frequently occur in the periphery) have been largely discounted. Thus the necessary apparatus is present within the brain for neurogenic regulation of the microvasculature. If such a function could be shown it would be, almost by definition, autonomic in nature.

PHYSIOLOGICAL AND PHARMACOLOGICAL EVIDENCE

A series of experiments were carried out in monkey which showed that chemical (carbamyl choline) stimulation of the main group of central noradrenergic neurons (locus coeruleus) resulted in a rapid alteration in two microvascular functions - cerebral blood flow and capillary permeability to water. Further, the alpha adrenergic antagonist phentolamine when administred intraventricularly reversed the effects of locus coeruleus stimulation. Later experiments showed that electical stimulation of locus could produce similar effects. These experiments were the first to show that the blood brain barrier was not a static constant property to the cerebral capillary, but a variable, alterable by pharmacological and even neurogenic means.

The above experiments strongly supported the fact that microvascular function could be altered by neurogenic mechanisms. It represented less conclusive evidence that the system involved was necessarily the central noradrenergic system. A series of pharmacological experiments were carried out in rat in which this issue was addressed. Tricyclic antidepressants are indirect central adrenergic agonists which act at the terminals by blocking reuptake of the biogenic amines. Administration of these drug mimicked the permeability effects of direct locus coeruleus stimulation. These effects could also be blocked by adrenergic antagonists or by the destruction of the central noradrenergic system with 6 OH-dopamine. On the other hand, destruction of the central serotonergic system, administration of dopamine antagonists, or antihistamines did not block the effect. These two sets of experiments indicated that the central noradrenergic system was capable of altering microvascular function when stimulated either directly or indirectly. Additional experiments were done to show that the vascular effects could be driven by physiological stimuli.

Increased mean arterial blood pressure whether produced by peripheral adrenergic agonists, angiotensin, or by transection of the vagus nerve distal to the carotid sinus all produced an increase in the cerebral extraction fraction of water similar in every way to that observed with central adrenergic stimulation.

Transection of the nerves relaying pressor information from the carotid sinus abolished these effects.

FURTHER ELUCIDATION OF THE CIRCUIT

A corollary of the hypothesis relating the central adrenergic system to the peripheral sympathetic system is that a major set of afferents to the locus coeruleus should be cholinergic (analogous to the cholinergic preganglionic neurons). Immunohistochemistry of choline acetyltransferase has demonstrated a group of cholinergic neurons that send axons into the locus. These neurons are part of a large loosely packed group of cholinergic cells which extend from the locus dorsally through the midbrain in a ventral-rostral direction to the substantia nigra. This pontomesencephalic cholinergic group also gives rise to axons which innervate the midbrain dopaminergic and serotonergic neurons. Its strategic location in the pons and midbrain tegmentum places it in a position to receive information from virtually all ascending and descending pathways. The organizational anatomy of this complex suggests that this group may function as a cholinergic "switchboard" which integrates incoming information and directs and coordinates the out-put to all three biogenic amines, orchestrating their functions according to the needs of the brain and the organism as a whole. The regulation of microvascular tone and permeability would be only a small part of its autonomic function. Its "special" functions would include establishing baseline tone of the various neuronal systems, regulating and modulating hormonal release, and mediating such higher "autonomic" functions as temperature control, the sleep-wake cycle, food intake, and concentration.

CONCLUSION

The evidence compiled not only by our group, but by many others conclusively shows that the brain contains neural circuits which modify the tone of non-neuronal structures. These systems are clearly autonomic in their function. The same systems, though not involved in carrying coded information themselves are involved in modulating the circuits that are involved in information processing. Thus, in the brain, as in the periphery, a clear understanding of function requires the separation of the two basic types of neural circuits. The first are the classical sensory-motor-information processing systems. The second are circuits which maintain organ homeostasis and influence the information processing systems in much the same way as a gain or sensitivity control effects a signal without actually carrying the signal itself.

These later circuits are best conceptualized as a central autonomic system.

DISCUSSION

Chaired by G.C. Palmer.

W. Meier-Ruge:

I have missed the differentiation between capillaries and larger vessels containing smooth muscle cells as far as the regulatory systems. Do you think that the stimulation of the locus coeruleus can influence the permeability of the capillaries independently on the peripheral sympathetic system?

B. Hartman:

I believe that the effect of central noradrenegic pathways on capillary permeability is independent on the sympathetic system. An evidence is represented by the fact that the experiments were carried out on sympathectomized animals.
In terms of flow, we have a number of cases in which no relation exists between changes in flow and water extraction. There may be the possibility that, within the central noradrenegic system, cerebral blood flow and capillary permeability are influenced by two different groups of neurons, both acting independently on the peripheral sympathetic innervation.
We do not know the mechanism by which blood flow is affected: one hypothesis is that the changes are mediated by alterations of glial volume. In the case of arterioles, the smooth muscle cells may have a direct influence by the central noradrenergic innervation.

W. Meier-Ruge:

In reference to the permeability, what do you think is increased after locus coerulens stimulation: water passage, K+ entry?

B. Hartman:

I realize that it is important to know and understand the mechanisms that result in a total increase of capillary permeability. However, what I would like to point out is that water permeability is simply a marker, a tool to study some very basic membrane functions and their dynamic modifications in response to neuronal manipulations. The changes in capillary permeability we have observed probably do not account for major fluxes of water, since the greatest contribution comes from differences in hydrostatic pressure. Nevertheless, although I do not know the physiological meaning of the changes in water permeability, I do believe that this parameter reveals a direct influence of the neurons on capillary function. Returning to your question, the mechanisms at the basis of this phenomenon do not seem to be based on ion passages.

G. Lucignani:

Are the changes in water extraction due to modifications in vascular permeability or to an increase in membrane surface area? This is a very critical issue since it implies methodological problems in the measure of CBF.

B. Hartman:

For our experiments we have followed two methods. The first (Raichle technique) is based on the use of a single isotope (15O-H2O) and an external detector: what we measure is the amount of radiolabelled water under the probe (mainly cortical) as a function of time. The units are expressed as ml per 100 g of perfused tissue per min.; this means that the number of capillaries does not influence the PS product as long as the surface area per unit capillary is not modified. Thus, the amount of water which permeates the brain over time depends on the total capillary surface (or number) and not on capillary recruitment. If the number of capillaries increase also PS product increases (since the diffusion of substances depends also on blood flow, the value ml/100 g tissue is divided per the wash out, which is also related to the flow). Thus, PS purely reflects the diffusion of the water into the brain.
An interesting confirmation of this concept is that in conditions of stroke the PS product is not altered, despite of the tremendous clustering of the capillaris (whereas blood flow is changed because the area is not perfused).
What we are not able to distinguish, however, is whether an elevation of water PS product reflects a capillary surface dilatation or on increase in microvessel number.
The extraction fraction method (Oldendorf technique) is affected by the flow. However, there are many reasons to believe that the observed values of water extraction represent changes in endothelial permeability. In fact, an increase in blood flow could explain a modification in extraction values, but in the absence of flow changes (due to the autoregulatory mechanisms in response to an elevation in blood pressure) these values must reflect the permeability capacity of the microvasculature.

V.J. Cunningham:

If the substance is not completely extracted,the changes in capillary number may alter the PS. Furthermore, we do not know if the extraction capacity is the same in all the capillary wall.

S. Rapoport:

In case of an increase in blood pressure, it takes about 1-2

min before the onset of the autoregulatory mechanisms (Hirsch and Korner, 1964; Rapoport, 1976); during this short period of time H_2O would flow into the brain, according to the Starling hypothesis, whereas after the establishment of autoregulation, there would be a transient outflux.
The movement of water, therefore, depends on both systemic and hydrostatic pressures. Not only would the H_2O extraction fraction be related to blood flow, but also to time and water fluxes occurring before and after autoregulation.

B. Hartman:

The uptake of ^{3}H-H_2O is different from the flux of H_2O. Prior the vascular injection of radiolabelled water there is no radioactivity in the brain; after the injection what we measure is simply a radioactive diffusion gradient. In this perspective and for our purposes it is irrelevant what the direction of the fluxes is.
I want to underline that the question is whether or not there is a neurogenic influence on capillary diffusion. If we lesion the vagus nerve the effect that we have observed is lost, while if the nervous system is intact we do see the effect: this clearly suggests a neurogenic mediation of the phenomenon. The purpose and the mechanisms at the basis of the changes in H_2O diffusion are not clear and constitute a separate issue of investigation.

- Hirsch H, Korner K, 1964. Pfleugers Arch Ges Physiol, 280:316- 325.
- Rapoport SI, 1976. Blood brain barrier in Physiology and Medicine, Raven Press, New York.

Innervation of the cerebral vasculature and its putative role in the coupling of flow to metabolism.

Hans C. Lou, M.D., Ph.D., The John F. Kennedy Institute, Dpt. of Neuropediatrics, Gl. Landevej 7, DK-2600 Glostrup, Denmark

Lars Edvinsson, M.D., Ph.D., Dpt. of Internal Medicine, University Hospitals Lund, S-22185 Lund, Sweden

Abstract

A tight coupling exists between brain function and cerebral perfusion in most situations. The Roy and Sherrington hypothesis has been widely accepted to account for the phenomenon: Increased neuronal metabolic activity will give rise to the accumulation of vasoactive catabolites, which decrease vascular resistance and thereby increase blood flow until normal homeostasis is reestablished. However, the hypothesis does not account for the disproportionate increase in flow that occurs in a number of circumstances. There are additional difficulties in reconciling more recent experimental data with the Roy and Sherrington hypothesis. In this review we direct attention toward the rich perivascular nerve supply to all parts of the cerebral circulation as possibly being an alternative control system allowing for rapid parallel changes in flow and neuronal activity. In particular the intrinsic polysynaptic cholinergic vasodilator system may serve as such a system.

Key words: brain metabolism, neurotransmitter, vasodilatation, cerebral blood flow.

NATO ASI Series, Vol. H 33
Regulatory Mechanisms of Neuron to Vessel Communication in the Brain. Edited by F. Battaini et al.

The coupling of neuronal activity and blood flow in normal brain has been increasingly evident as sophisticated techniques allowing a high-resolution spatial analysis of neuronal activity and flow have been developed. A very close correlation has been shown between local cerebral blood flow (lCBF), as measured with ^{14}C-iodoantipyrine, and local cerebral metabolic rate for glucose (lCMRGlc), as measured with ^{14}C-2-deoxyglucose. Greenberg and colleagues (1979) have refined the spatial topical characterization of the metabolism/flow association by the discrete vibrissal stimulation model in combination with the above-mentioned autoradiographic methods. The results showed almost identical spatial distribution of activation of flow and metabolism, limited to cylindrically shaped parts of the brain parenchyma with diameters of 375 to 500 μm (Greenberg et al. 1979). Silver (1978) has shown that localblood flow increases as soon as approximately 1 second after neuronal excitation, and that the increase is limited to 250 μm from the site of the increased neuronal activity. These results are particularly important because they indicate that flow can be adjusted very rapidly at the microvascular level according to metabolic demands (i.e., functional subunits or columns).

The nature of the coupling mechanism(s) has been the subject of much discussion. Originally Roy and Sherrington (1890) presented the hypothesis that metabolic products in the extracellular fluid acted as regulators of vascular tone. Many candidates for such a mediater have been proposed: O_2, Co_2, H^+, adenosine and "glycolytic intermediates". A simple homeostatic mechanism is, however, unable to explain a number of characteristics for the coupling mechanism: Several investigators have found that flow increases more than metabolism during activation (Silver 1978, Nilsson et al 1978, Fox and Raichle 1986). The speed of flow increase after stimulation (Silver 1978) also makes it difficult to assume a homeostatic mechanism based on the accumulation of metabolites. Of more interest is the elegantly presented hypothesis of K^+ siphonising from nerve depolarisation sites through astrocytes and endfeet to the microvessels. This hypothesis offers one acceptable mechanism for the coupling of flow to metabolism (Paulson and Newman 1987).

A number of phenomena do, however, illustrate that this model probably is complemented with neural coupling between flow and metabolism. This chapter will present some of the mostly, but not exclusively, circumstantial evidence, for the role of neuronal coupling.

It is evident that a number of neurotransmitters and neurotransmitter candidates in brain are vasoactive and capable of adjusting the tone of brain vessels (Table). A dense plexus of nerve fibers in the walls of cerebral vessels, forming a mini brain, has been documented. Therefore, neurons may activate arterioles as well as target neurons and could, in principle, constitute the coupling mechanism between metabolism and flow. To evaluate the possibility of neuronal adjustment of flow to metabolism, it is above all necessary to consider the vasomotor activity of some of these neurotransmitters, as well as some pertinent anatomical data.

Effects of Neurotransmitter on flow

ACETYLCHOLINE. In studies of coupling process between neural activity and blood flow in some physiological conditions (e.g., sleep-wake cycles (Pearce et al 1981), it has been observed that spontaneous shifts in electroencephalogram (EEG) from high to low voltage states are accompanied by increases in CBF and $CMRO_2$ of 26 + 7% and 22 + 4%, respectively (measured by a modification of the Kety-Schmidt technique). A detectable increase in CBF was noted within 30 seconds of the change in EEG even with this rather crude experimental approach (Pearce et al 1981). Of particular importance is that administration of the cholinergic blocker, scopolamine, abolished the increase in CBF without affecting the increase in $CMRO_2$ during the transition of the EEG from high to low voltage. As scopolamine has no known direct vascular effect of its own but may penetrate the blood-brain barrier, the result has been taken as evidence for the idea that cholinergic mechanisms mediate increases in CBF that accompany an increase in $CMRO_2$ and a decrease in EEG amplitude.

DOPAMINE. The presence of dopamine receptors D_1 and D_2) both in cerebral neurons and in the blood vessel walls suggests that dopaminergic neurons exert a direct influence on target neurons as well as on flow. In recent studies, administration of the D_1 receptor agonist SKF 38393 failed to change the overall CBF/glucose ratio (1.75 ml/μmole versus 1.77 ml/μmole), while the D_2 agonist LY 171555 provoked a marked increase in both blood flow and cerebral glucose use. The ratio of flow to glucose utilization increased slightly (from 1.75 ml/μmole to 1.97 ml/μmole)(Sharkey and McCulloch 1985a). The change was described as insignificant. However, in the ventral striatum, CBF was massively elevated (by more than 100%) whereas glucose use was only minimally altered (Sharkey and McCulloch 1985b). This could be due to a different temporal resolution of the two techniques, but the discrepancies between the findings in different regions strongly suggest a selective influence of dopamine on flow in the ventral striatum, a region richly innervated by dopaminergic neurons.

NORADRENALINE. Noradrenaline, as well as other adrenergic agents, may induce constrictions of isolated cerebral arteries in a concentration-dependent manner. This vasoconstrictor effect is antagonized by alpha-adrenoceptor antagonists (Edvinsson 1982, Edvinsson and Owman 1974). Conversely, preconstricted cerebral arteries are dilated by catecholamines through an interaction with beta-adrenoceptors. In vivo, however, the situation is more controversial. Some studies have shown an increase in cerebral perfusion after intraarterial or intravenous administration of catecholamines (McKenzie et al 1976, Purves and James 1969), others have shown no change in flow (Olesen 1972), and the most frequent result is a slight decrease in perfusion (Lluch et al 1973, Oberförster et al 1973). The diversity of the findings may be related to 1: variable numbers of alpha- and beta-adrenoceptors in different cerebral vascular regions and species, and 2: variable degrees of blood-brain barrier permeability of the catecholamines in different experimental models. For instance, Dahlgren and associates (1980) reported that intravenous infusion of adrenaline had no effect on $CMRO_2$ or CBF if the systemic pressor effect of epinephrine was prevented by a servo-controlled device. But, if the systemic pressor effect, which presumably opens the blood-brain barrier, was allowed to take place, $CMRO_2$ increased 1.6 to 2 fold, and cerebral blood flow 3 to 4 fold. A similar relationship was found with norepinephrine (MacKenzie et al 1976). Measurements of cerebral tissue metabolites failed to show alterations in cerebral cortical glycogen, lactate, adenosine triphosphate, adenosine disphosphate, AMP, or cyclic AMP concentrations (Berntman et al 1978). It can be concluded from these studies that systemic administration of catecholamines has only a modest direct influence on the net cerebral circulation via alpha- and beta-adrenoceptors in the vessel walls (Edvinsson 1982). However, if the blood-brain barrier is circumvented, a marked stimulatory effect on $CMRO_2$ is noted and is accompanied by an even larger increase in CBF, without change in the concentration of energy metabolites. The increase in CBF, which under these conditions is out of proportion to the increase in $CMRO_2$, cannot be totally explained as a direct vascular effect or as a result of increased $CMRO_2$. Stimulation of neurogenic control mechanisms via dilatory beta-adrenoceptors may be assumed to be at least partly responsible for the flow increase.

SEROTONIN. Like catecholamines, serotonin has both vasoconstrictor and vasodilator effects on cerebral vasculature (Harper and Mackenzie 1977a, Harper and MacKenzie 1977b). Serotonin applied directly to larger isolated cerebral vessels causes constriction (Edvinsson et al 1978, Toda and Fujita 1973), which may be important in adjusting autoregulatory range (Lou and Edvinsson 1986). The vasoconstriction can be blocked by classical serotonergic blockers such as methysergide, pizotifen, and cyproheptadine Edvinsson et al 1978).

After blockade of the vasoconstrictor effect, and provided that the pial arteries have been given an active tone, serotonin may produce concentration-dependent relaxation in vitro. This dilation can be inhibited by the beta-adrenoceptor blocker propranolol (Edvinsson et al 1978, Mena and Vidrio 1976). In the intact vascular bed in situ, perivascular microapplication of serotonin constricts the larger cerebral arteries (diameter greater than 100 μm), and induces dilation of smaller arteriolar resistance vessels (Harper and MacKenzie 1977). Since the dilator response to serotonin in situ was inhibited by propranolol, local serotonin release may be capable of decreasing arteriolar tone by interaction with betaadrenoceptors and causing local increases in flow.

GAMMA-AMINOBUTYRIC ACID. Gamma-aminobutyric acid (GABA) is the most widely distributed neurotransmitter in the human brain and has a predominantly inhibitory action of central neuronal pathways. Two different types of GABA receptors have been demonstrated to date: $GABA_A$, which has a postsynaptic, mainly vasodilatory action (Edvinsson and Krause 1979, Fujiwara et al 1975), although a biphasic response has been reported (Suzuki et al 1984); and a $GABA_B$ receptor, which has a presynaptic inhibitory effect on sympathetic nerve terminals. The global effects of GABA on CBF and metabolism are not known in detail. However, the effects of the systemic administration of muscimol, a GABA agonist that can penetrate the blood-brain barrier more freely than GABA itself, have been investigated. The results have been contradictory. Edvinsson and associates (1980) and Alborch and associates (1984) have demonstrated an increase in CBF following systemic administration, whereas Kelly and McCulloch (1983) observed a proportional decrease in local CBF and glucose utilization. The interpretation of these studies is hampered by the difficulty in separating a possible direct effect of the circulating agent on the brain vessel per se from a neuronally mediated effect. The latter may be either secondary to an effect on neuronal energy metabolism (the Roy and Sherrington hypothesis) or due to activation of neuro-nal control of the circulation. In humans, Roland and Friberg (1983) observed that intracarotid administration of the GABA agonist tetrahydroisoxazolopyridinol (THIP) decreased local rCBF in all regions examined. However, it was noted that local CBF decreased less in regions that had been activated by the performance of certain tasks.

A closely related, naturally occuring metabolite, gamma-hydroxybutyrate, which is a potent inhibitor of cerebral metabolism, has been the subject of some investigation (Snead 1977). Kuschinsky and colleagues 1985), using the 2-deoxyglucose and iodoantipyrine methods, reported global reductions (by 51%) in glucose utilization, but no change in lCBF. The authors interpreted this discrepancy as due to the observed increase in arterial pCO_2 (about 8 mm Hg) following the administration of gamma-hydroxybutyrate. It is, however, apparent from their local values of glucose utilization (lCGU) and blood flow (lCBF) that a pronounced heterogeneity exists: In cerebellar cortex the lCGU decreased from 61 + 8 to 31 + 12 ($p<0.01$), whereas lCBF doubled (from 102 + 22 to 197 + 12, $p<0.01$). A similar pattern was seen in other cerebellar structures, as well as in the visual cortex. On the other hand, lCGU and lCBF were both significantly decreased in structures such as substantia nigra, pons, and hypothalamus. This heterogeneity cannot be explained by changes in arterial pCO_2. On one hand, the results strongly contradict the notion that the metabolic rate of neurons is the sole determinant of flow. On the other hand, if gamma-hydroxybutyrate (or GABA) exerts a direct influence on the vasculature and flow, the heterogeneity could be due to differences in the density of gamma-hydroxybutyrate receptors in various cerebral structures and vessels.

Innervation of Cerebral Vessels and Coupling

During the 1930s a number of experiments were performed on the petrosal nerve and the geniculate ganglion showing that stimulation of either structure resulted in vasodilatation of cortex pial vessels. The fact that this response could be attenuated by atropine indicated that cholinergics could exert a direct influence on cerebral circulation (Chorobski and Penfield, 1932). The cerebral vasculature is supplied with acetylcholinesterase-containing and vasoactive intestinal peptide-containing (VIP) fibers (Edvinsson 1985, Edvinsson et al 1972). Although much research has been done on these fibers during the last two decades, their origin is still debated. Three sources have been proposed: 1: The fibers originate in microganglia located along several of the large cranial arteries (Gibbins et al 1984); 2: Cerebrovascular acetylcholinesterase-containing and VIP fibers originate in the sphenopalatine ganglion (Hara et al 1985); and 3: Cerebrovascular cholinergic and VIP fibers originate in cerebral cortex and are distributed to the perivascular fiber network (Eckenstein and Baughman 1984).

A rich supply of sympathetic nerves has been demonstrated in cerebral vessels of most species, including humans; these fibers originate in the superior cervical ganglion (Edvinsson 1982). The nerve fibers are located in the adventitia and at the outer border of the medial layer. The innervation extends down to the smalles arterioles, or possibly even to precapillaries (15 μm), and are regularly seen in cerebral veins. Sympathetic fibers contain, besides noradrenaline, the neurotransmitter candidate neuropeptide Y (NPY) (Edvinsson 1985). The cell bodies are primarily found in the superior cervical ganglion, but some perivascular fibers emanate from the stellate ganglion (nerves to the vertebral artery and the distal part of the basilar artery). A small contribution to the internal carotid artery nerve plexus may occur via the vidian nerve.

Perivascular meningeal projections have been demonstrated from the trigeminal ganglion (Mayberg et al 1981) and in 1983 several groups were able to show with tracer techniques that horseradish peroxidase applied perivascularly around the middle cerebral artery was transported to the ophthalmic (Keller et al 1985, Liu-Chen et al 1984) branch of the trigeminal ganglion. The transport material was mainly located ipsilaterally. A few molecules of horseradish peroxidase were, however, seen contralaterally. The neurotransmitter candidate in these fibers was initially thought to be substance P; however, recent experiments have shown that substance P is co-localized with neurokinin A and calcitonin gene-related peptide (Edvinsson 1985). Thus, this system may involve at least three neuropeptides.

In addition to sympathetic fibers that have a peripheral origin there is ultrastructural evidence for central adrenergic innervation of cerebral vasculature (Swanson et al 1977)(Table). Raichle and associates (1976) have shown in functional studies that chemical and electrical stimulation of the major cell group of the central adrenergic system, the locus coerulens, results in a pronounced effect on the intracerebral microvasculature of sympathetic ganglionectomized monkeys that changes the capillary permeability and sometimes the CBF. It is still an unresolved question whether there is a close functional and/or anatomical relationship between this intrinsic innervation and the extrinsic adrenergic innervations from the superior cervical ganglion.

The serotonergic innervation of cerebral vessels has also been investigated as a possible coupler. Serotonergic fibers emanating from brainstem raphe nuclei innervate pial arteries and arterioles (Edvinsson et al 1983), and cerebral microvessles (i.e., capillaries and venules (Reinhard et al 1979). These findings have, for the first time, provided us with an anatomical substrate for a possible serotonergic coupling of normal function and flow. The evidence is based to a large extent on neurochemical findings, since 1: Serotonergic fibers are fairly scarce and weakly immunofluorescent (Edvinsson et al. 1983), and 2: Sympathetic fibers may accumulate serotonin from circulating platelets.

Electrical Stimulation of Intact Brain and Coupling

For many years it has been known that the brainstem may influence global cerebral blood flow (Reis 1979). It has been assumed that such flow increases were the result of an increase in local brain metabolism. However, as early as 1976 McKee and associates demonstrated cerebral vasodilatation following electrical stimulation of the cerebellum, independent of any change in cerebral metabolism.

In a more recent and well-controlled study, Iadecola and associates (1983) demonstrated that electrical stimulation of the dorsal medullary reticular formation caused an increase in rCBF. This increase could not be attributed to an effect secondary to systemic factors, such as blood gases or arterial blood pressure, or to a global metabolic stimulation, since stimulation at adjacent sites in the brainstem failed to alter CBF. The demonstrated increase in flow was not coupled to EEG activation. Furthermore, transection experiments showed that the flow increase was the result of excitation of pathways intrinsic to the brain. There are indications from autoradiographic studies that stimulation of the dorsal medullary reticular formation is closely coupled to metabolism in certain cortical blood flow without any simultaneous change in cortical glucose utilization. This mechanism is polysynaptic and is probably dependend upon intrinsic cortical cholinergic neurons as the dilator response may be partly abolished by local application of atropine (Arneric et al 1987). Thus, an intrinsic cholinergic neural system exists that is capable of influencing flow. As the system is multisynaptic, it may have several functions, for instance: 1) A central, brain stem controlled, vasomotor regulator operating in different sleep-wake states and affecting the entire brain, and 2) A control mechanism ensuring coupling between function and flow at the local level.

Table

Cerebrovascular Neurotransmitter Candidates

Neurotransmitter	Origin	Sites of action	Cerebrovascular response (and receptor types)
Dopamine	Substantia nigra	A M	Dilation
Norepinephrine	Locus ceruleus Sympathetic ganglia	A M A V	Controversial Constriction (alpha-adrenoceptors) Dilation (beta-adrenoceptors)
Serotonin	Raphe nuclei	A M	Constriction (large vessels) Dilation (microvessels)
Acetylcholine	Cortex	A	Dilation
	Sphenopalatine ganglion VIIth cranial nerve	A V	
Neuropeptide Y	Sympathetic ganglia	A V	Constriction
Vasoactive intestinal peptide	Cortex sphenopalatine-ganglion	A V	Dilation
Substance P/ neurokinin	Trigeminal ganglion	A M V	Dilation
Calcitonin gene-related peptide	Trigeminal ganglion	A V	Dilation

A: Arteries; M: Microvessels; V: Veins.

References

Alborch E, Torregrosa G, Terrasa J, Estrada C (1984): GABA-receptors mediate cerebral vasodilatation in the unanesthetized goat. Brain Res 321:103-110

Arneric SP et al (1987): Local cholinergic mechanisms participate in the increase in cortical cerebral blood flow elicited by electrical stimulation of the fastigial nucleus in rat. Brain Res 411:212-225

Berntman L, Dahlgren N, Siesjö BK (1978): Influence of intravenously administered catecholamines on cerebral oxygen consumption and blood flow in the rat. Acta Physiol Scand 104:101-108

Chorobski J, Penfield W (1932): Cerebral vasodilator nerves and their pathway from medulla oblongata. With observations on the pial and intracerebral vascular plexus. Arch Neurol Psychiatry 28:1257-1289

Dahlgren N, Rosén J, Sakabe T, Siesjö BK (1980): Cerebral functional, metabolic and circulatory effects of intravenous infusion of adrenaline in the rat. Brain Res 184:143-152

Eckenstein F, Baughman RW (1984): Two types of cholinergic innervation in cortex, one co-localized with vasoactive intestinal polypeptide. Nature 309:153-154

Edvinsson L (1985): Functional role of perivascular peptides in the control of cerebral circulation. Trends Neurosci 8:126-131

Edvinsson L (1982): Sympathetic control of cerebral circulation. Trends Neurosci 5:425-428

Edvinsson L, Degueurce A, Duverger DD et al (1983): Central serotonergic nerves project to the pial vessels of the brain. Nature 306:55-57

Edvinsson L, Hardebo JE, Owman C (1978): A pharmacological analysis of 5-hydroxytryptamine receptors in intracranial and extracranial vessels of cat and man. Circ Res 42:143-151

Edvinsson L, Krause DN (1979): Pharmacological characterization of GABA receptors mediating vasodilation of cerebral arteries in vitro. Brain Res 173:89-97

Edvinsson L, Larsson B, Skärby T (1980): Effect of the GABA receptor agonist muscinol on regional cerebral blood flow in the rat. Brain Res 185:445-448

Edvinsson L, Nielsen KC, Owman C, Sporrong B (1972): Cholinergic mechanisms in pial vessels. Histochemistry, electron microscopy, and pharmacology. Z Zellforsch Mikroskop Anat 134:311-325

Edvinsson L, Owman C (1974): Pharmacologic characterization of adrenergic alpha and beta receptors mediating vasomotor response of cerebral arteries in vitro. Circ Res 35:835-849

Fox PT, Raichle ME (1986): Focal physiological uncoupling of cerebral blood flow and oxidative metabolism during somatosensory stimulation in human subjects. Proc Natl Acad Sci USA 83:1140-1144

Fujiwara M, Muramatsu I, Shibata S (1975): Gamma-aminobutyric acid receptor on vascular smooth muscle of dog cerebral arteries. Br J Pharmacol 55:561-562

Gibbins IL, Brayden JE, Bavan JA (1984): Distribution and origin of VIP-immunoreactive nerves in the cephalic circulation of the cat. Peptides 5:209-212

Greenberg J, Hand P, Sylvestro A, Reivich M (1979): Localized metabolic-flow couple during functional activity. Acta Neurol Scand 60, suppl 72:12-13

Hara H, Hamill GS, Jacobowitz DM (1985): Origin of cholinergic nerves to the rat major cerebral arteries: coexistence with vasoactive intestinal polypeptide. Brain Res Bull 14:179-188

Harper AM, MacKenzie ET (1977a): Cerebral circulatory and metabolic effects of 5-hydroxy tryptamine in anesthetized baboons. J Physiol (London) 271:721-733

Harper AM, MacKenzie ET (1977b): Effects of 5-hydroxytryptamine on pial arteriolar calibre in anesthetized cats. J Physiol (London) 271:735-746

Iadecola C, Arbit E, Nakai M et al (1982): Increased regional cerebral blood flow and metabolism elicited by stimulation of the dorsal medullary reticular formation in the rat: evidence for an intrinsic neural system in brain regulating cerebral metabolism. In: Heistad DD & Marcus ML (eds): Cerebral blood flow: effects of nerves and neurotransmitters. New York, Elsevier, pp 485-492

Iadecola C, Nakai M, Arbit E, Reis DJ (1983): Global cerebral vasodilatation elicited by focal electrical stimulation within the dorsal medullary reticular formation in anesthetized rat. J Cereb Blood Flow Metab 3:270-279

Keller JT, Beduk A, Saunders MC (1985): Origin of fibers innervating the basilar artery of the cat. Neurosci Lett 58:263-268

Kelly PT, McCulloch J (1983): The effect of the GABA-ergic agonist muscimol upon the relationship between local cerebral blood flow and glucose utilization. Brain Res 258:338-342

Kuschinsky W, Suda S, Sokoloff L (1985): Influence of gamme-hydroxybutyrate on the relationship between local cerebral glucose utilization and local cerebral blood flow in the rat brain. J Cereb Blood Flow Metab 5:58-64

Liu-Chen L, Gillespie SA, Norregaard TV, Moskowitz MA (1984): Co-localization of retrogradely transported wheat germ agglutinin and the putative neurotransmitter substance P within trigeminal ganglion cells projecting to cat middle cerebral artery. J Comp Neurol 225: 187-192

Lluch S, Reimann C, Glick G (1973): Evidence for the direct effect of adrenergic drugs on the cerebral vascular bed in un-anaesthetized gloat. Stroke 4:50-56

Lou HC, Edvinsson L (1986): Influence of serotonin and norepinephrine on flow capacity/pressure characteristics of feline isolated cerebral arteries. Stroke 17:91-94

MacKenzie ET, McCulloch J, Harper AA (1976): Influence of endogenous norepinephrine on cerebral blood flow and metabolism. Am J Physiol 231:489-494

Mayberg MA, Langer RS, Zervas NT, Moskowitz MA (1981): Perivascular meningeal projections from cat trigeminal ganglia: possible pathway for vascular headaches in man. Science 213:228

McKee JC, Denn MJ, Stone HL (1976): Neurogenic cerebral vasodilatation from electrical stimulation of the cerebellum in the monkey. Stroke 7:179-186

Mena MA, Vidrio H (1976): On the mechanism of the coronary dilator effect of serotonin in the dog. Eur J Pharmacol 36:1-5

Nakai M, Iadecola C, Ruggiero DA et al (1983): Electrical stimulation of cerebellar fastigial nucleus increases cortical blood flow without change in local metabolism: evidence for an intrinsic system in brain for primary vasodilation. Brain Res 260:35-49

Nilsson B, Rehncrona S, Siesjö BK (1978): Coupling of cerebral metabolism and blood flow in epileptic seizures, hypoxia and hyperglycemia. IN: Elliott K, O'Connor M (eds) Cerebral vascular smooth muscle and its control, Ciba Foundation Symposium 56, New York, Elsevier, pp 199-214

Oberförster G, Lang R, Zimmer R (1973): Direct effects of the alpha and beta sympathomimetic amines on the cerebral circulation of the dog. Pflügers Arch 340:145-160

Olesen J (1972): the effect of intra carotid epinephrine, norepinephrine, and angiotension ont he regional cerebral blood flow in man. Neurology 22:978-987

Paulson OB, Newman EA (1987): Does the release of potassium from astrocyte and feet regulate cerebral blood flow? Science 237:896-898

Pearce WJ, Screming OU, Sonnenschein RR, Rubinstein EH (1981): The electroencephalogram, blood flow, and oxygen uptake in rabbit cerebrum. J Cereb Blood Flow Metab 1:419-428

Purves MJ, James IM (1969): Observations on the control of cerebral blood flow in the sheep fetus and newborm lamb. Circ Res 25:651-667

Raichle ME, Grubb RL, Gado MH et al (1976): Correlation between regional cerebral blood flow and oxidative metabolism. Arch Neurol 33: 523-526

Reinhard JF, Leibmann JE, Schlosberg AJ, Moskowitz MA (1979): Serotonin neurons project to small blood vessels in the brain. Science 206:85-87

Reis DJ (1979): Nervous control of the central blood flow in normal health and in relationship to cerebrovascular disease. In: Castaigne P, Lhermitte F, Gautier JC et al (eds.) Maladie vasculaires cérébrales deuxieme conférénce de la salpetriere. Paris, Bailliere-Tindall, pp 197-224

Roland PE, Friberg L (1983): Are cortical rCBF increases during brain work in man due to synaptic excitation or inhibition? J Cereb Blood Flow Metab 3 (Suppl 1):244-245

Roy CW, Sherrington CS (1890): On the regulation of the blood supply of the brain. J Physiol (London) 11:85-108

Sharkey J, McCulloch J (1985a): Relationship between local cerebral blood flow and glucose utilization with selective dopamine receptor agonists. J Cereb Blood Flow Metab 5 (Suppl 1):537-538

Sharkey J, McCulloch J (1985b): Relationship between local cerebral blood flow and glucose utilization with selective dopamine receptor agonists. In Ingvar DH, Owman C, Siesjö BK (eds) Brain 85 Scientific Abstracts. Lund, Studenterlitteratur, p 250

Silver IA (1978): Cellular microenvironment in relation to local blood flow. In Elliott K, O'Connor M (eds) Cerebral vascular smooth muscle and its control, Ciba Foundation Symposium 56. New York, Elsevier, pp 49-61

Snead OC (1977): Gammahydroxybutyrate. Life Sci 20:1935-1944

Suzuki T, Chada T, Shibuya M et al (1984): Gamma-aminobutyricacid-induced contraction of the basilar artery. Pharmacology 29:24-30

Swanson LW, Connelly MA, Hartman BK (1977): Ultrastructural evidence for central monaminergic innervation of blood vessels in the paraventricular nucleus of the hypothalamus. Brain Res 136:166-173

Toda N, Fujita Y (1973): Responsiveness of isolated cerebral and perepheral arteries to serotonin, norepinephrine, and transmural electrical stimulation. Circ Res 33:98-104

DISCUSSION

Chaired by G.C. Palmer.

R. Pumain:

Concerning the pyramidal cells in the layer of the cortex, which release glutamate, is there any evidence that the vasculature in the cortex is sensitive to this neurotransmitter?

H. Lou:

I turn the question to Dr. Mc Kenzie.

E. Mc Kenzie:

Glutamate and aspartate are all devoid of direct vascular effect, there is no evidence for innervation. Some studies report that glutamate causes vascular constriction, but only at very high doses.

B. Hartman:

It is known that brain fibers exit to innervate the large vasculature. There are 3 sources, as you have shown, one of which is the trigeminal nerve. Can you speculate on what kind of information the trigeminus carries to the vascular system?

H. Lou:

The trigeminus carries mainly vasodilators mediated through substance P and related neuropeptides to the cerebral vasculature.

G.C. Palmer:

What is the source of VIP fibers innervating the vasculature?

H. Lou:

This is still under investigation: the cortex or the sfenopalatine ganglion seem to be the most probable sources.

J. Mc Culloch:

There is indeed a growing body of literature which describes anatomical tracer studies identifying the importance of the sfenopalatine ganglion as a source of cholinergic and VIP fibers to the blood vessels

G.C. Palmer:

What about the cortical source of VIP innervation?

J. Mc Culloch:

As Dr. Lou pointed out, VIP staining neurons are predominantly localized in layers II and III of the cortex. There is a number of studies showing that they innervate the blood vessels. The other focus of these studies was CCK innervation originating from the cortex.

PEPTIDERGIC INNERVATION OF THE CEREBRAL VASCULATURE AND ITS FUNCTIONAL SIGNFIICANCE

James McCulloch and Lars Edvinsson
Wellcome Neuroscience Group
University of Glasgow
Glasgow, G61 1QH, U.K.

INTRODUCTION

In the last decade, the number of neurotransmitters which have been proposed to be associated with the cerebral vasculature has increased to approximately 20. However, the dominant factors primarily responsible for the dynamic regulation of cerebral blood flow have long been considered to be vasoactive products of cellular metabolism, with perivascular hydrogen ions, potassium ions and adenosine being the most favoured candidates. Many cerebrovascular investigators would ascribe only a minor, if any, role in the regulation of cerebral blood flow to neurotransmitters. In the last five years, new anatomical evidence has rekindled great interest in neurogenic cerebrovascular influences. Cerebral blood vessels are not only innervated by nerve fibres originating in sympathetic ganglia and cranial nerve ganglia but there is considerable anatomical evidence that intracerebral neurones gives rise to process, some of which form synaptic contacts with other cortical neurones, and others which are intimately associated with the cerebral vasculature. Such anatomical arrangements could provide a mechanism by which neuronal firing and thus energy generation could be linked to local blood flow directly by a neurotransmitter. (For detailed review of cerebrovascular innervation, see Edvinsson and McCulloch, 1987).

In this chapter, we will focus on two neuronal systems whose influences on the cerebral vasculature are quite dissimilar; the trigemino-cerebrovascular system which originates in the trigeminal ganglion and the vasoactive intestinal polypeptide (VIP) containing cerebrovascular nerve fibres, a portion of which putatively originates in cortical perikarya.

TRIGEMINO-CEREBROVASCULAR SYSTEM

Anatomy. The fifth cranial nerve (the trigeminal) has long been accepted to be the principal afferent neuronal pathway for the transmission of pain

NATO ASI Series, Vol. H 33
Regulatory Mechanisms of Neuron to Vessel Communication in the Brain. Edited by F. Battaini et al.

of a cerebrovascular origin (Moskowitz, 1984). Although there are numerous, elegant descriptions by classical anatomists of nerves from the trigeminal ganglion contacting the pial vessels, only with the advent of modern axonal tracing techniques and immunocytochemistry has the anatomy and the constituent neurotransmitters of the trigemino-cerebrovascular system been established definitively. Retrograde transport of horseradish peroxidase, from the cerebral arteries, sinuses, and dura mater, to perikarya in the opthalmic division of the trigeminal ganglion is observed consistently (Mayberg et al. 1984; Liu-Chen et al. 1984; Keller et al. 1985; Arbab et al. 1985; O'Connor and Kooy, 1988). It should be emphasized that the trigeminal innervation of the cerebral vasculature is a major source of nerve fibres; the number of cells in the trigeminal ganglion labelled retrogradely from the proximal portion of the middle cerebral artery is equivalent to the number of labelled cells observed in the superior cervical ganglion (Mayberg et al. 1984). Moreover, only for the trigeminal ganglia and superior cervical ganglion is there unanimity among the modern neuroanatomists of cerebrovascular innervation in all species studied. The trigemo-cerebrovascular nerve fibres certainly contain substance P, and calcitonin gene-related peptide (CGRP), and probably contain neurokinin A and possibly somatostatin, cholecystokinin, dynrphin B and bombesin/gastrin releasing peptide. There has, as yet, been only one systematic attempt to establish whether each of these peptides is present in all trigeminal nerves or only on specific subpopulations, or to examine the relative levels of these peptides in the cerebrovascular bed (O'Connor and Kooy, 1988, vide infra). Moreover, functional interactions between the various putative constituents of the trigemino-cerebrovascular system remain to be explored adequately.

Nerve fibres containing substance P-immunoreactivity have been identified around pial arteries in every mammalian species examined hitherto, including man, cat, guinea pig, rabbit, gerbil, and rat (Edvinsson et al. 1981; Uddman et al. 1981; Edvinsson and Uddman, 1982; Yamamoto et al. 1983; Matsuyama et al. 1984). Although interspecies differences have not been systematically studied, cerebral arteries from the cat and guinea pig have an abundant supply of substance P fibres; porcine cerebral arteries have a moderate supply, with the cerebral vessels from humans, rabbits, and rats possessing only a few substance P-containing nerve fibres (Uddman et al. 1981). Substance P-immunoreactive nerve fibres are slightly more abundant in the cerebral arteries from the rostral portion of the Circle of Willis, such as the anterior cerebral artery (Edvinsson et al. 1981; Liu-Chen et al. 1983; Matsuyama et al. 1984). Surgical division of the trigeminal ganglion in cats effects a marked reduction (50%) in cerebrovascular substance P levels and an almost complete loss of substance P immunoreactivity in cerebral arteries (Liu-Chen et al. 1983; Uddman et al. 1985). Chronic treatment with capsaicin, which is known to deplete peptides from primary sensory afferent fibres, causes an almost complete loss of substance P from pial arteries of the guinea pig (Duckles and Buck, 1982). These two lines of evidence point to the presence of substance P in sensory fibres originating in the trigeminal ganglion which innervate the cerebral vasculature.

The tachykinin, neurokinin A, has recently been identified in cerebrovascular nerves (Edvinsson and Jansen, 1987). Neurokinin A and substance P are derived from a common precursor, β-prepro-tachykinin (Nawa et al. 1983), and equal amounts of neurokinin A and substance P are found in the cerebrovascular tissue from cat, cow, pig and man. Up to five times more neurokinin A than substance P is found in cerebral vessels of rodents, suggesting a difference in posttranslatonal processing of tachykinins in rodents. Neurokinin A is co-localized with substance P in cerebrovascular nerve fibres and in cell bodies in the trigeminal ganglion, which is the probable source of the cerebrovascular neurokinin A fibres.

Calcitonin gene-related peptide (CGRP) was identified originally from structural analysis of the expression products of the calcitonin gene. Alternative processing of RNA transcribed from the calcitonin gene leads to the production in neuronal tissue of a 37 amino acid peptide, CGRP (Amara et al. 1982; Rosenfeld et al. 1983). Calcitonin gene-related peptide immunoreactivity is present in perikarya in the trigeminal ganglia and in the cerebral vasculature of all species examined, including man. CGRP is frequently co-localized with substance P in trigeminal cell bodies, although the number of CGRP containing cells exceeds the number of substance P containing cells (Rosenfeld et al. 1983; Uddman et al. 1985; Lee et al. 1985; McCulloch et al. 1986; O'Connor and Kooy, 1988).

In animals in which the trigeminal nerve has been surgically divided 14 days before sacrifice, there was a complete absence of CGRP- and substance P-immunoreactive nerve fibres in cerebral blood vessels ipsilateral to the lesion (McCulloch et al. 1986). Cerebrovascular tissue contains substantial amounts of authentic CGRP, as demonstrated with HPLC; CGRP cannot be detected in cerebral arteries after unilateral lesions of the trigeminal nerve 14 days prior to sacrifice (Uddman et al. 1985).

In the most systematic study of co-localization of these peptides presently available, (O'Connor and Kooy, 1988), approximately 32% of trigeminal ganglion cells projecting to the cerebral vessels contain CGRP whereas 16% of trigeminal projection contains substance P and CCK, almost invariably co-localized with each other and with CGRP. A major portion of the trigeminal projection to the cerebral vasculature appears to contain only CGRP (O'Connor and Kooy, 1988). It remains to be established whether these patterns of co-localization in the cerebral vasculature of the rat are present in other species.

Physiology. The direct vasomotor effects of the neuropeptides present in the trigemino-cerebrovascular system have been characterised in details in cerebral vessels in vitro and in situ (Edvinsson et al. 1981; McCulloch et al. 1986; Edvinsson and Jansen, 1987; Hanko et al. 1985; McCulloch and Kelly, 1984; Uddman et al. 1983; Moskowitz et al. 1986). CGRP is a most potent dilator of cerebral blood vessels; the tachykinins (substance P, neurokinin A and B) also dilate cerebral vessels though less, potently than CGRP: and the other peptides appear to have minimal direct effects on cerebral vascular diameter (Fig. 1).

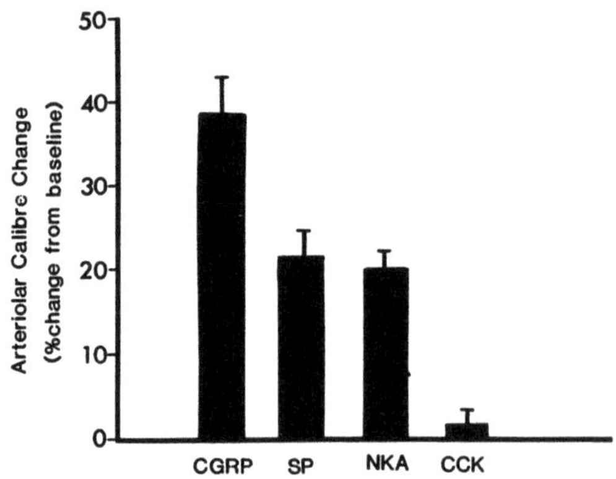

Fig. 1 - Vasomotor responses of feline pial arterioles in situ to perivascular microapplication of 4 neuropeptides, putatively contained in cerebrovascular nerbe fibres origintingin the trigeminal ganglion. The maximum dilatation (± SEM) induced by each peptide is shown.

Investigations of pial vascular responses in situ in animals with chronic trigeminal lesions have been particularly revealing of the vasomotor significance of the trigemino-cerebrovascular system. Chronic surgical lesions of the ipsilateral trigeminal nerve do not modify the magnitude of the pial arteriolar responses to perivascular microapplication of various vasoactive materials, including noradrenaline, alkaline CSF (pH 7.6), prostaglandin $F_{2\alpha}$ barium chloride. However, the durations of the vasoconstrictions were significantly prolonged in cats with trigeminal lesions (Fig. 2). The duration and magnitude of responses of vasodilator stimuli such as subarachnoid injections of acidic CSF (pH 6.8) were not altered by trigeminal nerve lesions (McCulloch et al. 1986). Thus, the trigemino-cerebrovascular system appears to be a component in a neuronal system that facilitates the restoration of vascular diameter after excessive vasoconstriction. It is unclear whether this mechanism is a local response, resulting from direct mechanical stimulation of sensory nerve endings, or whether it is part of a reflex arc with the trigeminal nerve constituting the afferent or efferent limb (or both). These equations remain for investigation. The potency of CGRP as a cerebrovascular dilator and its sustained action in vivo, point to the involvement of this peptide in the restoration of vascular calibre after vasoconstriction.

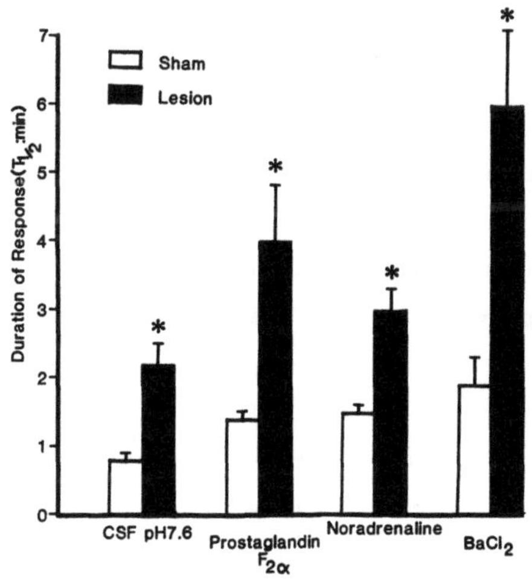

Fig. 2 - Vasomotor responses of feline cortical arterioles in situ. The duration of the vasoconstrictor responses to alkaline CSF, prostaglandin $F_{2\alpha}$, noradrenaline and barium chloride were significantly prolonged after trigeminal ganglia lesions compared to sham-operated controls. The duration of response was assessed as the time (mins) for half restoration of preinjection diameter. *$p<0.01$. Data from McCulloch et al. (1986).

The conceptual approach, applied to the cerebrovascular CGRP neuronal system, has been to examine whether this dilator peptide might provide a mechanism for preventing reductions in cerebral tissue perfusion. Amino acid residues with the attendant biosynthetic demands and its long lasting effect on the cerebral circulation would be involved in the moment-to-moment regulation regulation of cerebral blood flow (see however the role of VIP below). The failure of trigeminal lesion to alter local cerebral blood flow and its normal relationship to local glucose use provides support for such a veiw (McCulloch et al. unpublished observations).

In conclusion, the propostion that the trigemino-cerebrovascular system is a neurogenic mechanism capable of detecting and responding to emergency situations where excessive vasoconstriction of the larger cerebral arteries threatens the survival of the central nervous system, has important implications not just for cerebrovascular physiology but also in pathological states such as migraine and subarachnoid haemorrhage.

VASOACTIVE INTESTINAL POLYPEPTIDE (VIP) CONTAINING NERVE FIBRES

Although VIP was the first of the neuropeptides shown to innervate cerebral blood vessels (Larsson et al. 1976), elucidation of its functional significance has proved a most intractable problem. The anatomical organisation of the cerebrovascular VIP innervation has not been characterised definitively. VIP is colocalized with a structurally related peptide, peptide histidine isoleicine (PHI), whose functional significance is obscure. VIP dilates cerebral vessels but in vivo its effects on cerebral tissue perfusion are complicated by the effects of the peptide upon cerebral oxidative metabolism. In consequence, unlike the trigemino-cerebrovascular system (see above), no coherent view has yet emerged of the roles which VIP may play in cerebrovascular regulation despite extensive reserch effort.

Anatomy. Cerebral blood vessels (major cerebral arteries, arterioles and veins) of all mammalian species studied to date are invested with nerve fibres which contain VIP-like immunoreactivity (Larsson et al. 1976; Edvinsson et al. 1980, 1982; Hara et al. 1985; Gibbons et al. 1984). There are considerable inter-species variation in the perivascular levels of VIP with the highest levels being noted in feline cerebral vasculature (Fig. 3) (Edvinsson et al. 1980).

There appears to be multiple origins of the cerebrovascular VIP fibres. In the rat, the sphenopalatine ganglion appears to be a major source of VIP fibres to cerebral arteries of the anterior portion of the Circle of Willis (Hara et al. 1985; Suzuki et al. 1988). In the cat, cerebrovascular nerve fibres appear to originate, at least in part, from microganglia in the cavernous plexus and external rete (Gibbons et al. 1984) and similar "miniganglia" associated with the carotid vasculture contribute to the cerebrovascular VIP levels in the rat (Suzuki et al. 1988). The sources of cerebrovascular VIP fibres which has aroused

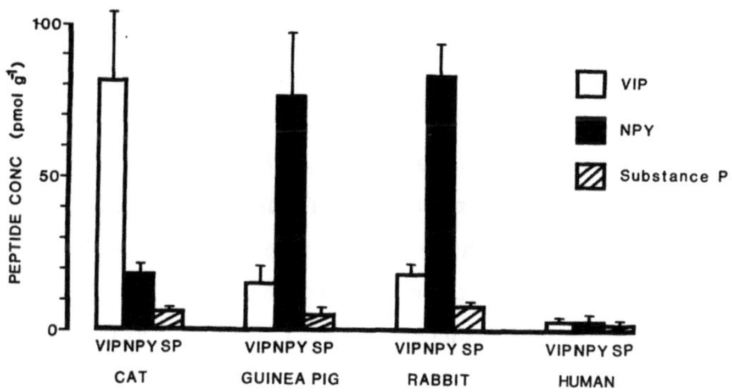

Fig. 3 - Interspecies differences in cerebrovascular neuropeptide concentrations. Data are from the published investigations of Edvinsson and his associates (see Edvinsson and McCulloch, 1987).

greatest interest are neurones within the cerebral cortex. There are a number of reports (Loren et al. 1979; Eckenstein and Baughman, 1984; Itakura et al. 1987). VIP containing perikarya within the neocortex, give rise to processes, some of which appear to form synaptic contacts with other neurones and others which have truly intimate contact with intracerebral arterioles. Such an anatomical arrangement provides a mechanism in which neuropeptides such as VIP could participate in linking local cerebral blood flow to neuronal excitability and energy generation and use (McCulloch, 1983).

Physiology. VIP and PHI relaxes cerebral arteries in vitro and dilate pial arterioles in situ (Larsson et al. 1976; McCulloch and Edvinsson, 1980,

Edvinsson and McCulloch, 1985; Suzuki et al. 1984; Duckles and Said, 1982; Wei et al. 1980). The maximum increases in feline pial arteriolar calibre elicited by VIP and PHI are 23 ± 5% and 23 ± 6% respectively which stands favourable comprison with peptides present in the trigeminal projection (see Fig. 1).

VIP increases cerebral blood flow provided it gains access to cerebral resistance vessels (i.e. after intracerebral administration or systemic administration with hypertonic blood-brain barrier disruption) (Fig. 4) (McCulloch and Edvinsson, 1980; Heistad et al. 1980; Wilson et al. 1981; Itakura et al. 1987). This VIP induced increase in blood flow is not attributable to the direct actions of the peptide on cerebral blood vessels. Increased cerebral oxidative metabolism parallel the increased CBF after VIP administration (Fig. 4) in all the studies in which cerebral metabolism has been adequately assessed. Autoradiographic investigation with ^{14}C-2-deoxyglucose after intracortical and intrastriatal injection of low concentration of VIP (20 pmoles) have emphasised the importance of the effects of the peptide upon metabolism and has allowed their anatomical complexity to be explored (McCulloch et al. 1983; McCulloch and Kelly, 1983).

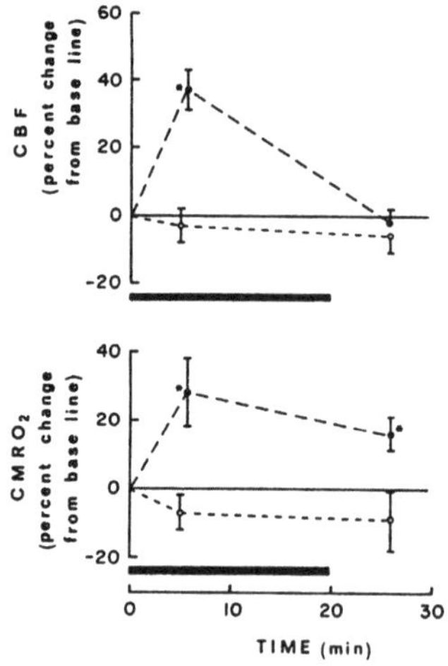

Fig. 4 - Effect of VIP upon cerebral blood flow (CBF) and cerebral oxygen consumption ($CMRO_2$) in anaesthetised baboons. VIP (10^{-11}mol/min) was administered as an intracarotid infusion from time 0 to 20 mins. Open circles: alterations that occurred under normal circumstances. Solid circle: alteration that occurred with VIP after blood brain barrier opening. *P<0.05. From McCulloch and Edvinsson, 1980.

One recent report Yaksh et al. (1987) has provided the first tentative evidence that VIP participates directly in coupling local cerebral blood flow to local neuronal activity. With a cranial window technique in cats, these investigators has stimultion of the mesencephalic reticulr formation which evokes an immediate dustained increase (80%) in pial arteriolar diameter. The presence of VIP antisera (to "block" the effects of endogenous VIP released during the stimultion) completely abolished the early phase of the pial arteriolar dilatation. It should be recognised that there are two main weakness in the study in relation the the role played by VIP in flow-metabolism coupling under normal circumstances. Firstly, the magnitude of the pial arteriolar response to reticular formation stimulation exceeds that observed under normal physiological activation; indeed, it exceeds the dilatation seen with drug induced seizures (Kuschinsky and Wahl, 1979). Secondly, the response attributed to VIP during the reticular formation stimulation greatly exceeds tht which VIP effects is normal, uncontracted vessels. Notwithstanding these comments, the strategy employed by Yaksh et al. (1978) provide a clear indication of the approach which is required if VIP involvement (if any) in flow metabolism coupling is to be unravelled.

REFERENCES:

Amara SG, Jonas V, Rosenfeld MG, Ong ES, Evans RM (1982). Alternative RNA processing in calcitonin gene expression generates mRNAs encoding different polypeptide products. Nature 298:240-244.

Arbab MA-R, Wilklund L, Svendgaard NAa (1985). Origin and distribution of the cerebral vascular innervation in rat investigated with retrograde and anterograde WGA-HRP. J Cereb Blood Flow Metab 5:S491-S492.

Duckles SP, Buck SH (1982). Substance P in the cerebral vasculature: depletion by capsaicin suggests a sensory role. Brain Research 245:171-174.

Duckles SP, Said SI (1982). Vasoactive intestinal polypeptide as a neurotransmitter in the cerebral circulation. Eur J Pharmacol 78:371-374.

Eckenstein F, Baughman RW (1984). Two types of cholinergic innervation in cortex, one co-localized with vasoactive intestinal polypeptide. Nature (London) 309:153-155.

Edvinsson L, Fahrenkrug J, Hanko J, Owman C, Sundler F, Uddman R (1980). VIP (vasoactive intestinal polypeptide)-containing nerves of intracranial arteries in mammals. Cell Tissue Res 208:135-142.

Edvinsson L, Jansen I (1987). Characterization of tachykinin receptors in isolated basilar arteries of guinea-pig. Br J Pharmac 90:553-559.

Edvinsson L, McCulloch J (1985). Distribution and vasomotor effects of peptide HI (PHI) in feline cerebral blood vessels in vitro and in situ. Regul Peptides 10: 345-356.

Edvinsson L, McCulloch J (eds) (1987). Peptidergic mechanisms in the cerebral circulation. Horwood, London.

Edvinsson L, McCulloch J, Uddman R (1981). Substance P:immunohistochemical localization and effect upon pial arteries in vitro and in situ. J Physiol 318:251-258.

Edvinsson L, McCulloch J, Uddman R (1982). Feline cerebral veins and arteries: comparison of autonomic innervation and vasomotor resposnes. J Physiol 325:161-173.

Edvinsson L, Uddman R (1982). Immunohistochemical localization and dilatatory effect of substance P on human cerebral vessels. Brain Research 232:466-471.

Gibbins IL, Brayden JE, Bevan JA (1984). Perivascular nerves with immunoreactivity to vasoactive intestinal polypeptide in cephalic arteries of the cat: Distribution, possible origins and functional implications. Neuroscience 13:1327-1346.

Hanko J, Hardebo JE, Kahrström J, Owman C, Sundler F (1985). Calcitonin gene-related peptide is present in mammalian cerebrovascular nerve fibres and dilates pial and peripheral arteries. Neuroscience Letters 57:91-95.

Hara H, Hamill GS, Jacobowitz DM (1985). Origin of cholinergic nerves to the rat major cerebral arteries: Coexistence with vasoactive intestinal polypeptide. Brain Res Bull 14:179-188.

Heistad DD, Marcus ML, Said SI, Gross PM (1980). Effect of acetylcholine and vasoactive intestinal peptide on cerebral blood flow. Am J Physiol 239:H73-H80.

Itakura T, Yokote H, Okuno T, Naka Y, Nakakita K, Kamel I, Nakai K, Imai H, Komai N (1987). Regulation of rCBF by intracortical vasoactive intestinal polypeptide-containing neurons. J Neurosurg 67:93-96.

Keller JT, Beduk A, Saunders MC (1985). Origins of fibers innervating the basilar artery of the cat. Neuroscience Letters 58:263-268.

Kuschinsky W, Wahl M (1979). Perivascular pH and pial arterial diameter during bicuculline induced seizures in cats. Pflügers Arch 382:81-85.
Lee Y, Kawai Y, Shiosaka S, Takami K, Kiyama H, Hillyard, CJ, Girgis S, MacIntyre I, Emson PC, Tohyama M (1985). Coexistence of calcitonin gene-related peptide and substance P-like peptide in single cells of the trigeminal ganglion of the rat: immunohistochemical analysis. Brain Reserch 330-194-196.
Larsson L-I, Edvinsson J, Fahrenkrug J, Hakanson R, Owman C, Schaffalitzky de Muckadell O, Sundler F (1976). Immunohistochemical localization of vasodilatory peptide (VIP) in cerebrovascular nerves. Brain Res 113:400-404.
Liu-Chen LY, Han DH, Moskowitz MA (1983). Pia arachnoid contains substance P originating from trigeminal neurons. Neuroscience 4: 803-838.
Liu-Chen LY, Gillespie SA, Norregaard TV, Moskowitz MA (1984). Co-localization of retrogradely transported wheat germ agglutinin and the putative neurotransmitter substance P within trigeminal ganglion cells projecting to cat middle cerebral artery. J Comp Neurol 225:187-192
Loren I, Emson PC, Fahrenkrug J, Björklund A, Alumets J, Hakanson R, Sundler F ((1979). Distribution of vasoactive intestinal polypeptide in the rat and mouse brain. Neuroscience 4:1953-1976.
Matsuyama T, Matsumoto M, Shiosaka S, Kayakawa T, Yoneda S, Kimura K, Abe H, Tohyama M (1984). Dual innervation of substance P-containing neuron system in the wall of the cerebral arteries. Brain Reserch 322:144-247.
Mayberg MR, Zervas NT, Moskowitz MA (1984). Trigeminal projection to supratentorial pial and dural blood vessels in cats demonstrated by horseradish perioxidase histochemistry. J Comp Neurol 223: 46-56.
McCulloch J (1983). Peptides and the microregulation of blood flow in the brain. Nature 304:120.
McCulloch J, Edvinsson L (1980). Cerebral circulatory and metabolic effects of vasoactive intestinal polypeptide. Am J Physiol 238:H449-H456.
McCulloch J, Kelly PAT (1983). A functional role for vasoactive intestinal polypeptide in anterior cingulate cortex. Nature 304:438-440.
McCulloch J, Kelly PAT (1984). Effects of cholecystokinin octapeptide on pial arteriolar diameter. J Cereb Blood Flow Metab 4:625-628.
McCulloch J, Kelly PAT, Uddman R, Edvinsson L (1983). Functional role for vasoactive intestinal polypeptide in the caudate nucleus: A 2-deoxy[^{14}C]glucose investigation. Proc Natl Acad Sci (USA) 80: 1472-1476.
McCulloch J, Uddman R, Kingman TA, Edvinsson L (1986). Calcitonin gene-related peptide: functional role in cerebrovascular regulation. Proc. Natl Acad Sci (USA) 83:5731-5735.
Moskowitz MA (1984). The neurobiology of vascular head pain. Ann Neurol 16:157-168.
Moskowitz MA, Brezina LR, Kuo C (1986). Dynorphin B-containing perivascular axons and sensory neurotransmitter mechnisms in brain blood vessels. Cephalalgia 6:81-86.
Nawa H, Hirose T, Takashima H, Inayam S, Nakanishi S (1983). Neucleotide sequence of cloned cDNAs for two types of bovine substance P precursor. Nature 306:32-36.
O'Conor TP, Kooy van der D (1988). Enrichment of a vasoactive neuropeptide (calcitonin gene related peptide) in the trigeminal sensory projection to the intracranial arteries. J Neurosci 8(7):2468-2476.
Rosenfeld MG, Mermod J-J, Armara SG, Swanson LW, Sawchenko PE, Rivier J, Vale WW, Evans RM (1983). Production of a novel neuropeptide encoded by the calcitonin gene by tissue-specific RNA processing. Nature 304:129-135.

Suzuki N, Hardebo JE, Owman C (1988). Origins and pathways of cerebrovascular vasoactive intestinal polypeptide-positive nerves in rat. J Cereb Blood Flow Metab 8:697-712.
Suzuki Y, McMaster D, Lederis K, Rorstad OP (1984). Characterization of the relaxant effects of vasoactive intestinal peptide (VIP) and PHI on isolated brain arteries. Brin Res 322:9-16.
Uddman R, Edvinsson L, Ekman R, Kingman T, McCulloch J (1985). Innervation of the feline cerebral vasculture by nerve fibers containing calcitonin gene-related peptide: trigeminal origin and co-existence with substance P. Neuroscience Letters 62:131-136.
Uddman R, Edvinsson L, Owman C, Sundler F (1981). Perivascular substance P: occurence and distribution in mammalian pial vessels. J Cereb Blood Flow Metab 1:227-232.
Uddman R, Edvinsson L, Owman C, Sundler F (1983). Nerve fibres containing gastrin-releasing peptide around pial vessels. J Cereb Blood Flow Metab 3:386-390.
Wei EP, Kontos HA, Said SI (1980). Mechanism of action of vasoactive intestinal polypeptide on cerebral arterioles. Am J Physiol 239: H765-H768.
Wilson DA, O'Neill JT, Said SI, Traystman RJ (1981). Vasoactive intestinal polypeptide and the canine cerebral circulation. Circ Res 48:138-148.
Yamamoto K, Matsuyama T, Shiosaka S, Inagaki S, Senba E, Shimizu Y, Ishimoto I, Hayakawa T, Matsumoto M, Tohyama M (1983). Overall distribution of substance P-containing nerves in the wall of the cerebral arteries of the guinea pig and its origins. J Comparative Neurology 215:421-426.
Yaksh TL, Wang J-Y, Go VLW, Harty GJ (1987). Cortical vasodilatation produced by vasoactive intestinal polypeptide (VIP) and by physiological stimuli in the cat. J Cereb Blood Flow Metab 7:315-326.

DISCUSSION

Chaired by Hartman B. and Racagni G.

B. Siesjo:

Did you observe denervation supersensitivity for the peptides that you have studied?

J. Mc Culloch:

We have looked for denervation supersensitivity by performing dose-response curve to CGRP or substance P in denervated animals but we have not found any evidence for it.
However, the vaso-constrictor response appears to be prolonged. The hypothesis is that in case of vasoconstriction after denervation the afferent fibers send the information to the brain, probably via the trigeminal system; the elaboration of these data leads to a local release of CGRP in order to restore the vascular caliber. This could explain why the level of sensitivity to peptides is still normal.

G. Racagni:

Is it possible that the mechanism at the basis of the lack of denervation supersensitivity is an allosteric modulation at the level of noradrenergic receptor sites?

J. Mc Culloch:

This is a possibility that could be taken into account.

C. Fieschi:

You have shown the effect of the trigeminal system on the time course of pial vessel constriction. Do you have any evidence on the effects of this system on intraparenchymal vessels?

J. Mc Culloch:

We have not investigated the effect of the trigeminal system on intracerebral vasculature; however, our feeling is that there may be a difference in response based mainly on the different patterns of innervation: the large extraparenchymal vessels are extensively innervated by CGRP and substance P fibers, whereas the intraparenchymal microvasculature seems to be sparsely innervated. This point may be of importance in the case of subarachnoid hemorrhage, occurring at the level of large vessels which are able to display a protective neurogenic mechanism.

E. Mc Kenzie:

Do subarachnoid hemorrhage patients who die of delayed

neurological deficits following the late vasospasm have drammatic reduction in CGRP levels?

J. Mc Culloch:

The delayed vasospasm is particularly interesting because many factors contribute to its development: one hyphothesis is to test whether this event is related to the time course of replenishing of the fibers with CGRP.

H. Lou:

Does CGRP reduce the constrictor response induced by K^+?

J. Mc Culloch:

We did not test the effect of CGRP or K^+ because the ion not only produces constriction but also induces transmitter release, so that we cannot distinguish the depolarization effect of K+ from the secondary reflex events.

H. Lou:

I have made this question since the relase of K^+ can contribute to vasoconstriction and its prolongation in certain conditions.

J. Mc Culloch:

I agree. Many factors, and not only CGRP, contribute to the duration of the vascular responses and the restoration of the caliber. For exemple, the intraluminal pressure is another very important mechanism in vascular homeostasis.

F. Joo:

Have you ever studied the effect of capsaicin, which selectively depletes the SP neurons, on the cerebral vasculature?

J. Mc Culloch:

The acute injection of capsaicin around the pial vessel leads to an enormeous vasoconstriction followed by a rebound vasodilatation; the repetition of the injection in the same vessels causes again vasoconstriction but no rebound vasodilatation. Probably the physiological mechanisms at the basis is the release of neurotransmitters causing vasodilatation. Interestingly, the animals with trigeminal lesion display only vasoconstriction without subsequent dilatation.

R. Orkand:

What's the relative contribution of glial cells and neurons

in the measure of glucose metabolism?

J. Mc Culloch:

All the anatomical elements contribute to the basal glucose metabolism: neuronal soma, terminals, glia etc. However, the dynamic alterations that we observe under certain circustances appear to be associated with modifications of electrochemical activities (firing rate), suggesting the involvment of the neuronal terminals.

B.Hartman:

What's the meaning of the various types of receptors located at vascular level? How do you study the specific function of each one?

J.Mc Culloch:

The problem can be approached by several points of view and methodologies: anatomy and biochemistry are certainly great sources of information; however a further step into the insight of the physiology of the system is to investigate the in vivo situations, using as tools agonist or antagonists, stimulations or lesions of the neuronal pathways.

NEUROTRANSMITTER COUPLED RESPONSES IN THE MICROVASCULATURE OF THE BRAIN UNDER NORMAL AND PATHOLOGICAL CONDITIONS

Gene C. Palmer, Sally A. McCreedy and Louis R. Freedman

Pharmacology Department, Pennwalt Corporation, 755 Jefferson Road, Rochester, New York 14623 USA

ABSTRACT

The central microvasculature consisting of capillaries and pia arachnoid is innervated by adrenergic and peptide neurons acting to regulate local blood flow, promote transport of substances into the brain, control permeability of small molecules and induce metabolism. These processes may be controlled by cyclic AMP and cyclic GMP which are formed from adenylate cyclase and guanylate cyclase upon appropriate neurohumoral stimulation. The sensitivity of the receptors mediating enzyme activation is altered during neuropathological states, namely, hypertension, diabetes, catecholamine depletion, irradiation, stroke, aging and Parkinsonism.

KEY WORDS

Capillaries, Pia-arachnoid, Adrenergic Receptors, Adenylate Cyclase, Guanylate Cyclase, Hypertension, Diabetes, Stroke

INTRODUCTION

The blood-brain barrier (BBB) is composed of several mechanisms which limit the passage of substances from the peripheral circulation into the brain: 1) tight junctions between endothelial cells; 2) highly specialized transport and carrier systems; 3) a reduced rate of pinocytosis; 4) enzymes which inactivate circulating neurohumoral substances; 5) a system of lipid membranes to prevent entry of polar molecules; 6) astrocytic processes; and 7) nerve endings which control capillary permeability to small molecules and perhaps regulate local blood flow (Hardebo and Owman, 1980; Goldstein and Betz, 1983; Palmer, 1986, Pardridge, 1988). With regard to neuronal control of permeability across central capillaries, Hartman and coworkers (1972) using dopamine-beta hydroxylase as a marker localized adrenergic nerve endings on microvessels. Ablation of the superior cervical ganglia did not result in a disappearance of the nerve endings whose source was later described to arise from the locus ceruleus. Physiological experiments indicated that central capillary permeability to small molecules and water was under central adrenergic control (Raichle et al., 1975; Hartman et al.,

NATO ASI Series, Vol. H 33
Regulatory Mechanisms of Neuron to Vessel Communication in the Brain. Edited by F. Battaini et al.

1980; Preskorn et al., 1980). Westergaard (1975) indicated that intraarterial administration of norepinephrine (NE) and cyclic AMP would enhance the transport of large molecules across the BBB. These studies suggested that noradrenergic neurons might control capillary permeability via a receptor action coupled to cyclic AMP mechanisms. At this time Joo and coworkers (1972, 1975a) showed that cyclic AMP increased the rate of pinocytosis in central capillaries. With the advent of techniques to isolate sufficient capillaries for biochemical study these workers demonstrated a histamine-sensitive adenylate cyclase (Joo et al., 1975a). Later we (Baca and Palmer, 1978) reported the presence of a catecholamine-responsive enzyme. Shortly thereafter three groups of investigators using ligand binding techniques showed the presence of adrenergic receptors on cerebral microvessels and pia-arachnoid (pia) from different species (Friedman and Davis, 1980; Harik et al., 1980; Peroutka et al., 1980). Using a pathological model, the spontaneously hypertensive rat, I was able to show an attenuated sensitivity of adenylate cyclase to catecholamines within the cerebral capillaries (Palmer, 1979). Further studies by us along with an extensive series of receptor ligand binding experiments conducted by Kobayashi, Magnoni and coworkers have shown alterations in adrenergic receptor coupled processes in the central microvasculature in response to diabetes (Magnoni et al., 1985b), hypertension (Magnoni et al., 1983), monoamine depletion (Magnoni et al., 1986), aging (Kobayashi et al., 1982c), development (Kobayashi et al., 1982a) and stroke (Magnoni et al., 1985a, 1986). In more recent work receptors and associated coupling to biochemical and physiological events have been described for other monoamines (e.g.,serotonin), acetylcholine, adenosine and a variety of peptide hormones (Huang and Drummond, 1979; Huang and Rorstad, 1983; Reinhard, 1979; Grammas et al., 1983, 1985; Edvinsson et al., 1985; Steardo and Nathanson, 1987).

The goal of this communication is to review principally the investigations involving receptor linked mechanisms associated with biochemical (cyclic nucleotides) and physiological mechanisms of the microvasculature (capillaries and pia vessels) of the brain, especially where such events are correlated to neuropathological situations. Where appropriate, our studies of the past 12 years have been summarized and new or unpublished findings are presented.

ADENYLATE CYCLASE-PHOSPHODIESTERASE

Overview

Adenylate cyclase (EC 4.6.1.1) has been shown to be a receptor-coupled enzyme spanning the cell membrane and consisting of several subcomponents described as follows: 1) The receptors face the external milieu of the cell and recognize specific classes and subclasses of hormonal agents; 2) Activation of a receptor is coupled to a GTP regulator or transducer subunit which in turn consists of either or both stimulatory (Gs) and inhibitory (Gi) guanine nucleotide regulatory proteins. Each Gs/Gi moiety contains alpha, beta and gamma proteins. The beta moieties appear similar but the alpha components bind guanine nucleotides and are substrates for NAD-dependent ADP phosphorylation by bacterial toxins. The hydrolysis of GTP by high affinity GTPases is necessary to complete and terminate the coupling of the GTP transducer site to; 3) The catalytic component which promotes the synthesis of cyclic AMP from ATP. The catalytic component is highly sensitive to stimulation by Mn++. Agents such as forskolin and NaF activate the catalytic component, but the process apparently requires GTP. In some preparations of adenylate cyclase, Ca++-calmodulin may enhance receptor-induced enzyme stimulation. Other enzyme preparations contain relatively large amounts of Gi transducer proteins which act to inhibit adenylate cyclase. The ratio of the Gs/Gi proteins may determine the ultimate response of the cell. In the case of dopamine (DA), for example, DA-1 receptors activate Gs and stimulate adenylate cyclase while DA-2 receptors in some cases activate Gi and inhibit the enzyme. In other situations, DA-2 receptors have no effect. Once formed, cyclic AMP activates specific protein kinases which influence key metabolic events. Cyclic AMP is degraded to 5'AMP by a family of phosphodiesterases some of which require Ca++-calmodulin for full expression of activity (for further discussions see: Palmer, 1983; Palmer et al., 1986).

Histochemical investigations have localized adenylate cyclase in smooth muscle and the luminal, abluminal, and basement membranes of capillaries in rat brain (Joo et al., 1975b; Szumanska et al., 1984; Vorbrodt et al., 1984). The biochemical activity of the enzyme in microvessels was first described by Joo et al., (1975a) and these authors reported the enzyme in Wistar rats to be activated by histamine.

Stefanovich (1979) established the presence of cyclic AMP and cyclic GMP dependent phosphodiesterases in microvessels from bovine cerebrum. High Km and low Km cyclic AMP phosphodiesterases were present in rat cerebral capillaries. The enzyme was insensitive to Ca++-calmodulin, despite the presence of calmodulin in capillaries. Enzyme activity was found in both the particulate and supernatant fractions and appeared capable of cyclic AMP metabolism over a wide substrate range. Moreover, the pial enzyme possessed specific activities similar to the capillaries and was likewise insensitive to Ca++-calmodulin. Phosphodiesterase was present at birth in both pia and capillaries and the activity did not change during development (Palmer, 1980; Palmer and Palmer, 1983; Palmer et al., 1983).

Protein kinases were first characterized in bovine brain capillaries by Pardridge and coworkers (1985). Protein phosphorylation was localized exclusively to the capillary plasma membrane and was nearly comparable to the activity of synaptic membranes. The major phosphoprotein was synapsin I and a 50-55K triplet protein appeared exclusive to capillaries. Cyclic AMP dependent phosphorylation could not be detected in this preparation. However, in the rat microvessel preparation Ca++-calmodulin dependent protein kinase was the major phosphorylating enzyme. Furthermore, the activity of cyclic GMP dependent protein kinase was higher than the cyclic AMP dependent enzyme (Joo et al., 1988).

Adrenergic-Coupled Cyclic AMP Response in Microvessels and Pia-Arachnoid (Pia).

Following the work of Joo et al., (1975a) in which a capillary fraction from Wistar rat brain was found to contain an adenylate cyclase responsive to histamine-2 receptor activation, we initiated studies to examine the nature of these receptors in greater detail. Our work principally focused on the microvasculature of the rat and later the rabbit, where regional studies were performed. Tables 1 and 2 summarize the results of these investigations which indicate that, under the conditions of our laboratory, beta-2 receptors predominate in both microvessels and pia in their ability to activate adenylate cyclase in the rat. To a lesser extent, DA agonists and adenosine were the only other classes of neurohumoral agents that appeared active. Hormonal responses were magnified by GTP, especially by the nonhydrolyzable analog, Gpp(NH)p. Forskolin was highly active (ED50's for capillaries = 3 um and pia = 4 um) (Table 3) (Baca and Palmer, 1978; Palmer and Palmer, 1978, 1983; Palmer, 1980). The capillaries from rabbit brain contain an adenylate cyclase sensitive to a variety of adrenergic, DA and histamine responses. These mixed responses were evident throughout the major regions, except cerebellum where only NE was effective. Pia vessels were responsive only to beta adrenergic agonists (Palmer et al., 1980; Palmer, 1981a). Other investigators have shown the presence of beta adrenergic mediation of adenylate cyclase in swine pia (Friedman and Davis, 1980), cat microvessels (beta-2 receptors) (Nathanson, 1980), rat microvessels (Herbst et al., 1979; Szumanska et al., 1984), guinea pig microvessels (Huang and Drummond, 1979), cultured rat endothelia (beta-2 and alpha-2 receptors) (Karnushina et al., 1982), and cultured rat cerebral vascular smooth muscle cells (beta-2 receptors were more potent than alpha-1 receptors) (Wroblewska et al., 1984).

The presence of the adrenergic-coupled adenylate cyclases in cerebral microvasculature is strengthened by the extensive work using ligand binding techniques which revealed: Beta-2 and alpha receptors in swine pia (Friedman and Davis,1980), alpha-2 receptors in bovine pia (Tsukahara et al., 1983), alpha and beta receptors in bovine cerebral microvessels (Peroutka et al., 1980), beta receptors in cerebral microvessels from rat

Table 1: Adenylate Cyclase: Summary of Agonist Responses in Rat Cerebral Microvessels and Pia-arachnoid

	Percent Maximal Stimulation			
	Microvessels		Pia-arachnoid	
Agonist	Alone	+Gpp(NH)p	Alone	+Gpp(NH)p
Adrenergic Agonists				
Isoproterenol	75	173	50	81
Epinephrine	75		43	102
1'Norepinephrine	66	176	34	77
Fenoterol			37	63
Metaproternol	45		18	58
Salbutamol	39		19	37
Terbutaline	40			
Dobutamine	29		12	49
Tazolol	45		11	32
Phenylephrine			18	47
Methoxamine	IA	IA		
Dopamine Agonists				
Dopamine	46	174	IA	
M-7	44			
Pyribedil	14			
S-585	14			
Apomorphine	IA	IA		
Bromocriptine	IA			
Histamine Agonists				
Histamine	IA	IA	IA	
2-Me-histamine	IA	IA		
4-Me-histamine	IA	IA		
Prostaglandins				
E1	IA	IA		
E2	IA	IA		
F2 alpha	IA	IA		
Serotonin	IA	IA	IA	
Adenosine	23	43	IA	

Gpp(NH)p = 10 um

IA = inactive

N = 4-17

Table 2: Adenylate Cyclase: Summary of Adrenergic Agonist/Antagonist Responses in Rat Cerebral Microvessels and Pia-arachnoid

Concentration (um)

	Microvessels			Pia-arachnoid		
Agent	ED50	IC50	Ki	ED50	IC50	Ki
Norepinephrine	5.0			1.0		
+propranolol		0.2			1.4	0.08
+butoxamine		70.0			8.0	0.48
+practolol		800.0			100.0	5.94
+phentolamine					27.0	1.60
Isoproterenol	4.0			6.0		
+propranolol		5.4	0.5			
+butoxamine		165.0	11.0			
+practolol		1100.0	496.0			
+tazolol		50.0				
Epinephrine	1.5			1.0		
Fenoterol				5.0		
Metaproterenol	2.5			9.0		
Salbutamol	0.9			7.5		
+propranolol		0.6				
+butoxamine		0.9				
+practolol		500.0				
Terbutaline	4.0					
+butoxamine		3.6				
Dobutamine	0.7			2.2		
+propranolol		0.7				
+butoxamine		100.0				
+practolol		1.1				
Tazolol	0.07			10.0		
+propranolol		1.0				
+butoxamine		10.4				
+practolol		5.4				

Propranolol is racemate
ED50 - concentration of agonist to elicit half maximal stimulation.
IC50 - concentration of antagonist to inhibit maximal stimulation by 50%.

and pig (Harik et al., 1980), predominantly beta-2 receptors in rat cerebral microvessels (Kobayashi et al., 1981ab), and beta receptors in microvessels from human cortex (Kobayashi et al., 1982b).

During development, from the third through the tenth postnatal days in rats, brain capillaries and pia display a low level of adenylate cyclase sensitivity to NE. In addition, the capillaries are insensitive to Gpp (NH)p while in the pia the response is half that of the adult. In contrast, neonatal microvasculature is highly sensitive to forskolin (Table 3). In support of these data Kobayashi et al., (1982a) found that rat cortical capillaries contained fewer adrenergic receptors at postnatal days 10 and 20 than observed in the adult. In aged, 24 month-old, rats the magnitude of NE-induced formation of cyclic AMP correlated with the decrease in capillary beta receptors (Kobayashi et al., 1982c). The authors postulated that diminished numbers of beta receptors during development and aging could be associated with the lower capacity for the newborn and aged to regulate cerebrovascular mechanisms (Kobayashi et al., 1982ac).

Table 3: Development of Adenylate Cyclase
in Cerebral Microvessels and Pia-Arachnoid of Rat Brain

		Percent Stimulation over basal activity			
Tissue	Age Postnatal	NE 0.1mM	Gpp(NH)p 0.01mM	NE+ Gpp(NH)p	Forskolin 0.01mM
Capillaries	3 day	17	5	20	160
Pia	3 day	10	24	34	103
Pia	6 day	12	20	37	
Capillaries	8 day	10	1	15	80
Pia	8 day	10	10	17	73
Capillaries	10 day	15	5	25	90
Pia	10 day	10	25	29	88
Capillaries	adult	58	69	118	100
Pia	adult	42	44	78	125

See Palmer and Palmer (1983) for procedural details
N = 3, except adults N = 6.

What functional roles do adrenergic-coupled cyclic AMP systems play with regard to the function of the brain microvasculature? With regard to NE control of capillary permeability the following data are known: 1) Vesicular transport is enhanced in mouse brain following perfusion of the ventricles with either cyclic AMP or NE (Westergaard, 1975); 2) Intracarotid administration of dibutyryl cyclic AMP increases the rate of pinocytosis in rat capillaries (Joo et al., 1972, 1975a); 3) NE fibers arising from the locus ceruleus control the permeability of capillaries to water and possibly maintain a constant regional cerebral volume in the face of changes in systemic blood pressure and osmolarity (Raichle et al., 1975; Hartman et al., 1980; Preskorn et al., 1980); and 4) Lesion of the locus ceruleus followed by induction of either hypertension or seizures results in an increase in the permeability of the BBB, suggesting that adrenergic innervation of microvessels helps preserve the integrity of the barrier during pathological states (Harik and McGunigal, 1984).

Other roles for adrenergic innervation of the central microvasculature involve beta receptor-mediated relaxation and alpha receptor-induced constriction of smooth muscle cells associated with precapillary arterioles and postcapillary venules. Cyclic AMP most likely is involved in the beta adrenergic response (Ulrich et al., 1982; for review see: Edvinsson, 1982).

Beta adrenergic stimulation of cyclic AMP in the cerebral microvasculature would be expected to exert a modulatory role upon specific metabolic processes via an interaction with the protein kinase phosphorylation systems (Joo et al., 1988). Cultures of rat brain capillary endothelia or smooth muscle cells respond to NE with an enhanced breakdown of glycogen. On the other hand, serotonin stimulated synthesis of glycogen (Spatz et al., 1986). Glycogen breakdown is a result of increased requirement for energy and perhaps these two monoamines play opposing roles in energy metabolism. Other biochemically related cyclic nucleotide processes in microvessels have yet to be described. Endothelial cells from rat brain contain actin-like filaments (LeBeux and Willemot, 1978) whose function has not been determined. An intriguing possibility might be a cyclic nucleotide mediated phosphorylation process whereby these filaments contract and cause a separation of endothelial tight junctions resulting in leakage of substances across the BBB.

Dopamine (DA)

It is not known for certain whether DA nerve endings are present in cerebral microvessels, however, one study did reveal DA nerve endings on capillaries in the striatum (Braun et al., 1980). Dopamine apparently constricts cerebral vessels and pia via an interaction with alpha adrenergic receptors. If tone is applied to the vessels DA mediates vasodilation via specific DA receptors (Altura et al., 1980; Edvinsson and MacKenzie, 1977; McCulloch and Harper, 1977; Brodde, 1982). Pial arterioles of rats vasodilate when exposed to small concentrations of DA (Altura et al., 1980), however, adenylate cyclase in pial vessels from rat, rabbit, and cat were only marginally sensitive to DA (Nathanson and Glasser, 1979; Palmer et al., 1980; Palmer and Palmer, 1983). On the other hand, DA readily stimulated the enzyme in microvessels and cultured endothelia from rat, cat and different regions of the rabbit brain, except cerebellum (Baca and Palmer, 1978; Palmer and Palmer, 1978; Palmer, 1981; Karnushina et al., 1982; Nathanson and Glasser, 1979). In these studies DA activation of adenylate cyclase was usually less than that observed for NE, unless GTP analogs were included in the assay (Palmer and Palmer, 1978; Palmer, 1981). The receptor mediating the DA response in the rat was distinct from that for NE and isoproterenol (Baca and Palmer, 1978). The use of radioligand binding techniques has revealed limited numbers of DA-2 receptors only in bovine cerebral microvessels (Peroutka et al., 1980). This receptor is usually coupled to adenylate cyclase in a negative fashion. The presence of the DA-1 receptor coupled to adenylate cyclase in a positive manner has not yet been determined with ligand binding techniques.

In a few instances, DA-responsive adenylate cyclases have been shown to undergo alteration following neuropathological conditions. In spontaneously hypertensive rats (SHR) both cerebral cortical tissue and associated microvessels displayed a reduced sensitivity to DA when compared to normotensive WKY rats (Palmer, 1979). The brain of SHR contains abnormally high levels of DA (Versteeg et al., 1976) which could, over a period of time, lead to subsensitivity of DA receptors. Alternatively, when rats received subchronic injections of reserpine, which depletes central stores of DA, the sensitivity of adenylate cyclase to the monoamine was enhanced (Palmer and Palmer, 1983), an indication of functional denervation supersensitivity of the receptor. Postnatal X-irradiation directed at the cerebral cortex of rats did not result in any change in DA-adenylate cyclase in the associated microvessels (Palmer et al., 1982). Injections of the DA releasing agent, amphetamine, promoted an increase in the permeability of the BBB in the rat, an event inhibited by the neuroleptic, pimozide (Braun et al., 1980). Moreover, amphetamine caused greater changes in the barrier permeability in SHR (Domer et al., 1980). Taken together, the data with DA-adenylate cyclase suggest, albeit tenuously, that this monoamine plays a role in the control of vasomotor responses, as well as in the control of permeability of substances from blood to brain.

Serotonin (5-HT)

Central capillaries, the choroid plexus, and pial vessels receive 5-HT innervation via the raphe nuclei (Reinhard et al., 1979; Edvinsson et al., 1983). Receptor ligand binding investigations with bovine microvessels did not reveal prominent binding sites for 5-HT (Peroutka et al., 1980). Moreover, 5-HT-sensitive adenylate cyclase was not present in isolated central capillaries or cultured endothelia (Baca and Palmer, 1978; Karnushina et al., 1982; Spatz and Mrsulja, 1982; Huang and Rorstad, 1983). Westergaard (1975) demonstrated an enhanced permeability of the BBB following perfusion of either 5-HT or cyclic AMP into the ventricles of rats, however, other workers failed to confirm this finding (Hardebo et al., 1981; Gulati et al., 1984). The role of 5-HT in pial arteries is complex. Normally 5-HT constricts pial vessels, but at high doses or in the presence of trauma or ischemia, 5-HT may dilate cerebral vessels indirectly by evoking release of NE (Edvinsson and MacKenzie, 1977; Marin et al., 1979). Spatz and coworkers (1986) demonstrated that 5-HT acted in opposition to NE in cultured brain endothelia and smooth muscle cells to preserve glycogen stores.

Histamine - Prostaglandins

A histamine-2 receptor activation of adenylate cyclase has been reported for microvessels from guinea pig brain (Huang and Drummond, 1979; Karnushina et al., 1980a). Mixed H-1/H-2 histamine receptors mediate stimulation of adenylate cyclase in central microvessels of Wistar rats (Joo et al., 1975a, 1984) but not in cultured endothelia or microvessels and pia from other strains of rats (Karnushina et al., 1982; Herbst et al., 1979; Palmer and Palmer, 1978). Mixed histamine receptor coupled adenylate cyclases were likewise present in microvessels from different regions of the rabbit brain except cerebellum. Alternatively, the pial vessels were insensitive to histamine (Palmer et al., 1980; Palmer, 1981a). Histamine increases the permeability of the BBB (Gross et al., 1981). An interesting series of experiments revealed that after prolonged exposure of rats to hypobaric-hypoxia the permeation of serum albumin into endothelial cells increased in conjunction with the number of pinocytotic vesicles. Intracarotid injection of histamine stimulated this process even further via an H-2 receptor mechanism. Histamine did not, however, modify the structure of endothelial junctions in normoxic animals. In addition, during hypobaric-hypoxia capillary adenylate cyclase activity increased, an occurrence also linked to pinocytosis (Joo et al., 1981; Dux et al., 1981, 1982, 1984). Moreover, histamine stimulated the synthesis of a variety of prostanoids in normal brain microvessels, an outcome that was augmented following exposure to chronic hypoxia (Joo et al., 1981; Dux et al., 1982). These results are intriguing because prostaglandins of the E series and prostacyclin activate adenylate cyclase in both capillaries and cerebrovascular muscle cells (Huang and Drummond, 1979; Karnushina et al., 1982; Huang and Rorstad, 1983; Spatz, 1986). Prostacyclin dilates microvessels and inhibits aggregation of platelets as a protective response to hypoxic/ischemic situations and has been tried with limited success to treat victims of stroke (Miller et al., 1984; Huczynski et al., 1985). In addition to PGI2 other prostaglandins (D2 and E1) were shown to protect mice from cerebral hypoxia/anoxia (Masuda et al., 1986). The histamine/prostaglandin interactions in stroke disorders are contradictory in that histamine itself appears to act in a deleterious manner by promoting brain edema, while at the same time evoking the synthesis of beneficial species of prostaglandins which, along with histamine, serve to promote vasodilatation (Dux and Joo, 1982; Wahl and Kuschinsky, 1979; Vane et al., 1987). In related work, H-2 antagonists inhibited brain edema resulting from exposure of dogs to irradiation (Joo et al., 1976). Moreover, production of edema in rats by surgical decompression was associated with an increased sensitivity of microvascular adenylate cyclase to histamine (Joo et al., 1984).

Other Neurohumoral Agents

Adenosine-During ischemia adenosine is released in quantity and mediates cerebrovasodilatation thereby enhancing blood flow to the affected brain region (Winn et al., 1981). One source of adenosine is the rapid breakdown of ATP following an ischemic episode. Human cerebral capillaries possess a four-fold greater capacity for adenosine uptake than found in the pia or cortical tissue (Kalaria and Harik, 1988). Moreover, rat and bovine microvessels have high affinity adenosine uptake mechanisms (Wu and Phillis, 1982; Stefanovich, 1983). Adenosine binds specifically to A-2 type receptors in human microvessels and pia (Kalaria and Harik, 1988), and has been shown to activate adenylate cyclase in cat pial vessels at a potency comparable to its ability to elicit relaxation (Edvinsson and Fredholm, 1983). Alternatively, we observed a lower specific binding of 2-Cl-adenosine to rat pia in conjunction with little or no capability to activate adenylate cyclase. In contrast, adenosine binding was greater in rat cerebral microvessels where, in common with guinea pig cerebral microvessels, the presence of an adenosine-sensitive adenylate cyclase has been demonstrated (Palmer and Ghai, 1982; Huang and Drummond, 1979; Schutz et al., 1982). Physiological studies conducted by Hardebo et al., (1987) localized an A-2 type receptor (activates adenylate cyclase) on P-1 receptors of microvascular smooth muscle. This receptor caused muscle relaxation while the P-2 receptor promoted constriction in smooth muscle; but dilation of endothelia. More data are needed to determine the functional significance of the diverse adenosine receptors associated with the microvascular elements of the brain.

Peptide Hormones-Parathyroid hormone potently activates adenylate cyclase in microvessels isolated from rat, guinea pig, cattle and swine brains (Huang et al.,1983; Huang and Rorstad, 1984). Parathyroid hormone induces relaxation of cerebral arteries from different species and this process appears to be mediated by cyclic AMP (Suzuki et al., 1983).

Vasoactive intestinal polypeptide (VIP) is another peptide capable of stimulating cyclic AMP in isolated cerebral microvessels. In the rat the process requires GTP to obtain full expression of enzyme activity. Microvessels from other species (guinea pig, cattle and pig) were considerably less responsive to VIP, while the peptide was inhibitory in cultured rat endothelial cells. On the other hand, homogenates of bovine cerebral arteries displayed a high degree of sensitivity to VIP (Huang and Rorstad, 1984; Karnushina et al., 1982). VIP induces a potent dilatation of cat cerebral vascular arteries accompanied by a modest rise in cyclic AMP (Edvinsson et al., 1985). Relaxation of these arteries does not require endothelial relaxing factor which would obviate any role for acetylcholine-cyclic GMP contributions (Duckles and Said, 1982). Moreover, VIP nerve fibers have been immunohistochemically localized in the adventitia-media border of pial arteries in cats and in the perivascular region of canine extracranial arteries (Larsson et al., 1976; Abrams et al., 1982; for review see: Edvinsson, 1985).

Another peptide namely, calcitonin gene-related peptide (CGRP), has been identified in fine varicose nerve fibers that invest all major arterioles and arteries of the cat brain. Exogenous CGRP was a potent relaxant of cat cerebral arteries. Of all the peptides identified in perivascular nerve fibers, CGRP was the most potent in effecting cerebrovascular dilation (McCulloch et al., 1986). Vasodilatory CGRP nerve fibers were present in the adventitia of cerebral arteries of guinea pig, rabbit, and mouse. The source of the CGRP nerve cell bodies is the trigeminal ganglion where it is often co-localized with substance-P and neurokinin-A. Human pial arteries also dilate in response to CGRP (McCulloch et al., 1986; Edvinsson et al., 1987, 1988). The pronounced relaxation of rat cerebral vessels by CGRP occurs simultaneously with a marked stimulation of cyclic AMP. On the other hand, substance-P did not evoke cyclic AMP formation and its ability to dilate cerebral vessels apparently requires endothelial relaxing factor (Edvinsson et al., 1985). Presently, it is unknown whether substance-P or neurokinin-A mediate smooth muscle relaxation via an influence on cyclic nucleotide formation.

Other peptide hormones affect permeability of central microvessels and in some instances promote vasodilation: 1) Angiotensin increases pinocytotic activity (Nag et al., 1979), permeability to water (Grubb and Raichle, 1981), binds with high affinity to dog microvessels (Speth and Harik, 1985), acts to constrict larger cerebral vessels (Wei et al., 1978), and has either no action on (Huang and Rorstad, 1983) or inhibits cyclic AMP formation (Karnushina et al., 1982); 2) Neuropeptide-Y constricts cerebral vessels and inhibits adenylate cyclase (Fredholm et al., 1985); 3) Vasopressin receptors are present in the cerebral microvessels of swine (Pearlmutter et al., 1988) and the peptide has been shown to alter the permeability of microvessels to water (Grubb and Raichle, 1981), but apparently has no influence on cyclic AMP mechanisms (Karnushina et al., 1982); and 4) Gamma amino butyric acid inhibits cyclic AMP in cultured endothelia (Karnushina et al., 1982) and acts to dilate pial vessels (for reviews see: Edvinsson, 1985; Palmer, 1986).

It appears that a variety of neurohumoral agents, all act in common to relax vascular smooth muscle by formation of cyclic AMP, however, each is mediated via a different receptor.

GUANYLATE CYCLASE

Guanylate cyclase [EC 4.6.1.2] has been localized histochemically to the luminal and basal membranes of capillaries isolated from rat brain (Karnushina et al., 1980b). Biochemical determinations using disrupted cellular preparations revealed that pia contained an 8-fold greater enzyme activity than capillaries. In the presence of low Mn^{++}, Ca^{++} activated the enzyme 2-4 fold using either soluble fractions (100,000 x g) or homogenates as enzyme sources from the two tissues. In the presence of Mg^{++} (3 mM) and catalase (30 ug), a NO forming activator, hydroxylamine, enhanced guanylate cyclase 3-4 fold. Neither monoamines nor muscarinic agonists influenced the enzyme (Karnushina et al., 1980b; Palmer, 1981b; Palmer et al., 1983).

Intracarotid injection of cyclic GMP analogs into rats resulted in an enhanced transport of albumin across the BBB, as well as increased numbers of pinocytotic vesicles in the central capillaries (Joo et al., 1983). This latter action was observed earlier with cyclic AMP (Joo et al., 1972, 1975a) and indicates that the two nucleotides might act in concert to regulate central pinocytotic mechanisms. It is not known whether pinocytosis is mediated via cyclic nucleotide dependent protein kinases. Recent work by Joo et al.,(1988) revealed that rat cerebral capillaries contained higher levels of cyclic GMP dependent protein kinase than of cyclic AMP dependent protein kinase.

The work of Grammas et al., (1983, 1985) indicated the presence of muscarinic receptors in rat cerebral capillaries that inhibited catecholamine-elicted formation of cyclic AMP. The critical study would be to determine whether muscarinic agonists have the capability to form cyclic GMP using intact preparations of microvessels. In interesting recent work, atrial natriuretic peptide was shown to activate guanylate cyclase-cyclic GMP in both intact and disrupted capillaries from rabbit brain. The process did not require NO, but was a direct stimulation of the enzyme. The pial enzyme was only marginally influenced by atrial natriuretic peptide while the choroid plexus, especially the epithelial cells therein, was highly sensitive to the hormone (Steardo and Nathanson, 1987). The rat choroid plexus contained receptors to atrial natriuretic peptide that were coupled to cyclic GMP and increased in number following experimental hydrocephalus (McCarty and Plunkett, 1986; Tsutsumi et al., 1987, 1988). Atrial natriuretic peptide inhibited cerebrospinal fluid production in the rabbit choroid plexus (Steardo and Nathanson, 1987) and may serve to counteract the effects of agents such as cholera toxin which elevate cerebrospinal fluid levels by a process presumably linked to cyclic AMP synthesis (Epstein et al., 1977). Norepinephrine which is also coupled to cyclic AMP serves, like atrial natriuretic peptide, to inhibit formation of cerebrospinal fluid (Lindvall and Owman, 1980). With respect to capillary function, NE has been shown to increase permeability to water (Preskorn et al., 1980). Perhaps atrial natriuretic peptide acting through cyclic GMP might inhibit this process and thereby limit the production of cerebral edema.

Pial vessels and arterioles contain small amounts of smooth muscle which form cyclic AMP in response to beta adrenergic agonists (Wroblewska et al., 1984; Palmer, 1983). The effect of beta adrenergic-linked cyclic AMP is vasodilation while alpha adrenergic agonists promote constriction (Edvinsson, 1982). Perhaps muscarinic-induced relaxation of smooth muscle via endothelial relaxing factor (NO) activation of guanylate cyclase- cyclic GMP (Vane et al., 1987) is another process whereby the two cyclic nucleotides work in concert within the cerebral microvasculature.

NEUROPATHOLOGICAL CONDITIONS

Hypertension

When arterial pressure increases, cerebral vessels maintain constant blood flow by vasoconstriction. Marked increases in pressure exceed this autoregulatory capacity resulting in BBB disruption and cerebral edema (Edvinsson, 1982; Heistad, 1982). Spontaneously hypertensive rats (SHR) represent a genetic model of hypertension in which experimental findings can be readily compared to the normotensive parent, WKY strain. A loss of adrenergic control of the BBB has been demonstrated in SHR. For example, agents such as NE or amphetamine cause greater barrier disruption in SHR (Domer et al., 1980; Ben-Menachem et al., 1982). When the brain of normal rats is depleted of NE innvervation the integrity of the BBB is disrupted during NE-induced hypertension (Harik and McGunigal, 1984), an indication that an intact adrenergic nervous system is required to protect against hypertension (Edvinsson, 1982). Cerebrovascular hypertrophy is characteristic of stroke-prone SHR and disappears following removal of sympathetic innervation. The hypertrophy serves to protect cerebral vessels from stress (Edvinsson, 1982; Werber and Heistad, 1984).

The electron microscope has revealed fewer pinocytotic vesicles, canalization of tight junctions and thickened basal lamina in central capillaries from SHR (Singh et al., 1981). Pial vessels from stroke prone SHR have been described as having a "sausage string" appearance (Werber and Heistad, 1984).

With respect to receptor-coupled actions, adenylate cyclase sensitivity to catecholamines (NE, DA, isoproterenol, salbutamol and dobutamine) was significantly reduced in cerebral vessels and cortex of SHR (Palmer, 1979). A similar lack of adenylate cyclase responses have been demonstrated repeatedly in peripheral vascular and myocardial tissue from SHR (Bhalla and Ashley, 1978; Asano et al., 1988). Magnoni and coworkers (1983) reported a decrease in the number of beta adrenergic receptors in cerebral microvessels from genetic (SHR) or experimentally-induced hypertensive rats. Coupled to the finding that vascular and parenchymal catecholamine content was greater in SHR brain (Versteeg et al., 1976; also See Palmer, 1986), the data indicate that adrenergic receptors on central microvessels possess the capability of "down regulation" in the face of excessive catecholamine input. Further work of this nature reported increased alpha-1 receptors (Kobayashi et al., 1985) and reduced activity of angiotensin-converting enzyme (Kobayashi et al., 1984) in microvessels of SHR. The increase in alpha-1 receptors may represent a compensatory mechanism of the SHR to control blood flow through arterioles while the beta receptor reduction might limit cyclic AMP-induced edema via pinocytotic processes.

Monoamine Depletion

Depletion of central catecholamines by chemical or surgical methods has been shown repeatedly to produce a denervation sensitivity of adrenergic receptor-coupled events leading to cyclic AMP synthesis (for review see: Palmer, 1983). Likewise, we observed that subchronic injections of reserpine resulted in increased adenylate cyclase responses to NE and DA in the rat cerebral capillaries. Pial responses following reserpine were even greater, plus the NE and isoproterenol stimulation was further augmented in the presence of Gpp(NH)p, an occurrence not present in the capillaries (Palmer and Palmer, 1983). These studies show that microvessels have the capability to "up regulate" in response to environmental changes. In one corollary investigation, Preskorn et al., (1984) used the tetrabenazine model of depression in rats to assess the influence of NE on central capillary permeability to water. Administration of CO_2 normally enhances the adrenergic firing rate of the locus ceruleus to increase capillary permeability to water. This regulatory control mechanism was blocked following monoamine depletion by tetrabenazine.

Irradiation

Irradiation of the brain is followed by a breakdown in the BBB resulting in edema due to damage of the capillary endothelium. In addition, increased thickening of the pial vessels and injury to nerve terminals occurs. Perikarya and glia, however, appear somewhat resistant to irradiation. The damage is postulated to involve free radical mechanisms (Remler and Marcussen, 1981; Clemente et al., 1960; Cockerham et al., 1988). Immediately following irradiation significant elevations in the cerebrospinal fluid levels of cyclic AMP and cyclic GMP were observed in monkeys (Catravas et al., 1979). In chronic studies, neonatal rats received X-irradiation of the posterior cortex. Neuropathology was evident in the dentate gyrus of the adult hippocampus concomitant with a reduced NE sensitivity of adenylate cyclase. The capillaries of the adult cortex were undamaged, but the pia-arachnoid displayed hyperplasia as well as diminished responses of adenylate cyclase to NE and/or Gpp(NH)p (Palmer et al., 1982). Further investigations involving neonatal X-irradiation of the cerebellum in rats have resulted in the following observations for the adult: 1) Decreased cyclic AMP stimulation by NE (Hoffer et al., 1976; Turck et al., 1981); 2) An increase in Type I cyclic AMP dependent protein kinase and a decrease in cyclic GMP dependent protein kinase activity (Dolphin et al., 1983); and 3) A decrease in the total number of beta and specifically beta-2 adrenergic receptors (Turck et al., 1981; Minneman et al., 1981). In the rat striatum and cerebrum irradiation respectively reduced DA receptors and forskolin-adenylate cyclase (Andersen and Nielsen, 1987), as well as alpha-2 receptors (Nyquist-Battie et al., 1983). Schlegel et al., (1980) provided evidence that the free radicals produced by irradiation inactivated the GTP-transducer site on adenylate cyclase. When free radicals were produced under in vitro conditions and presumably formed following post ischemic reperfusion in vivo, it was shown that GTP mediated coupling of adenylate cyclase was especially vulnerable to damage (Palmer et al., 1986; Palmer, 1987).

Diabetes

The role of the microvasculature during diabetes is to maintain constant cerebral blood flow despite the associated pathology which in the streptozotocin model of the rat is manifested by degeneration of microvessels with enlarged vacuoles. The arterioles possess irregularly shaped muscle cells, swollen endoplasmic reticula, vacuoles and decreased staining of cytoplasmic proteins. Moreover, there is a loss in the total number of venules (Bohlen, 1987). These rats also display an increase in the permeability of the BBB (Stauber et al., 1981). Retinal microvessels likewise are especially susceptible to diabetes-induced damage resulting in a host of microscopic changes including endothelial proliferation, dilation and microaneurysms. The small hemorrhages contribute to retinal degeneration and eventually blindness (Yanoff, 1969; Sosula et al., 1972). We looked at cyclic nucleotide enzymes in microvessels, pia, cerebrum and retina of rats made diabetic with streptozotocin. Receptor sensitivities of NE-and-DA-adenylate cyclase were significantly depressed in the cerebrum and retina of diabetic rats. All modalities of adenylate cyclase stimulation [NE, Gpp(NH)p, forskolin] were drastically reduced in the cerebral capillaries, while pial responses were scarcely affected (Palmer et al., 1983). These studies were in contrast to results obtained following neonatal X-irradiation of the rat brain, where the pial enzyme suffered the principal damage and the retinal enzyme became hyperresponsive to DA (Palmer et al., 1982). Neither phosphodiesterases nor guanylate cyclases were changed in the studies of the diabetic rat brain. In support of our work Magnoni and coworkers (1985b) reported a significant reduction in the number of beta adrenergic receptors in cerebral microvessels from genetic and streptozotocin-induced diabetic rats. The reduced responsiveness of the microvessels may contribute to the neuropathology of diabetes and could additionally contribute to the lowered capacity of diabetic vessels for metabolism (Morisaki et al., 1982). Peripheral tissues from diabetic animals have likewise been shown to possess attenuated beta receptor coupled adenylate cyclases (Atkins et al., 1985).

Stroke

The biochemical sequelae following stroke-related disorders (anoxia, ischemia, post ischemic reperfusion, vasospasm etc..) are complicatd by a host of metabolic events namely, energy metabolism, transmitter release, ionic fluxes especially calcium, breakdown of phospholipids with release of free fatty acids, prostanoid synthesis and free radical formation. Capillary damage occurs, small vessels lose their capability for autoregulation (Ong et al., 1987), and edema ensues (Raichle, 1983). Free radicals damage complex molecules especially those associated with cell membranes. Thus, complex membrane enzymes such as Na+,K+-ATPase and adenylate cyclase are highly vulnerable (Palmer, 1987). The GTP-transducer site mediating the synthesis of cyclic AMP (Gs protein) is particularly sensitive to the consequences of post ischemic reperfusion (Palmer et al., 1986). Despite the extensive ongoing investigations in the stroke field only a few studies have involved receptor-coupled cyclic nucleotide events within the microcirculation. In the gerbil following global ischemia (no recirculation) the sensitivity of adenylate cyclase to NE, Gpp(NH)p and forskolin in the pial vessels was unchanged (Palmer, unpublished). After common carotid occlusion of either the right or the left artery, the cerebral microvessels in gerbils and rats contained a decreased density of beta adrenergic receptors, especially localized to the left side of the brain (Magnoni et al., 1985b). The decreased receptors on the left did not occur if the rats were treated with 6-hydroxydopamine or the corpus callosum was sectioned prior to stroke (Magnoni et al., 1986). Subsequent to subarachnoid hemorrhage in humans, high affinity presynaptic alpha-2 receptors disappeared from cerebral arteries. The authors attributed the receptor loss to the increased vulnerability of patients to delayed vasospasm following a cerebrovascular accident (Tsukahara et al., 1988). Dux and coworkers (1981, 1982, 1984) reported an interesting series of events in rat microvessels following prolonged hypoxia or hypobaric-hypoxia. These conditions increase adenylate cyclase activity and release histamine which promotes the synthesis of E type prostaglandins and prostacyclin which then act in concert to stimulate production of cyclic AMP. The cyclic nucleotide enhances pinocytotic activity resulting in edema observed in the swollen processes of astrocytes. Pinocytosis has also been observed in other species following ischemia (Shibata et al., 1988). Histamine may additionally worsen edema by an H-2 receptor action to open the tight junctions between endothelial cells (Joo et al., 1976). A proliferation of capillary endothelium following stroke has been observed by us (unpublished) in dog brain and by Garcia et al., (1981) in the rat. Immature capillaries lack tight junctions and could likewise contribute to delayed edema (Goldstein and Betz, 1983).

The series of events described in the preceding paragraph could be postulated to contribute to the scheme of metabolic activity described in Figure 1. The processes of cyclic nucleotide action contribute in a beneficial manner to relax smooth muscle and increase cerebral blood flow. Associated events such as prostacyclin formation would prevent platelet aggregation and along with synthesis of E type prostaglandins and release of adenosine would dilate cerebral arterioles. Both adenosine and cyclic AMP act to dampen central firing of neurons and would presumably preserve crucial metabolic substrates. The enhanced pinocytotic activity observed after stroke could be a cyclic nucleotide mediated process whereby additional substrates are supplied to depleted nerve cells. Conversely, the process mediated by both cyclic nucleotides could be deleterious resulting in edema. Via an activation of protein kinase the large increases in cyclic nucleotides observed following a stroke would likewise promote key metabolic processes and energy metabolism within the affected neurons. Events acting in opposition would include inhibition of cyclic AMP synthesis by substances that activate guanylate cyclase. The thrombin-Ca^{++}-phospholipase-induced release of arachidonate from capillary membranes would act beneficially to synthesize cyclic GMP, however, the process forms thromboxanes which promote vasoconstriction and platelet aggregation leading to decreased cerebral blood flow. Furthermore, thromboxane synthesis along with calcium-induced conversion of xanthine dehydrogenase to xanthine oxidase forms free radicals. The free radicals attack macromolecules, lysosomes and endothelial membranes resulting in worsening of edema and cellular damage. Unfortunately, little information is available with regard to the ratios of these opposing neuropathological events or as to the sequences in which they occur. (For relevant discussions see: Raichle, 1983; Palmer, 1986; Palmer et al., 1987; Vane et al., 1987).

Figure 1: Cylic Nucleotide Related Events in Stroke and Hypoxia.

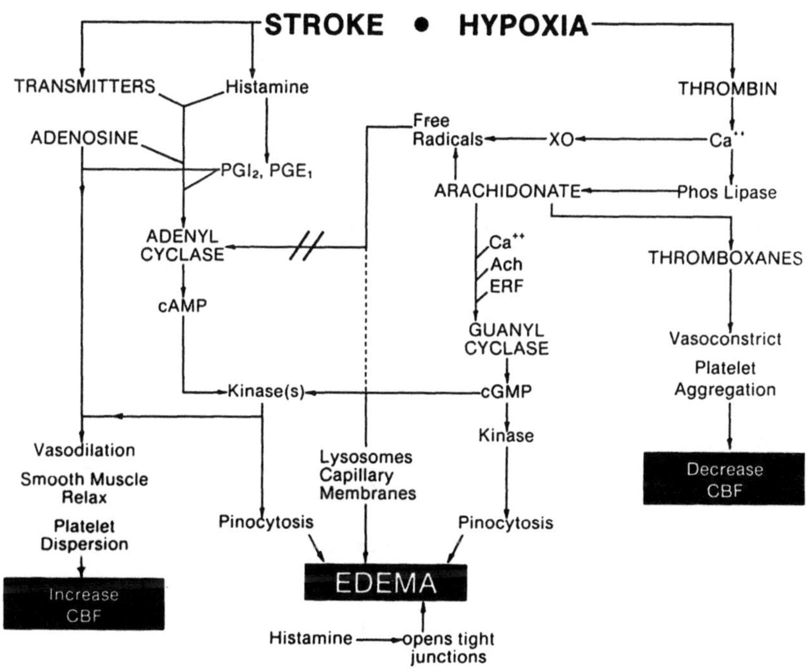

Figure 1 Legend

Abbreviations: Ach = Acetylcholine, adenyl cylase = adenylate cyclase, cAMP = cyclic AMP, CBF = cerebral blood flow, cGMP = cyclic GMP, ERF = endothelial relaxing factor, guanyl cyclase = guanylate cyclase, phoslipase = phospholipase, and XO = Xanthine Oxidase.

Other CNS Disorders

In Parkinson patients adrenergic receptor binding studies were performed using alpha-1, alpha-2 and beta ligands. The microvessels from the putamen were found to have decreased numbers of alpha-1 binding sites. The authors felt that receptor changes might be reflected in alterations in cerebral blood flow reported in these patients (Cash et al., 1985).

If noradrenergic control of cerebral microvessels was destroyed, chemically induced seizures resulted in a greater leakage of albumin across the blood-brain barrier (Harik and McGunigal, 1984). The sequelae associated with seizure activity include rapid energy depletion, edema, and anoxia; conditions not unlike stroke. Using a rat kindling model of epilepsy, subthreshold electrical stimulation (8 mA x 2 sec, 2 x day) was applied via corneal electrodes until all rats displayed grade 4 seizure activity (6 days). On day 7 the pia-arachnoid and capillaries were removed (Head et al., 1980) and adenylate cyclase sensitivity to neurohumoral agents was subsequently determined in the frozen tissue homogenates.

The freezing/thawing cycle destroyed the sensitivity of adenylate cyclase to NE (0.1mM) in the presence of GTP (0.01mM), however, a prominent enzyme response to added forskolin (0.01 mM) was seen. Kindling-type seizures significantly elevated basal adenylate cyclase in the capillaries but only a tendency was evident in the pia. The percent activation of the enzyme by forskolin was depressed, albeit to a nonsignificant degree in both capillaries and pia. These unpublished findings are presented in Table 4.

Table 4: Adenylate Cyclase in Rat Cerebral Microvessels and Pia-arachnoid Following Development of Kindling Seizures

			Adenylate Cyclase	
Tissue	N	Stimulus	Control	Epilepsy
Microvessels	4	basal	25 ± 5	35 ± 6 (p<.003)
Microvessels	4	forskolin	602 ± 97%	512 ± 67%
Pia	5	basal	23 ± 3	27 ± 4
Pia	5	forskolin	375 ± 27%	302 ± 23%

Basal adenylate cyclase is expressed as p moles cyclic AMP formed/min/mg protein. Forskolin activity is percent stimulation over basal. Values are Mean ± SE.

Perhaps the anoxic conditions produced by the kindling seizures elevated the basal activity of the capillary preparation an occurrence in

Table 5: Influence of Neuropathological Conditions on Receptor Coupled Systems in the Central Microvasculature

Disease and Species	Tissues	Adenyl Cyclase Sensitivity	Receptor Binding	Reference
Hypertension	MV	decrease	decrease beta	1,2
(rats)	MV		increase alpha-1	3
Diabetes	MV	decrease	decrease beta	4,5
(rats)	pia	no change		4
Monoamine depletion	MV	increase	decrease beta	6,9
(rats)	pia	increase		6
Irradiation	MV	no change		7
(rats)	pia	decrease		7
Stroke (gerbil)	pia	no change		8
(gerbil and rat)	MV		decrease beta	9
(rats)	MV	increase		10
Parkinsonism (human)	MV		decrease alpha	11
Seizures	MV	basal ↑		8
	pia	none		8
Development	MV	decrease	decrease alpha, beta	12, Table 3
	pia	decrease		Table 3
Aging	MV	decrease	decrease beta	13

MV - Microvessels, Adenylate cyclase sensitivity is in response to monoamines

References:

1)Palmer, 1979
2)Magnoni et al., 1983
3)Kobayashi et al., 1985
4)Palmer et al., 1983
5)Magnoni et al., 1985b
6)Palmer and Palmer, 1983
7)Palmer et al., 1982
8)Palmer et al., (unpublished)
9)Magnoni et al., 1985a, 1986
10)Dux et al., 1982, 1984
11)Cash et al., 1985
12)Kobayashi et al., 1982a
13)Kobayashi et al., 1982c

common with the findings of Dux et al., (1984) following exposure of rats to prolonged hypobaric/hypoxia.

CONCLUSION

The receptors mediating the stimulation of adenylate cyclase in the microvasculature of the central nervous system are sensitive to environmental and neuropathologic conditions. The principal work has involved the role of the adrenergic input to the microvasculature and these findings are summarized in Table 5. Recent work has focused on the roles of peptide innervation, as well as the nature of the diverse types of adenosine receptors. Future work with regard to neuropathological conditions hopefully will be directed to the role of these neurohumoral agents. Additional work is needed with respect to the metabolic control exerted by the cyclic nucleotides under both normal and disease states.

ACKNOWLEDGEMENTS

The authors wish to thank Kathy M. Montesano for typing the manuscript, Fred C. Kaiser for photography, Rick Marciszewski for art work, George Garske for proof reading, Dr. J.C. Blosser for technical advice and use of neurochemical laboratory, and Dr. J.C. Strand for computer assistance in formatting the manuscript.

REFERENCES

Abrams GM, Recht LD, Haldar J, Zimmerman EA (1982) Vasoactive intestinal peptide (VIP): Localization and pharmacologic effects on canine extracranial arteries. Neurol 32:A110-A111

Altura BM, Gebrewold A, Lassoff S (1980) Biphasic responsiveness of rat pial arterioles to dopamine: Direct observations on the microcirculation. Br J Pharmacol 69:543-544

Andersen PH, Nielson M (1987) Irradiation inactivation studies of the dopamine D-1 receptor and dopamine-stimulated adenylate cyclase. Neurosci 83:167-172

Asano M, Masuzawa K, Matsuda T (1988) Evidence for reduced beta-adrenoceptor coupling to adenylate cyclase in femoral arteries from spontaneously hypertensive rats. Br J Pharmacol 94:73-86

Atkins FL, Dowell RT, Love S (1985) Beta-adrenergic receptors, adenylate cyclase activity, and cardiac dysfunction in the diabetic rat. J Cardiovasc Pharmacol 7:66-70

Baca GM, Palmer GC (1978) Presence of hormonally-sensitive adenylate cyclase receptors in capillary-enriched fractions from rat cerebral cortex. Blood Vessels 15:286-298

Ben-Menachem E, Johansson BB, Svensson TH (1982) Increased vulnerability of the blood-brain barrier to acute hypertension following depletion of brain noradrenaline. J Neural Trans 53:159-167

Bhalla RC, Ashley T (1978) Altered function of adenylate cyclase in

myocardium of the spontaneously hypertensive rat. Biochem Pharmacol 27:1967-1971
Bohlen HG (1987) Cerebral microvascular control in normal, hypertensive and diabetic conditions. In: McDonagh PF (ed). Microvascular Perfusion and Transport in Health and Disease. Karger Publishing, New York, pp 80-112
Braun U, Braun G, Sargent T III (1980) Changes in blood-brain permeability resulting from d-amphetamine, 6-hydroxydopamine and pimozide measured by a new technique. Experientia 36:207-209
Brodde OE (1982) Vascular dopamine receptors: Demonstration and characterization by in vitro studies. Life Sci 31:289-306
Cash R, Lasbennes F, Sercombe R, Seylaz J, Agid Y (1985) Adrenergic receptors on cerebral microvessels in control and Parkinsonian subjects. Life Sci 37:531-536
Catravas GN, Wright SJ Jr, Trocha PJ, Takenaga J (1979) Cyclic AMP, cyclic GMP and amino acid levels in the cerebrospinal fluid of the primate: Effects of radiation. Fed Proc 38:231
Clemente CD, Yamazaki JN, Bennett LR, McFall RA (1960) Brain radiation in newborn rats and differential effects of increased age. Neurology 10:669-675
Cockerham LG, Arroyo CM, Hampton JD (1988) Effect of 4-hydroxypyrazolo (3,4-d) pyrimidine (allopurinol) on postirradiation cerebral blood flow: Implications of free radical involvement. Free Radical Biol Med 4:279-284
Dolphin AC, Detre JA, Schlichter DJ, Nairn AC, Yeh HH, Woodward DJ, Geengard P (1983) Cyclic nucleotide-dependent protein kinases and some major substrates in the rat cerebellum after neonatal X-irradiation. J Neurochem 40:577-581
Domer FR, Sankar R, Cole S, Wellmeyer D (1980) Dose-dependent, amphetamine-induced changes in permeability of the blood-brain barrier of normotensive and spontaneously hypertensive rats. Exp Neurol 70:576-585
Duckles SP, Said SI (1982) Vasoactive intestinal peptide as a neurotransmitter in the cerebral circulation. Eur J Pharmacol 78:371-374
Dux E, Joo F, Hideg J, Dux L, Gecse A, Mezei Z (1981) Prostacyclin and prostaglandin synthesis in rat brain microvessels: Characteristic changes in hypoxaemia. Acta Biol Acad Sci Hungary 32:291-300
Dux E, Joo F (1982) Effects of histamine on brain capillaries. Exp Brain Res 47:252-258
Dux E, Joo F, Gecse A, Mezei Z, Dux L, Hideg J, Telegdy G (1982) Histamine-stimulated prostaglandin synthesis in rat brain microvessels. Agents Actions 12:146-148
Dux E, Temesvari P, Joo F, Adam G, Clementi F, Dux L, Hideg J, Hossman K-A (1984) The blood brain barrier in hypoxia: Ultrastructural aspects and adenylate cyclase activity of brain capillaries. Neurosci 12:951-958
Edvinsson L, MacKenzie ET (1977) Amine mechanisms in the cerebral circulation. Pharmacol Rev 28:275-348
Edvinsson L (1982) Sympathetic control of cerebral circulation. Trends Neurosci 5:425-429
Edvinsson L, Degueurce A, Duverger D, MacKenzie ET, Scatton B (1983) Central serotonergic nerves project to the pial vessels of the brain. Nature 306:55-57
Edvinsson L, Fredholm BB (1983) Characterization of adenosine receptors in isolated cerebral arteries of cat. Br J Pharmacol 80:631-637
Edvinsson L (1985) Functional role of perivascular peptides in the control of cerebral circulation. Trends Neurosci 8:126-131
Edvinsson L, Fredholm BB, Hamel E, Jansen I, Verrecchia C (1985) Perivascular peptides relax cerebral arteries concomitant with

stimulation of cyclic adenosine monophosphate accumulation or release of an endothelium-derived relaxing factor in the cat. Neurosci Lett 58:213-217
Edvinsson L, Ekman R, Jansen I, McCulloch J, Uddman R (1987) Calcitonin gene-related peptide and cerebral blood vessels: Distribution and vasomotor effects. J Cereb Blood Flow Metab 7:720-728
Edvinsson L, Brodin E, Jansen I, Uddman R (1988) Neurokinin A in cerebral vessels: Characterization, localization and effects in vitro. Reg Peptides 20:181-197
Epstein MH, Feldman AM, Brusilow SW (1977) Cerebrospinal fluid production: Stimulation by cholera toxin. Science 196:1012-1013
Fredholm BB, Jansen I, Edvinsson L (1985) Neuropeptide Y is a potent inhibitor of cyclic AMP accumulation in feline cerebral blood vessels. Acta Physiol Scand 124:467-469
Friedman AH, Davis JN (1980) Identification and characterization of adrenergic receptors and catecholamine-stimulated adenylate cyclase in hog pial membranes. Brain Res 183:89-102
Garcia JH, Conger KA, Briggs L (1981) Brain capillaries in acute and subacute regional ischemia. In Cervos-Navarro J, Fritschka E (eds). Cerebral Microcirculation and Metabolism. Raven Press, New York, pp 83-91
Goldstein GW, Betz AL (1983) Recent advances in understanding brain capillary function. Ann Neurol 14:389-395
Grammas P, Diglio CA, Marks BH, Giacomelli F, Wiener J (1983) Identification of muscarinic receptors in rat cerbral cortical microvessels. J Neurochem 40:645-651
Grammas P, Diglio CA, Giacomelli F, Weiner J (1985) Cholinergic-adrenergic receptor interactions in cerebral microvessels. J Neurochem 44:1732-1735
Gross PM, Teasdale GM, Angerson WJ, Harper AM (1981) H-2 Receptors mediate increase in permeability of the blood-brain barrier during arterial histamine infusion. Brain Res 210:396-400
Grubb RL Jr, Raichle ME (1981) Intraventricular angiotensin II increases brain vascular permeability. Brain Res 210:426-430
Gulati A, Agarwal SK, Shukla R, Srimal RC, Dhawan BN (1984) Evidence for the lack of serotonergic mechanisms in the regulation of the blood-brain barrier. Pharmacol Res Comm 16:181-188
Hardebo JE, Owman C (1980) Barrier mechanisms for neurotransmitter monoamines and their precursors at the blood-brain interface. Ann Neurol 8:1-11
Hardebo JE, Owman C, Wiklund L (1981) Influence of neurotransmitter monoamines and neurotoxic analogues on morphologic blood-brain barrier function. In: Cervos-Navarro J, Fritschka E (eds). Cerebral Microcirculation and Metabolism. Raven Press, New York, pp 177-180
Hardebo JE, Kahrstrom J, Owman C (1987) P1- and P2- purine receptors in brain circulation. Eur J Pharmacol 144:343-352
Harik SI, Sharma VK, Wetherbee JR, Warren RH, Banerjee SP (1980) Adrenergic receptors of cerebral microvessels. Eur J Pharmacol 61:207-208
Harik SI, McGunigal T Jr (1984) The protective influence of the locus ceruleus on the blood-brain barrier. Ann Neurol 15:568-574
Hartman BK, Zide D, Udenfriend S (1972) The use of dopamine beta-hydroxylase as a marker for the central noradrenergic nervous system in rat brain. Proc Nat Acad Sci USA 69:2722-2726
Hartman BK, Swanson LW, Raichle ME, Preskorn SH, Clark HB (1980) Central adrenergic regulation of cerebral microvascular permeability and physiologic evidence. Adv Exp Med Biol 131:113-126
Head RJ, Hjelle JT, Jarrott B, Berkowitz B, Cardinale G, Spector S (1980) Isolated brain microvessels: Preparation, morphology, histamine and catecholamine contents. Blood Vessels 17:173-186

Heistad DD (1982) Effects of nerves on cerebral vessels in stroke, cerebral edema, and hypertension. J Lab Clin Med 99:139-147
Herbst TJ, Raichle ME, Ferrendelli JA (1979) Beta-adrenergic regulation of adenosine 3',5'-monophosphate concentration in brain microvessels. Science 204:330-332
Hoffer BJ, Freedman R, Woodward DJ, Daly JW, Skolnick P (1976) Beta-adrenergic control of cyclic AMP- generating systems in cerebellum: Pharmacological heterogeneity confirmed by destruction of interneurons. Exp Neurol 51:653-667
Huang M, Drummond GI (1979) Adenylate cyclase in cerebral microvessels: Action of guanine nucleotides, adenosine, and other agonists. Molec Pharmacol 16:462-472
Huang M, Hanley DA, Rorstad OP (1983) Parathyroid hormone stimulates adenylate cyclase in rat cerebral microvessels. Life Sci 32:1009-1014
Huang M, Rorstad OP (1983) Effects of vasoactive intestinal polypeptide, monoamines, prostaglandins, and 2-chloroadenosine on adenylate cyclase in rat cerebral microvessels. J Neurochem 40:719-726
Huang M, Rorstad OP (1984) Cerebral vascular adenylate cyclase: Evidence for coupling to receptors for vasoactive intestinal peptide and parathyroid hormone. J Neurochem 43:849-856
Huczynski J, Kostka-Trabka E, Sotowska W, Bieron K, Grodzinska L, Dembska-Kiec A, Pykosz-Mazur E, Peczak E, Gryglewski RJ (1985) Double-blind controlled trial of the therapeutic effects of prostacyclin in patients with completed ischemic stroke. Stroke 16:810-814
Joo F (1972) Effect of N6O2-dibutyryl cyclic 3',5'-adenosine monophosphate on the pinocytosis of brain capillaries of mice. Experientia 28:1470
Joo F, Rakonczay Z, Wollemann M (1975a) cAMP-mediated regulation of the permeability in the brain capillaries. Experientia 31:582-583
Joo F, Toth I (1975b) Brain adenylate cyclase: Its common occurrence in the capillaries and astrocytes. De Naturwissenschaften 8:397
Joo F, Szucs A, Csanda E (1976) Metiamide-treatment of brain edema in animals exposed to 90 yttrium irradiation. J Pharm Pharmacol 28:162-163
Joo F, Dux E, Karnushina IL, Halasz N, Gecse A, Ottlecz A, Mezei Z (1981) Histamine in brain capillaries. Agents Actions 11:129-134
Joo F, Temesvari P, Dux E (1983) Regulation of the macromolecular transport in the brain microvessels: The role of cyclic GMP. Brain Res 278:165-174
Joo F, Mihaly A, Temesvari P, Dux E (1984) Basic molecular events underlying transendothelial transport in brain capillaries. In: Go KG, Baethman A (eds). Recent Progress in the Study of Brain Edema. Plenum Press, New York, pp 107-116
Joo F, Olah Z, Novak R, Lengyel I, Dux E (1988) Second messenger effects on protein phosphorylation in brain microvessels. Am Soc Neurochem 19:99
Kalaria RN, Harik SI (1988) Adenosine receptors and the nucleoside transporter in human brain vasculature. J Cereb Blood Flow Met 8:32-39
Karnushina IL, Palacios JM, Barbin G, Dux E, Joo F, Schwartz JC (1980a) Studies on a capillary-rich fraction isolated from brain: Histamine components and characterization of the histamine receptors linked to adenylate cyclase. J Neurochem 34:1201-1208
Karnushina I, Toth I, Dux E, Joo F (1980b) Presence of the guanylate cyclase in brain capillaries: Histochemical and biochemical evidence. Brain Res 189:588-592
Karnushina IL, Spatz M, Bembry J (1982) Cerebral endothelial cell culture I. The presence of beta-2 and alpha-2-adrenergic receptors linked to adenylate cyclase activity. Life Sci 30:849-858
Kobayashi H, Maoret T, Ferrante M, Spano PF, Trabucchi M (1981a) Subtypes of beta-adrenergic receptors in rat cerebral microvessels. Brain Res 220:194-198
Kobayashi H, Memo M, Spano PF, Trabucchi M (1981b) Identification of

beta-adrenergic receptor binding sites in rat brain microvessels, using [125I] Iodohydroxybenzylpindolol. J Neurochem 36:1383-1388
Kobayashi H, Cazzaniga A, Spano P, Trabucchi M (1982a) Ontogenesis of alpha-and-beta receptors located on cerebral microvessels. Brain Res 242:358-360
Kobayashi H, Frattola L, Ferrarese C, Spano PF, Trabucchi M (1982b) Characterization of beta adrenergic receptors on human cerebral microvessels. Neurology 32:1384-1387
Kobayashi H, Maoret T, Spano PF, Trabucchi M (1982c) Effect of age on beta-adrenergic receptors on cerebral microvessels. Brain Res 244:374-377
Kobayashi H, Take K, Wada A, Izumi F, Magnoni MS (1984) Angiotensin converting enzyme activity is reduced in brain microvessels of spontaneously hypertensive rats. J Neurochem 42:1655-1658
Kobayashi H, Wada A, Izumi F, Magnoni MS, Trabucchi M (1985) Alpha-adrenergic receptors in cerebral microvessels of normotensive and spontaneously hypertensive rats. Circ Res 56:402-409
Larsson L-I, Edvinsson L, Fahrenkrug J, Hakanson R, Owman CH, Schaffalitzky, de Muckadell O, Sundler F (1976) Immunohistochemical localization of a vasodilatory polypeptide (VIP) in cerebrovascular nerves. Brain Res 113:400-404
LeBeux YJ, Willemot J (1978) Actin-like filaments in the endothelial cells of adult rat brain capillaries. Exp Neurol 58:446-454
Lindvall M, Owman C (1980) Evidence for the presence of two types of monoamine oxidase in rabbit choroid plexus and their role in breakdown of amines influencing cerebrospinal fluid formation. J Neurochem 34:518-522
Magnoni MS, Kobayashi H, Cazzaniga A, Izumi F, Spano PF, Trabucchi M (1983) Hypertension reduces the number of beta-adrenergic receptors in rat brain microvessels. Circ 67:610-612
Magnoni MS, Kobayashi H, Frattola L, Spano PF, Trabucchi M (1985a) Effect of common carotid occlusion on beta-adrenergic receptor function in cerebral microvessels. Stroke 16:505-509
Magnoni MS, Kobayashi H, Trezzi AC, Spano PF, Trabucchi M (1985b) Beta-adrenergic receptor function in brain microvessels of diabetic rats. Life Sci 36:1-6
Magnoni MS, Kobayashi H, Govoni S, Battaini F, Pasinetti G, Trabucchi M (1986) Partial reversal of asymmetry in microvessel neurochemical changes after ischemia by corpus callosum section. Circ 73:784-788
Marin J, Salaices M, Sanchez CLF (1979) Release of [3H]-noradrenalin induced by 5-hydroxytryptamine from cat pial arteries. J Pharm Pharmacol 31:818-821
Masuda Y, Ochi Y, Ochi Y, Karasawa T, Hatano N, Kadokawa T, Shimizu M (1986) Protective effect of prostaglandins D2, El and I2 against cerebral hypoxia/anoxia in mice. Naunyn-Schmideberg's Arch Pharmacol 334:282-289
McCarty R, Plunkett LM (1986) Binding site for atrial natriuretic factor (ANF) in brain: Alterations in Brattleboro rats. Brain Res Bull 17:767-772
McCulloch J, Harper AM (1977) Cerebral circulation: Effect of stimulation and blockade of dopamine receptors. Am J Physiol 233:H222-H227
McCulloch J, Uddman R, Kingman TA, Edvinsson L (1986) Calcitonin gene-related peptide: Functional role in cerebrovascular regulation. Proc Natl Acad Sci USA 83:5731-5735
Miller VT, Coull BM, Yatsu FM, Shah AB, Beamer NB (1984) Prostacyclin infusion in acute cerebral infarction. Neurology 34:1431-1435
Minneman KP, Pittman RN, Yeh HH, Woodward DJ, Wolfe BB, Molinoff PB (1981) Selective survival of beta-1-adrenergic receptors in rat cerebellum following neonatal X-irradiation. Brain Res 209:25-34
Morisaki N, Saito Y, Kumagai A (1982) Fatty acid oxidation of rat brain

microvessels in hypertension, aging and experimental diabetes. Atherosclerosis 42:221-227
Nag S, Robertson DM, Dinsdale HB (1979) Quantitative estimate of pinocytosis in experimental acute hypertension. Acta Neuropath 46:107-116
Nathanson JA, Glaser GH (1979) Identification of beta-adrenergic-sensitive adenylate cyclase in intracranial blood vessels. Nature 278:567-569
Nathanson JA (1980) Cerebral microvessels contain a beta2-adrenergic receptor. Life Sci 26:1793-1799
Nyquist-Battie C, Fortney R, Gochee A (1984) Adrenoreceptor development in rat cerebrum and the effects of ionizing radiation. Brain Res Bull 12:1-4
Ong BY, Kettler JJ, Bose D (1987) Alteration of pial vessel responses to blood pressure changes in rats after hypoxia. Can J Physiol Pharmacol 65:2265-2268
Palmer GC, Palmer SJ (1978) 5'-Guanylyl-imidodiphosphate actions on adenylate cyclase in homogenates of rat cerebral cortex plus neuronal and capillary fractions. Life Sci 23:207-216
Palmer GC (1979) Diminished adenylate cyclase responses in frontal cortex and cerebral capillaries of spontaneously hypertensive rats. Biochem Pharmacol 28:2847-2849
Palmer GC (1980) Beta adrenergic receptors mediate adenylate cyclase responses in rat cerebral capillaries. Neuropharmacology 19:17-23
Palmer GC, Chronister RB, Palmer SJ (1980) Adenylate cyclase responses to neurohumoral agonists in microvascular elements of the rabbit brain. J Neurobiol 11:503-508
Palmer GC (1981a) Regional distribution of neurohumoral responsive adenylate cyclase in rabbit brain microvessels. Neurosci Lett 21:207-210
Palmer GC (1981b) Distribution of guanylate cyclase in rat cerebral cortex: Neuronal, glial, capillary, pia-arachnoid and synaptosomal fractions plus choroid plexus. Neurosci 6:2547-2553
Palmer GC, Chronister RB, Christie-Pope BC, Gerbrandt L (1982) Adenylate cyclase responses in rat brain after unilateral postnatal X-irradiation. Exp Neurol 78:629-643
Palmer GC, Ghai G (1982) Adenosine receptors in capillaries and pia-arachnoid of rat cerebral cortex. Eur J Pharmacol 81:129-132
Palmer GC, (1983) Effects of psychoactive drugs on cyclic nucleotides in the central nervous system. Prog Neurobiol 21:1-133
Palmer GC, Palmer SJ (1983) Adenylate cyclase sensitivity to catecholamines and forskolin in rat pia-arachnoid and cerebral microvessels. Neuropharmacology 22:213-219
Palmer GC, Wilson GL, Chronister RB (1983) Streptozotocin-induced diabetes produces alterations in adenylate cyclase in rat cerebrum, cerebral microvessels and retina. Life Sci 32:365-374
Palmer GC (1986) Neurochemical coupled actions of transmitters in the microvasculature of the brain. Neurosci Biobehav Rev 10:79-101
Palmer GC, Jones DJ, Palmer SJ, Christie-Pope BC, Poulakos L (1986) Further probes into the molecular sites of damage to cerebral adenylate cyclase following post ischemic reperfusion. Neurochem Path 5:1-24
Palmer GC (1987) Free radicals generated by xanthine oxidase-hypoxanthine damage adenylate cyclase and ATPase in gerbil cerebral cortex. Metab Brain Dis 2:243-257
Pardridge WM, Yang J, Eisenberg J (1985) Blood-brain barrier protein phosphorylation and dephosphorylation. J Neurochem 45:1141-1147
Pardridge WM (1988) Recent advances in blood-brain barrier transport. Ann Rev Pharmacol Toxicol 28:25-39
Pearlmutter AF, Szkrybalo M, Kim Y, Harik SI (1988) Arginine vasopressin receptors in pig cerebral microvessels, cerebral cortex and hippocampus. Neurosci Lett 87:121-126
Peroutka SJ, Moskowitz MA, Reinhard JF Jr, Snyder SH (1980) Neurotransmitter

receptor binding in bovine cerebral microvessels. 208:610-612
Preskorn SH, Hartman BK, Raichle ME, Swanson LW, Clark HB (1980) Central adrenergic regulation of cerebral microvascular permeability and blood flow: Pharmacologic evidence. Adv Exp Med Biol 131:127-138
Preskorn SH, Kent TA, Glotzbach RK, Irwin GH, Solnick JV (1984) Cerebromicrocirculatory defects in animal model of depression. Psychopharmacol 84:196-199
Raichle ME, Hartman BK, Eichling JO, Sharpe LG (1975) Central noradrenergic regulation of cerebral blood flow and vascular permeability. Proc Nat Acad Sci USA 72:3726-3730
Raichle ME (1983) The pathophysiology of brain ischemia. Ann Neurol 13:2-10
Reinhard JF Jr, Liebmann JE, Schlosberg AJ, Moskovitz MA (1979) Serotonin neurons project to small blood vessels in the brain. Science 206:85-87
Remler MP, Marcussen WH (1981) Time course of early delayed blood-brain barrier changes in individual cats after ionizing radiation. Exp Neurol 73:310-314
Schlegel W, Cooper DMF, Rodbell M (1980) Inhibition and activation of fat cell adenylate cyclase by GTP is mediated by structures of different size. Arch Biochem Biophys 201:678-682
Schutz W, Steurer G, Tuisl E (1982) Functional identification of adenylate cyclase-coupled adenosine receptors in rat brain microvessels. Eur J Pharmacol 85:177-184
Shibata S, Tsutsumi K, Inoue M, Fukushima M, Mori K (1988) Experimental cerebral infarction in the dog: Electron microscopic studies of the microvasculature. Neurosurgy 22:669-675
Singh DNP, Jeria M, Melax H, Cookson F, Yoshida O (1981) Ultrastructural changes of capillaries in the cerebral cortex of spontaneously hypertensive rats. IRCS Med Sci 9:485-486
Sosula L, Beaumont P, Hollows FC, Jonson KM (1972) Dilatation and endothelial proliferation of retinal capillaries in streptozotocin-diabetic rats: Quantitative electron microscopy. Invest Opthal 11:926-935
Spatz M, Mrsulja BB (1982) Progress in cerebral microvascular studies related to the function of the blood-brain barrier. Adv Cell Neurobiol 3:311-337
Spatz M, (1986) Recent advances in the study of cerebrovascular receptors. In: Baethmann A, Go KG, Unterberg A (eds). Mechanisms of Secondary Brain Damage. Plenum Publishing, New York, pp 283-294
Spatz M, Mrsulja BB, Wroblewska B, Merkel N, Bembry J (1986) Modulation of glycogen metabolism in cerebromicrovascular smooth muscle and endothelial cultures. Biochem Biophys Res Comm 134:484-491
Speth RC, Harik SI (1985) Angiotensin II receptor binding sites in brain microvessels. Proc Natl Acad Sci USA 82:6340-6343
Stauber WT, Ong S-H, McCuskey RS (1981) Selective extravascular escape of albumin into the cerebral cortex of the diabetic rat. Diabetes 30:500-503
Steardo L, Nathanson JA (1987) Brain barrier tissues: End organs for atriopeptins. Science 235:470-473
Stefanovich V (1979) Cyclic 3',5'-adenosine monophosphate phosphodiesterase (cAMP PDE) and cyclic 3',5'-guanosine monophosphate phosphodiesterase (cGMP PDE) in microvessels isolated from bovine cortex. Neurochem Res 4:681-687
Stefanovich V (1983) Uptake of adenosine by isolated bovine cortex microvessel. Neurochem Res 8:1459-1469
Suzuki Y, Lederis K, Huang M, LeBlanc FF, Rorstad OP (1983) Relaxation of bovine, porcine and human brain arteries by parathyroid hormone. Life Sci 33:2497-2503
Szumanska G, Palkama A, Lehtosalo JI, Uusitalo H (1984) Adenylate cyclase in the microvessels of the rat brain. Acta Neuropath 62:219-224

Tsukahara T, Taniguchi T, Fujiwara M, Handa H (1983) Characterization of alpha adrenoceptors in pial arteries of the bovine brain. Naunyn-Schmiedeberg's Arch Pharmacol 324:88-93

Tsukahara T, Taniguchi T, Miwa S, Shimohama S, Fujiwara M, Nishkawa M, Handa H (1988) Presynaptic and postsynaptic alpha 2-adrenergic receptors in human cerebral arteries and their alteration after subarachnoid hemorrhage. Stroke 19:80-83

Tsutsumi K, Niwa M, Kawano T, Ibaragi M, Ozaki M, Mori K (1987) Atrial natriuretic polypeptides elevate the level of cyclic GMP in the rat choroid plexus. Neurosci Lett 79:174-178

Tsutsumi K, Niwa M, Himeno A, Kurihara M, Kawano T, Ibaragi M, Ozaki M, Mori K (1988) Alpha atrial natriuretic peptide binding sites in the rat choroid plexus are increased in the presence of hydrocephalus. Neurosci Lett 87:93-98

Turck M, Yeh H, Woodward DJ, Schultz JE (1981) Beta-adrenergic receptors in rat cerebellum after neonatal X-irradiation: Effect of prolonged imipramine and lithium treatment. Neurosci Lett 27:357-362

Ulrich K, Auer LM, Kuschinsky W (1982) Cat pial venoconstriction by topical microapplication of norepinephrine. J Cereb Blood Flow Metab 2:109-111

Vane JR, Gryglewski RJ, Botting RM (1987) The endothelial cell as a metabolic and endocrine organ. TIPS 8:491-496

Versteeg DHG, Palkovits M, van der Gugtern J, Wijnen HLJM, Smeets GWM, de Jong W (1976) Catecholamine content of individual brain regions of spontaneously hypertensive rats (SH-rats). Brain Res 112:429-434

Vorbrodt AW, Szumanska G, Dobrogowska DH (1984) Cytochemical studies of adenylate cyclase in the choroid plexus and brain vessels of rat and mouse. J Histochem Cytochem 32:275-284

Wahl M, Kuschinsky W (1979) The dilating effect of histamine on pial arteries of cats and its mediation by H-2 receptors. Circ Res 44:161-165

Wei EP, Kontos HA, Patterson JL Jr (1978) Vasoconstrictor effect of angiotensin on pial arteries. Stroke 9:487-489

Werber AH, Heistad DD (1984) Effects of chronic hypertension and sympathetic nerves on the cerebral microvasculature of stroke-prone spontaneously hypertensive rats. Circ Res 55:286-294

Westergaard E (1975) The effect of serotonin, norepinephrine and cyclic AMP on the blood-brain barrier. J Ultrastruct Res 50:383

Winn HR, Rubio GR, Berne RM (1981) The role of adenosine in the regulation of cerebral blood flow. J Cereb Blood Flow Metab 1:239-244

Wroblewska B, Spatz M, Merkel N, Bembry J (1984) Cerebrovascular smooth muscle culture II. Characterization of adrenergic receptors linked to adenylate cyclase. Life Sci 34:783-791

Wu PH, Phillis JW (1982) Uptake of adenosine by isolated brain capillaries. J Neurochem 38:687-690

Yanoff M (1969) Ocular pathology of diabetes mellitus. Am J Opthal 67:21-38

DISCUSSION

Chaired by Hartman B. and Racagni G.

F. Joo:

I have a comment on the capillary response to stroke: the endothelial cells are highly enriched, as compared to the neurons, in xantine oxidase, indicating the possibility of a local formation of free radicals. Thus, in case of hypoxic challenge the endothelial reactivity is immediate. Recent results suggest that allopurinol, which is an inhibitor of the enzyme, is able to limit to a great extent the consequences of experimental ischemia.
Concerning the development of the BBB, do yo have evidence that the adenylate cyclase system develops in parallel with the beta adrenergic receptors?

G.C.Palmer:

I did not study the receptor ontogenesis; it is possible that the two parameters are completed during development, although not always is there a parallel in the maturation of the different systems. Sometimes receptor density at birth is the same as the adults, while for other systems it may be rather very low or even higher than in adults. Another point of interest could be the comparison of receptor and adenylate cyclase maturation in neuron and glial cells.

L. Betz:

What's the role of adrenergic receptors located on brain capillaries? According to the current hypothesis they seem to be related to functional changes of the BBB; however, I wonder whether they have a role in the general metabolism of the endothelial cells, as other cell types.

G.C. Palmer:

This is a possibility. A relation between capillary adenylate cyclase activation and pinocytosis has been observed. On the other hand, an interesting study has been published by Spatz on the glycolytic activity of norepinephrine in cultures of endothelial cells and the positive effect of serotonin on glycogen synthesis. This work opens new perspectives for the understanding of capillary physiology.

L. Betz:

I think there is a big gap between binding studies and adenylate cyclase stimulation and the interpretation of the function of receptor - stimulated process -.
There is another potentially significant point in the study of the biochemical parameters of isolated microvessels. The preparation of isolated capillaries are not constituted

purely by endothelial cells, but also contain pericytes in a ratio 2:1 respectively. This must be taken in account in the interpretation of the results.

G.C. Palmer:

I agree that such factors should be considered when interpreting the results.

PHYSIOLOGICAL AND PATHOLOGICAL EVENTS MEDIATED BY CAPILLARY NEUROTRANSMITTER RECEPTORS

Ferenc Joo*

Laboratory of Neuropathology and Neuroanatomical Sciences,
Institute of Neurological and Communicative Disorders
and Stroke, National Institutes of Health,
Bldg.: 36; Room: 4D04, Bethesda,
MD 20892, USA.

ABSTRACT

The view that the cerebral endothelial cells represent the cellular analogue of the blood-brain barrier has been generally accepted. The regulation of transport processes operating in the cerebral endothelial cells is of great current interest. From biochemical and electron microscopic studies it became evident that, contrary to the classical view, histamine is a potent activator of the pinocytosis and the albumin transport in the cerebral endothelial cells. This action of histamine is mediated mainly by histamine H_2-receptors, which are linked to the capillary adenylate cyclase. Cyclic nucleotides, produced locally upon environmental stimuli, can phosphorylate specific proteins in the cerebral endothelium, some of which are connected to the endothelial contractile system. Specific histamine H_2-receptor blockers proved to be able to prevent brain edema formation in different experimental models.

KEY-WORDS

Blood-brain barrier, permeability, receptors, cyclic nucleotides, histamine, protein phosphorylation, calcium.

NATO ASI Series, Vol. H 33
Regulatory Mechanisms of Neuron to Vessel Communication in the Brain. Edited by F. Battaini et al.

INTRODUCTION

Regulation of solute exchange between blood and brain extracellular fluids takes place in the cerebral endothelial cells, which comprise, in most cases, the "blood-brain barrier" (BBB, Bradbury, 1979; Joo, 1986). It was demonstrated in electron microscopic studies (Brightman & Reese, 1969) that the barrier, which prevented the passing of proteins from blood to brain, consisted of tight junctions between cerebral endothelial cells. Another equally important feature of the cerebral endothelial cells, in contrast to those of most other tissues, is the relative lack of pinocytotic vesicles. From a physiological point of view, the cerebral endothelial cells represent a very tight cellular barrier with high (1900Ω /cm^2) membrane resistance (Crone & Olesen, 1982). This potent restraint delays the penetration of solutes to the brain at the capillary endothelium. On the other hand, a number of polar metabolic substrates cross the plasma membranes of cerebral endothelium via highly stereospecific, self-saturable and inhibitable transport systems (Bradbury, 1985; Oldendorf, 1977; Pardridge, 1983). These particular structural and functional characteristics of cerebral capillaries are determined by the surrounding central nervous tissue, especially by the influence deriving from astrocytes (Arthur & al., 1987; Janzer & Raff, 1987; Stewart & Hayakawa, 1987; Stewart & Willey, 1981; Tao-Cheng & Brightman, 1988; Tao-Cheng & al., 1987).

REGULATION OF ENDOTHELIAL TRANSPORT PROCESSES

Water permeability

The brain vasculature is not freely permeable to water, in fact, it exhibits a permeability similar to those of membranes known to regulate water permeability (Peachy & Rasmussen, 1961; Raichle & Grubb, 1978). Brain vascular water transport is influenced by **hemodynamic** and **biochemical** factors, alike. Raichle & Grubb (1978) observed originally that centrally released **vasopressin** increased the permeability to water of brain capillaries. Confirming the validity of this finding, experiments carried out in Brattleboro rats with diabetes insipidus showed that lack of vasopressin altered indeed the course of brain edema formation after subarachnoid hemorrhage (Doczi & al., 1984). In addition to vasopressin, the stimulation of bilateral sympathetic chain (Grubb & al., 1977) or locus coeruleus (Raichle & al., 1976) also resulted in increased permeability for water indicating the importance of **noradrenergic** innervation in the regulatory process. On the other hand, **corticosteroids** reduce the bulk flow of water across the BBB (Olson & al., 1988). It has been shown recently (Doczi & al., 1987; Nathanson, 1988) that **atrial natriuretic peptide (ANP)** acts directly on the cerebral capillaries via specific receptors (Ibaragi & al., 1988) and reduces the water permeability, also. These data lend support to the concept that a central neuroendocrine system regulates brain volume homeostasis.

Recent results of Armstrong & al. (1988) show that, after osmotic opening of the BBB, slits of approximately 40 um in width can be formed in the cerebral endothelium through which water-soluble compounds of differing molecular size are transported in a size-dependent manner.

Regulation of brain electrolyte contents

Na^+/K^+-ATPase, a membrane-bound enzyme of brain microvessel endothelial cells, plays a major role in the maintenance of electrolyte (and water) homeostasis in the brain. The **sodium** exchange takes place presumably by means of at least two specific Na^+-transport system confined to the luminal membrane, whereas an energy-dependent carrier is located only on the abluminal membrane of the brain endothelial cells (Betz, 1983a; 1983b). In ischemic edema, the clearance of sodium and water is impaired due to damage to the endothelial cells (Lo & al., 1987).

Potassium is actively transported from brain to blood by Na^+/K^+-ATPase of brain microvessels and a high sodium-potassium pump activity is present in brain endothelial cells (Goldstein, 1979). According to the results of Jakobsen & al. (1987) an impaired Na^+/K^+-ATPase pump can lead in diabetes to insufficient maintenance of electrolyte homeostasis across the BBB and to the development of edema during treatment of diabetic ketoacidosis.

Calcium influx into the central nervous system is not regulated by a saturable mechanism that is sensitive to acute changes in plasma $[Ca]_i$ (Tai & al., 1986). It has been suggested by Nag (1988) that calcium is transported through the cerebral endothelium in the pinocytotic vesicles and that Ca^{2+}-ATPase plays a part in the transcellular Ca^{2+} transport.

CHARACTERISTICS OF TRANSENDOTHELIAL MACROMOLECULAR TRANSPORT

The operation of the transcytotic pathway for a blood-borne protein, **lectin wheat germ agglutinin** conjugated to the enzymatic tracer horseradish peroxidase, has been evidenced recently (Broadwell & al., 1988). The entry of **transferrin** into the brain is also thought to be transported by transcytosis (Fishman & al., 1987). For a long time, the pathways, forces, and mechanisms involved in the **albumin** transport from the blood plasma to the interstitial fluid were issues on which there was no general agreement. Recent data, however, clearly indicate that albumin is transported by transcytosis across the capillary endothelium (Milici & al., 1987). Alterations in the BBB permeability to albumin play an important role in the pathogenesis of brain edema. When the barrier is breached, as occurs in a wide variety of diseases, albumin together with extravasated fluid of varying composition enters the brain and accumulates in the extracellular space or occasionally in neurons and glial cells. As electron microscopic studies showed (Joo, 1972; Joo & al., 1983; Noble & Wrathall, 1987) the primary cellular mechanism for BBB breakdown appears to be the activation in the cerebral endothelium of the pinocytotic vesicular transport rather than the interendothelial passage across compromised tight junctions.

THE IMPORTANCE OF HISTAMINE IN THE OPENING OF THE BBB AND ACTIVATION OF ALBUMIN TRANSPORT IN THE BRAIN CAPILLARIES

It has been shown (Majno & al., 1969) that, in non-neuronal tissues, histamine promoted cellular contraction and separation of endothelial cells. In the endothelia of cerebral blood vessels, however, capillary permeability to protein was found not to increase in response to histamine, acetylcholine or epinephrine (Ashton & Cunha-Vaz, 1965; Broman & Lindberg-Broman, 1945). The lack of response to vasoactive substances was explained by the low content of contractile proteins in brain endothelia (Rapoport, 1976). The notion that histamine is uneffective in the cerebral endothelium has been quoted in text-books (Rapoport, 1976; Bradbury, 1979) and interpreted as one of the differences existing between microvessels of neuronal and non-neuronal origin. The first observation questioning the validity of this statement was published by Foldes & Kelentei (1954). Later, Wolff & al. (1975) reported on the cerebrovascular ultrastructural alterations after intraarterial infusions of histamine. In other studies (Dux & Joo, 1982), histamine was found to increase the transendothelial transport of albumin together with an induction of pinocytosis in the brain endothelium. In addition, histamine was detected in a fraction enriched in brain capillaries (Dux & al., 1987; Jarrott & al., 1979; Karnushina & al., 1980a). It seems probable that, in pathological circumstances, when histamine is released from the circulating blood (Beaven, 1976), intracerebral nerve endings (Takagi & al., 1986) or mast cells (Dropp, 1972) surrounding the capillaries, it can increase not only the local cerebral blood flow (Gross & al., 1981) but also the endothelial permeability.

MOLECULAR MECHANISMS OF HISTAMINE EFFECTS

With the elaboration of a procedure, by means of which a fraction enriched by functionally and metabolically intact cerebral microvessels could be obtained, direct biochemical investigations of the intraparenchymal microvasculature has become available (Joo, 1985; Joo & Karnushina, 1973).

Presence of histamine and other receptors in the cerebral endothelial cells

Description of the presence of **histamine H_2-receptors** linked to adenylate cyclase in the cerebral endothelium was the first notion (Joo & al., 1975; Karnushina & al., 1980a) in the literature, which suggested that these cells may have receptors for vasoactive substances similar to those of peripheral vessels. Subsequent studies revealed that in fact a mixed population of receptors coupled to adenylate cyclase is present in the cerebral endothelium (Joo, 1986) consisting of α_1- (Peroutka & al., 1980), β_2- (Herbst & al.,

1979; Kobayashi & al., 1981; Nathanson, 1980, Nathanson & Glaser, 1979) and weak β_1-**adrenergic receptors** (Peroutka & al., 1980), as well as, **dopamine** (Baca & Palmer, 1978), **adenosine** (activatory type, Huang & Drummond, 1979; Schutz & al., 1982) **prostacyclin** (Huang & Drummond, 1979), **vasoactive intestinal polypeptide** (Huang & Rorstad, 1983), **parathyroid hormone** (Huang & Rorstad, 1983), **GABA** (Krause & al., 1980; Rausch & al., 1983) and **muscarinic cholinergic** (Estrada & Krause, 1982; Grammas & al., 1983; Harik & al., 1981) receptors. All, with the exception of α-adrenergic receptors, mediate a vasodilatory response. The ligands for these receptors, i.e. the vasoactive catabolites can be released either from the brain tissue or from the rich perivascular nerve supply deriving from central and peripheral sources alike (for review see Lou & al., 1987) allowing the effective coupling of cerebral blood flow to neuronal metabolic activity. In addition to mediating vasoregulatory responses, histamine H_2-receptors were shown to be involved in the induction of macromolecular transport leading eventually to the formation of brain edema (Dux & Joo, 1982; Dux & al., 1984;1987).

Main features of cyclic nucleotide-metabolism in the cerebral endothelium

Both synthesizing (Baca & Palmer, 1978; Herbts & al., 1979; Huang & Drummond, 1979; Joo & al., 1975; Karnushina & al., 1980a and b; Kobayashi & al., 1981; Nathanson, 1980; Nathanson & Glaser, 1979; Palmer, 1981a; Palmer, 1981b; Schutz & al., 1982) and degrading (Stefanovich, 1979) enzymes of cAMP and cGMP were detected in the cerebral endothelial cells indicating that these second messenger molecules are produced and metabolized locally. Recently a correlation was found (Adam & al., 1987) between the activation of adenylate cyclase in cerebral endothelial cells and the induction of transcapillary albumin transport, suggesting that elevated cAMP may trigger macromolecule permeation through the microvessels. It would seem probable that in the cerebral endothelium cyclic nucleotides can apparently influence the rate of transcytosis. Indeed, in addition to an increase in the number of pinocytotic vesicles, a rise in the frequency of occurrence of the so-called bristle-coated vesicles, an organelle characteristic of protein transport, was observed (Joo, 1972; Joo & al., 1983) in the brain endothelium after dibutyryl cyclic nucleotide treatment coincident with the activation of albumin transport.

Modulation of protein phosphorylation by second messengers at the cerebral endothelium

The fact that protein phosphorylation occurs in the brain microvessels has been evidenced (Pardridge & al., 1985; Weber & al., 1987), but initial attempts to modulate the phosphorylation pattern of proteins by second messengers or certain vasoactive substances were negative (Pardridge & al., 1985). However, by improving the isolation procedure and by providing optimal conditions for incubation it was possible to reveal (Olah & al., 1988) the occurrence of Ca^{2+}-calmodulin-, Ca^{2+}/phospholipid-, cGMP- and cAMP-dependent protein phosphorylation in the cerebral endothelial cells. Several phosphoproteins were detected as common substrates for all four studied phosphorylating activities (Fig. 1).

Figure 1.

Protein substrates of Ca^{2+}-calmodulin system, Ca^{2+}/phospholipide-cGMP- and cAMP-dependent protein kinases

Ca^{2+}-calmodulin cGMP cAMP		cGMP	PK-C		cGMP cAMP	
P50 P55	P58(57)	P30 P33	P40	P45	P17	P66 P68
β- and α- subunits of CAM-dependent protein kinase II	α- and β-subunits of tubulin	DRAPP-32 (dopamine- and cAMP-regulated phosphoprotein)	Not known	Catalytic fragment of protein kinase C (PK C)	Myosin light chain	δ- and γ-subunits of nicotinic acetylcholine receptor

However, the phosphorylation kinetics of these proteins were different and depended on the second messenger added to the incubation. Generally, it can be concluded that those proteins which were phosphorylated by cyclic nucleotide-dependent protein kinases reached their maximal phosphorylation states very rapidly and preserved them longer than those being phosphorylated by the Ca^{2+}-dependent enzymes. It can be hypothesized that when endothelial cells are stimulated by vasoactive substances released either from circulating platelets or directly from the endothelial cytoplasm the generation of cyclic nucleotides is increased, which may activate cAMP- or cGMP-dependent protein kinases, and thereby increase transendothelial transport. In addition, increases of cytosolic free Ca^{2+} observed in endothelial cells in response to histamine (D'Amore & Shepro, 1977) and under conditions of in brain oedema elicited by kainic acid (Sztriha & Joo, 1986), can activate Ca^{2+}-calmodulin-dependent protein kinase, which proved to be the major phosphorylating enzyme in the cerebral microvessels (Olah & al., 1988). As reversible phosphorylation of membrane-bound proteins or of regulatory subunits is involved in carrier regulation of neuronal cell membranes (Greengard, 1978), it can be suggested that similar mechanisms may serve in the cerebral endothelium to control some of the transport systems operating across the blood-brain barrier under physiological and pathological conditions.

Characteristics of arachidonic acid metabolism in the cerebral endothelium

Prostaglandins (PG), prostacyclin and thromboxane (TX), the active metabolites of arachidonic acid, are modulators of vascular tone and reactivity. The presence of PGH_2 -E_2 isomerase and prostacyclin synthetase was detected first in the brain microvessels (Gerritsen & al., 1980). In other studies (Maurer & al., 1980), the synthesis of prostacyclin, $PGF_{2\alpha}$ and PGF_2 and TXA_2 was detected. A comparison, carried out by Goehlert & al. (1981) of the synthesis of prostaglandin 6-keto-$F_{1\alpha}$, E_2 and $F_{2\alpha}$ showed that 6-keto-prostaglandin $F_{1\alpha}$ was the major prostaglandin formed in the microvessels. Prostaglandin D_2 was not detected in the study of Maurer & al. (1980). In contrast, Gecse & al. (1982) reported that the main PGs synthesized by guinea pig microvessels were PGD_2 and PGE_2. Significantly less $PGF_{2\alpha}$ or prostacyclin stable metabolite, 6-oxo-$PGF_{1\alpha}$ were found.

Vasoactive substances have been shown to interfere with the metabolism of PGs in brain microvessels. Thus, norepinephrine (Birkle & al., 1981; Gecse & al., 1982) and histamine (Joo & al., 1981) were found to activate PG synthesis in brain cerebral microvessels. Hypoxemia, as a pathological condition, was also shown (Asano & al., 1985; Dux & al., 1981; Koide & al., 1985) to affect greatly the PG synthesis in brain microvessels, especially those of PGF_2, PGD_2, PGE_2 and TXB_2. The fact that the pattern of arachidonic acid cascade was seen to be changed in response to vasoactive substances and hypoxia (Moskowitz & al., 1984) strongly suggested that PG metabolites may participate in the pathogenesis of brain edema formation. Indeed, as evidenced by Chan & Fishman (1984), polyunsaturated fatty acids were able to produce brain edema in cortical slices. Leukotrienes, formed by intracranial tumors, were regarded to be important in the formation of vasogenic edema surrounding tumors (Black & al., 1986). It should be mentioned that, in the supply of arachidonic acid, the phospholipases play an important role.

PREVENTION OF BRAIN EDEMA FORMATION BY HISTAMINE RECEPTOR ANTAGONISTS

Edematous swelling, i.e. abnormal accumulation of fluid within the parenchyma intra- or extracellularly, or both, accompanies frequently a wide variety of cerebral diseases, such as tumors, trauma, infections, hypoxic, toxic and metabolic disorders. Because of the frequent occurrence, and life threatening severity, elucidation of cellular and molecular mechanisms leading to formation of brain edema is of obvious importance in order to find ways of effective therapeutic treatments.

The sequence of histamine-mediated events taking place in the cerebral endothelial cells and leading eventually to brain edema formation is summarized in Fig. 2. Supporting the validity of observations made on in vitro fractions of cerebral microvessels, the application of histamine H_2-receptor blockers proved to be effective in the prevention of brain edema formation in vivo, too. As the results of earlier studies indicated, the blockade of histamine H_2-receptors prevented the formation of brain edema accompanying radiation injury (Joo & al., 1976), kainic acid induced epilepsy (Sztriha & al., 1987), or acute hypoxia (Dux & al., 1987).

Figure 2.

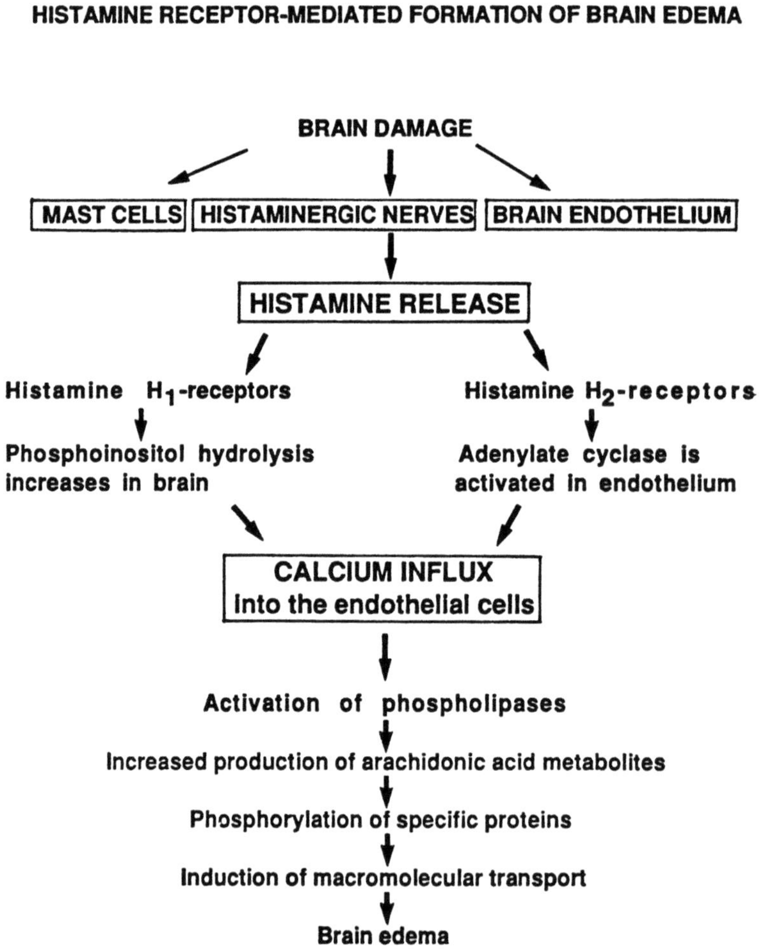

CONCLUDING REMARKS

Brain endothelial cells have a mixed-population of receptors mediating the effects of several vasoactive substances. A great number of the specific receptors is coupled to second messenger-generating enzymes, mainly to adenylate cyclase. Histamine is a strong activator of transcapillary albumin transport and adenylate cyclase in brain microvessels. The cellular basis of the induced macromolecular transport is transcytosis. Protein kinases, activated by second messenger molecules including cAMP, cGMP, Ca^{2+} and diacylglycerol have specific protein substrates in the cerebral endothelium, which can be phosphorylated. Histamine H_2-receptor blockers can prevent brain edema formation and be recommended for use in preoperative medication.

* **Acknowledgement.** The author is a recipient of a Fogarty Visiting Scientist fellowship. Permanent address: Laboratory of Molecular Neurobiology, Institute of Biophysics, Biological Research Center, 6701-Szeged, P.O. Box 521, Szeged, Hungary.

REFERENCES

Adam G & al. (1987) : Effects of acute hypoxia on the adenylate cyclase and Evans blue transport of brain microvessels. Neurochem Int, 10: 529-532.

Armstrong BK & al. (1987) : Size-dependent blood-brain barrier opening demonstrated with [^{14}C]sucrose and a 200,000-Da [^{3}H]dextran, Exp Neurol, 97: 686-696.

Arthur FE & al. (1987) : Astrocyte-mediated induction of tight junctions in brain capillary endothelium: an efficient in vitro model. Dev Brain Res, 36:155-159.

Asano T & al. (1985) : Ischemic brain edema following occlusion of the middle cerebral artery in the rat. II: Alterations of the eicosanoid synthesis profile of brain microvessels. Stroke, 16: 110-113.

Ashton N & Cunha-Vaz JG (1965) : Effect of histamine on the permeability of ocular vessels. Arch Ophtalmol, 73:211-223.

Baca GM & Palmer GC (1987) : Presence of hormonally-sensitive adenylate cyclase receptors in capillary-enriched fractions from rat cerebral cortex. Blood Vessels, 15: 286-297.

Betz AL (1983a) : Sodium transport in capillaries isolated from rat brain. J Neurochem, 41:1150-1157.

Betz AL (1983b) : Sodium transport from blood to brain: Inhibition by furosemide and amiloride. J Neurochem, 41: 1158-1164.

Birkle DL & al. (1981) : Prostaglandin levels in isolated brain microvessels and in normal and norepinephrine stimulated cat brain homogenates. Prostaglandins, 21: 856-877.

Black KL & al. (1986) : Increased leukotriene C_4 and vasogenic edema surrounding brain tumors in humans. Ann Neurol, 19:592-595.

Bradbury MWB (1979) : The Concept of a Blood-Brain Barrier. John Wiley, Chichester.

Bradbury MWB (1985) : The blood brain barrier. Transport across the cerebral endothelium. Circ Res, 57: 213-222.

Beaven MA (1976) Histamine. N Eng J Med, 294:320-325.

Brightman MW & Reese TS (1969) : Junctions between intimately apposed cell membranes in the vertebrate brain. J Cell Biol, 40: 648-677.

Broadwell RD & al. (1988) : Transcytotic pathway for blood-borne protein through the blood-brain barrier. Proc Natl Acad Sci USA, 85: 632-636.

Broman T & Lindberg-Broman AM (1945): An experimental study of disorders in the permeability of the cerebral vessels ("the blood-brain barrier") produced by chemical agents. Acta Physiol Scand, 10:102-125.

Chan PH & Fishman RA (1984) : The role of arachidonic acid in vasogenic brain edema. Fedn Proc, 43: 210-213.

Crone C & Olesen S-P (1982) : Electrical resistance of brain microvascular endothelium. Brain Res, 241: 49-55.

D'Amore PA & Shepro D (1977) : Stimulation of growth and calcium influx in cultured bovine aortic endothelial cells by platelets and vasoactive substances. J Cell Physiol, 92:117-184.

Doczi T & al. (1987) : Regulation of water and electrolyte contents: The possible involvement of central natriuretic factor. Neurosurgery, 21:454-458.

Doczi T & al. (1984) : Involvement of vasopressin in brain edema formation: Further evidence obtained from the Brattleboro diabetes insipidus rat with experimental subarachnoid hemorrhage. Neurosurgery, 14: 436-441.

Dropp JJ (1972) : Mast cells in the central nervous system of several rodents. Anat Rec, 174:227-238.

Dux E & Joo F (1982) : Effects of histamine on brain capillaries: fine structural and immunohistochemical studies after intracarotid infusion. Expl Brain Res, 47: 252-258.

Dux E & al. (1981) : Prostacyclin and prostaglandin synthesis in rat brain microvessels: characteristic changes in hypoxaemia. Acta Biol Hung, 32:291-300.

Dux E & al. (1984) : The blood-brain barrier in hypoxia: ultrastructural aspects and adenylate cyclase activity of brain capillaries. Neuroscience, 12: 951-958.

Dux E & al. (1987) : Protective effect of antihistamines on cerebral oedema induced by experimental pneumothorax in newborn piglets. Neuroscience, 22: 317-321.

Estrada C & Krause DN (1982) Muscarinic cholinergic receptor sites in cerebral blood vessels. J Pharm exp Ther, 221:85-90.

Fishman JB & al. (1987) : Receptor-mediated transcytosis of ferritin across the blood-brain barrier. J Neurosci Res, 18:299-304.

Foldes I & Kelentei B (1954) : Studies on the haemato-encephalic barrier. II. The effects of histamine with special reference to the passage of antibiotics. Acta Physiol Hung, 5: 149-162.

Gecse A & al. (1982) : Prostacyclin and prostaglandin synthesis in isolated brain capillaries. Prostaglandins, 23: 287-297.

Gerritsen ME & al. (1980) : Prostaglandin endoperoxidase metabolism by bovine cerebral microvessels. Biochim Biophys Acta, 619:

196-206.

Goehlert UG & al. (1981) : Biosynthesis of prostacyclin in rat cerebral microvessels and the choroid plexus. J Neurochem, 36:1192-1201.

Goldstein GW (1979) : Relation of potassium transport to oxydative metabolism in isolated brain capillaries. J Physiol, 1 286: 185-195.

Grammas P & al. (1983) : Identification of muscarinic receptors in rat cerebral cortical microvessels. J Neurochem, 40:645-651.

Greengard P (1978) : Phosphorylated proteins as physiological effectors. Science, 199:146-152.

Gross PM & al. (1981) : H_2-receptors mediate increases in permeability of the blood-brain barrier during arterial histamine infusion. Brain Res, 210:396-400.

Grubb RL Jr & al. (1977) : Effects of subarachnoid hemorrhage on cerebral blood volume, blood flow, and oxygen utilization in humans. J Neurosurg, 46: 446-453.

Harik SI, Sharma VK, Wetherbee JR, Warren RH and Banerjee SP (1981) Adrenergic and cholinergic receptors of cerebral microvessels. J Cereb Blood Flow Metab, 1:329-338.

Herbst TJ & al. (1979) : Raichle,M.E. and Ferrendi,J.A. β-Adrenergic regulation of adenosine 3',5'-monophosphate concentration in brain microvessels. Science, 204: 330-332.

Huang M & Drummond GI (1979) : Adenylate cyclase in cerebral microvessels: Action of guanine nucleotides, adenosine and other agonists. Molec Pharmac, 16: 462-472.

Huang M & Rorstad OP (1983) : Effects of vasoactive intestinal polypeptide, monoamines, prostaglandins and 2-chloroadenosine on adenylate cyclase in rat cerebral microvessels. J Neurochem, 40: 719-726.

Ibaragi M & al. (1988) : Rat atrial natriuretic polypeptide binding sites in rat brain microvessels. Neurochem Res, 13: 275-276.

Jakobsen J & al. (1987) : Cation permeability of the blood-brain barrier in streptozotocin-diabetic rats. Diabetologia, 30: 409-413.

Janzer RC & Raff MC (1987) : Astrocytes induce blood-brain barrier properties in endothelial cells. Nature, 325:253-257.

Jarrott B & al. (1979) Association of histamine with cerebral microvessels in regions of bovine brain. Brain Res, 168:323-330.

Joo F (1972) : Effect of N^6,O^6-dibutyryl cyclic 3',5'-adenosine monophosphate on the pinocytosis of brain capillaries in mice. Experientia, 28:1470-1471.

Joo F (1985) : The blood-brain barrier <u>in vitro</u>: Ten years of research on microvessels isolated from the brain. Neurochem Int, 7:1-25.

Joo F (1986) : New aspects to the function of the cerebral endothelium. Nature, 321: 197-198.

Joo F & al. (1981) : Histamine in brain capillaries. Agents and Actions, 11:129-134.

Joo F & Karnushina I (1973) : A procedure for the isolation of capillaries from rat brain. Cytobios, 8: 41-48.

Joo F & al. (1975) : cAMP-mediated regulation of the permeability in the brain capillaries. Experientia, 32: 582-583.

Joo F & al. (1976) : Metiamide-treatment of brain oedema in animals exposed to ^{90}-Yttrium irradiation. J Pharm Pharmacol, 28: 162-163.

Joo F & al. (1983) : Regulation of the macromolecular transport in the brain microvessels: the role of cyclic GMP. Brain Res, 278:165-174.

Karnushina I & al. (1980a) : Studies on a capillary-rich fraction isolated from brain: histaminic components and characterization of the histamine receptors linked to adenylate cyclase. J Neurochem, 34:1201-1208.

Karnushina I & al. (1980b) : Presence of the guanylate cyclase in brain capillaries: histochemical and biochemical evidences. Brain Res, 189: 588-592.

Kobayashi H & al. (1981) : Subtypes of β-adrenergic receptors in rat cerebral microvessels. Brain Res, 220: 194-198.

Koide T & al. (1985) : Alterations of the eicosanoid synthetic capacity of rat brain microvessels following ischemia: relevance to ischemic brain edema. J Neurochem, 44: 85-93.

Krause DN & al. (1980) : GABA receptors in bovine cerebral blood vessels: Binding studies with [^{3}H]muscimol. Brain Res, 185:51-57.

Lo WD & al. (1987) : Transport of sodium from blood to brain in ischemic brain edema. Stroke, 18:150-157.

Lou HC & al. (1987) : The concept of coupling blood flow to brain function: Revision required? Ann Neurol, 22: 289-297.

Majno G & al. (1969) : Endothelial contraction induced by histamine type mediators. J Cell Biol, 42:617-672.

Maurer P & al. (1980) : The synthesis of prostaglandins by bovine cerebral microvessels. Prostaglandins Med, 4:153-161.

Milici AJ & al. (1987) : Transcytosis of albumin in capillary endothelium. J Cell Biol, 105: 2603-2612.

Moskowitz MA & al. (1984) : Synthesis of compounds with properties of leukotrienes C_4 and D_4 in gerbil brains after ischemia and reperfusion. Science, 224: 886-889.

Nag S (1988) : Localisation of calcium-activated adenosine-triphosphatase (Ca^{2+}-ATPase) in intracerebral arterioles in acute hypertension. Acta Neuropathol, 75: 547-553.

Nathanson JA (1980) : Cerebral microvessels contain a β_2-adrenergic receptor. Life Sci, 26:1793-1799.

Nathanson JA (1988) : Direct application of a guanylate cyclase activator lowers intraocular pressure. Eur J Pharmacol, 147:155-156.

Nathanson JA & Glaser GH (1979) : Identification of β_2-adrenergic-sensitive adenylate cyclase in intracranial blood vessels. Nature, 278: 567-569.

Noble LJ & Wrathall JR (1987) : The blood-spinal cord barrier after injury: pattern of vasculat events proximal and distal to a transection in the rat. Brain Res, 424: 177-188.

Olah Z & al. (1988) : Kinetics of protein phosphorylation in microvessels isolated from rat brain: Modulation by second messengers. J Neurochem, 51: 49-56.

Oldendorf WH (1977) : The blood brain barrier. Expl Eye Res, 25: Suppl., 177-190.

Olson JJ & al. (1988) : Methylprednisolone reduces the bulk flow of water across an in vitro blood-brain barrier. Brain Res, 439:259-265.

Palmer GC (1981a) : Regional distribution of neurohumoral responsive adenylate cyclase in rabbit brain microvessels. Neurosci Lett, 21: 207-210.

Palmer GC (1981b) : Distribution of guanylate cyclase in rat cerebral cortex: neuronal, glial, capillary, pia-arachnoid and synaptosomal fractions plus choroid plexus. Neuroscience, 6:2547-2553.

Pardridge WM (1983) : Brain metabolism: A perspective from the blood brain barrier. Physiol Rev, 63:1481-1535.

Pardridge WM & al. (1985) : Blood-brain barrier protein phosphorylation and dephosphorylation. J Neurochem, 45:1141-1147.

Peachy LD & Rasmussen H (1961) : Structure of the toad's urinary bladder as related to its physiology. J Biophys Biochem Cytol, 10: 529-553.

Peroutka SJ & al. (1980) : Neurotransmitter receptor binding in bovine cerebral microvessels. Science, 208: 610-612.

Raichle ME & Grubb RL (1978) : Regulation of brain water permeability by centrally-released vasopressin. Brain Res, 143: 191-194.

Raichle M & al. (1976) : Central noradrenergic regulation of brain microcirculation. In Dynamics of Brain Edema (Eds.: H.M. Pappius and W. Feindel), Springer, Berlin Heidelberg New York, pp. 11-16.

Rapoport SI (1976) : Blood-Brain Barrier in Physiology and Medicine. Raven Press, New York.

Rausch WD & al. (1983) : Studies on the neurotransmitter binding to pig brain microvessels. J Neural Transm, (Suppl.) 18:33-44.

Schutz W & al. (1982) : Functional identification of adenylate cyclase-coupled adenosine receptors in rat brain microvessels. Eur J Pharmac, 85:177-184.

Stefanovich V (1979) : Cyclic 3',5'-adenosine monophosphate phosphodiesterase (cAMP PDE) and cyclic 3',5'-guanosine monophosphate phosphodiesterase (cGMP PDE) in microvessels isolated from bovine vessels. Neurochem Res, 4: 681-687.

Stewart PA & Hayakawa EM (1987) : Interendothelial junctional changes underlie the developmental "tightening" of the blood-brain barrier. Dev Brain Res, 32: 271-281.

Stewart PA & Wiley MJ (1981) : Developing nervous tissue induces formation of blood-brain barrier characteristics in invading endothelial cells: A study using quail-chick transplantation chimera. Devl Biol, 84: 183-192.

Sztriha L & Joo F (1986) : Intraendothelial accumulation of calcium in the hippocampus and thalamus of rats after systemic kainic acid administration. Acta Neuropathol, 72:111-116.

Sztriha L & al. (1987) : Histamine H_2-receptors participate the formation of brain oedema induced by kainic acid in the thalamus of rat. Neurosci Letters, 75:334-338.

Tai C-Y & al. (1986) : Calcium influxes into brain and cerebrospinal fluid are linearly related to plasma ionized calcium concentration. Brain Res, 385:227-236.

Takagi H & al. (1986) : Histaminergic axons in the neostriatum and cerebral cortex of the rat: a correlated light and electron microscopic immunocytochemical study using histidine decarboxylase as a marker. Brain Res, 364:114-123.

Tao-Cheng JH & Brightman MW (1988) : Development of membrane interactions between brain endothelial cells and astrocytes in vitro. Int J Devl Neurosci, 6:25-37.

Tao-Cheng JH & al. (1987) : Tight junctions of brain endothelium in vitro are enhanced by astroglia. J Neurosci, 7:3293-3299.

Weber M, Mehler M and Wollny E (1987) Isolation and partial characterization of a 56,000-Dalton phosphoprotein phosphatase from the blood-brain barrier. J Neurochem, 49:1050-1056.

Wolff JR & al. (1975) : Cerebrovascular ultrastructural alterations after intra-arterial infusions of oubain, scilla-glycosides, heparine and histamine. Acta Neuropathol, 31:45-58.

DISCUSSION

Chaired by Betz L.A. and Frattola L.

G.C.Palmer:

How relevant is the component of histamine neurons versus mast cells in histamine release in the CNS vasculature?

F. Joo:

The major sources of histamine in the brain are the mast cells. Their distribution within the brain is not uniform. The majority is confined to the thalamus. Interestingly, it has been observed that kainic acid causes cerebral edema in thalamic regions and that H2 antagonists are able to prevent this damaging effect. However, there is new evidence indicating the existence of a rich histaminergic innervation, originating from thalamic and hypothalamic regions.

G.C. Palmer:

Are there clinical observations on the benefical effects of H2 antagonists on cerebral edema?

F. Joo:

So far, H2 receptor antagonists have not yet been used in clinics to limit the consequences of ischemia, in particular brain edema. The data in literature have been obtained by pretreatment experiments on rats.

A. Pecile:

Do the endothelial cells store histamine?

F. Joo:

Yes, one of the questions is: why are endothelial cells able to store vasoactive substances?
Not only histamine but other transmitters are taken up by endothelial cells. No specific granules for storage have been so far identified in the capillaries: they are probably free in the cytoplasm from where they are released under depolarization and therefore act on the external surface of the endothelium by bindings to specific receptors; this may represent a way of self-regulation of the binding sites.

W. Meier-Ruge:

Do you believe that H2 antagonists may became a therapeutic tool in the treatment of stroke?

F. Joo:

I wish it would. The results of our experiments show a preventive effects of these drugs on the development of cerebral edema. Nevertheless, we are aware that the clinical situation is always more complicated. One of our future goals is to investigate the effect of these compounds given post ischemia, during recirculation.
It is however encouraging that several surgeous use H2 antagonists in the premedication before surgery. The main reasons are based on the observation that the patients show a better recovery and on the capacity of these drug to reduce the risk of stress gastric ulcer.

F. Sacchi:

As a clinicam I'm impressed by the results on the preventive effect of H2 blockers on brain edema. Are you aware of studies suggesting a potentiation between H2 antagonists and currently used therapeutic tools as glycerol, hydrocortison, hyperosmotic agents?

F. Joo:

This is an extremely interesting and important point. The effect on cerebral edema of the combination of different pharmacological tools is under investigation.
One main point of our future studies is to investigate the activity of lipooxygenase and cyclooxygenase blockers which inhibit leukotriens and PGs synthesis. Another possibility is to investigate the combination of H2 antagonists with calcium antagonists.

S.I. Rapoport:

Early data back in 1945 (Broman and Lindberg-Broma), showed that the application of histamine to the surface of the brain did not induce any change in capillary permeability. Do you have any measure of increase in permeability in the normal brain independent of the changes of the blood pressure after application of histamine at the surface of the capillaries or in the lumen?

F. Joo:

The results by this old work report the classical concept of the endothelium as resistant to circulating histamine and other neurotransmitters, but the most recent experiments argue against this view.

- Broman T, Lindberg-Broman, 1945. Acta Physiol Scand, 10:102- 125.

However, we do not have independent means to distinguish BBB responses to the direct application of histamine from changes in blood pressure.

S.I. Rapoport:

It has been shown that the movement of radiolabelled tracers across the BBB depends upon the local hydrostatic pressure at the capillary level. Can your results be interpreted in terms of changes of local pressure rather than changes in vesicular transport?

F. Joo:

The vasogenic edema formation may be partially associated with local changes in hydrostatic pressure, but several lines of evidence suggest that the increase in permeability due to macromolecular vesicular transport is greatly involved.

R. Orkand:

What's the specificity of response of the endothelial cells versus the astrocytes in controlling BBB function?

F. Joo:

The unique feature of cerebral endothelium is undoubtly the BBB function. However, the biochemical machinery of the capillary cells is influenced by the microenvironment, as the glia. It has been observed that glial cells display a suppressive action on the endothelial biochemical process, regulating many of the macromolecular transport mechanisms. In certain pathological conditions like brain edema, these events are of extreme relevance. The determinant influence of glial cells on capillary function is confirmed by experiments using co-cultures of glial and endothelial cells which lead to the in vitro expression of barrier characteristics.

J. Cervos Navarro:

The influence of glia on the capillary characteristics is certainly relevant, but in the blood liquor barrier the glial cells are not present in meninges, suggesting that the interactions between the two systems may not occur in every case.

R. Pumain:

You have mentioned the effect of kainic acid on thalamic regions, attenuated by H2 antagonists. However, this substance applied in any area would induce large variation in the interstitial space volume, via ionic changes between intra and extraneuronal compartment. Do you have any comment?

F. Joo:

Certainly the first action of kainic acid is to induce neuronal depolarization which finally leads to a rythmic and extended electical discharge of the neurons and an increase in metabolism and O2 consumption which may induce tissue hypoxia. However, also other elements such as the mast cells are depolarized and release transmitters like histamine. Thus, although the sites of primary lesion are the neurons and glial cells, the transmitters released can act in a second place at microvascular level.

BIOCHEMICAL CHANGES OF BRAIN MICROVESSELS DURING AGING

Magnoni M.S., Govoni S.*, Battaini F.**, Trabucchi M.**

Institute of Pharmacological Sciences,
Via Balzaretti 9
University of Milan,
20133 Milan

* Department of Pharmacobiology,
University of Bari,

** Department of Experimental Medicine and
Biochemical Sciences,
Chair of Toxicology
University of Rome,
Italy.

The capillary network, with its unique morphological and enzymatic properties, forms a dynamic interface between the blood and the cerebral parenchyma and actively participates to the regulation of the neuronal microenvironment.

The knowledge of the mechanisms controlling the blood brain barrier (BBB) may have potential clinical implications in physiological and pathological states, such as aging, cerebrovascular disorders, hypertension.

Several lines of evidence indicate that the capillaries respond to stimuli of various nature, as circulating hormones, local metabolic factors and neuronally-released transmitters.

Recent work has been aimed to the research on the existence of a neuronal regulation of microcirculation. The adrenergic innervation is so far the best investigated, on the basis of anatomical, physiological and biochemical studies. In fact, ultrastructural observations have demonstrated that

NATO ASI Series, Vol. H 33
Regulatory Mechanisms of Neuron to Vessel Communication in the Brain. Edited by F. Battaini et al.

intraparenchymal small vessels and capillaries are closely associated with adrenergic fibers originating from the Locus Coeruleus (Swanson et al., 1977; Rennels and Nelson, 1975; Hartman et al., 1972).

The relation between vascular function and neuronal activity seems to be mediated by adrenergic receptors (alfa 1, alfa 2, beta 2) located on cerebral endothelial cells, which have been identified and characterized by means of radioligand techniques (Kobayashi et al., 1981a; Kobayashi et al. 1982a; Kobayashi et al., 1981b).

Table I shows the characteristics of beta-adrenergic receptors in human and rat brain capillaries.

TABLE I

ICYP binding parameters to microvessels isolated from human and rat brain

	Bmax (fmol/mg prot.)	Kd (pM)
Human	85 ± 9	88 ± 7
Rat	119 ± 11	65 ± 9

Values are means $\pm$ SD of 4 experiments.
Cerebral microvessels from rat or human cortices were isolated by albumin flotation and glass bead filtration technique according to Kobayashi et al. (Kobayashi et al., 1981a). Beta-adrenergic receptors were measured using the specific radioligand 125I-iodocyanopindolol (ICYP).
The maximun number of binding sites (Bmax) and the dissociation constant (Kd) were extrapolated with a Scatchard analysis.

The functional role of the receptors is not completely known. However, evidence based on the experimental manipulation of the Locus coeruleus indicates the involvement of adrenergic innervation on the regulation of vascular permeability (Raichle et al., 1975; McKenzie et al., 1976; Bates et al., 1977). This is further confirmed by recent data suggesting that the adrenergic innervation is important in regulating the capacity of the BBB to actively transport Na^+ and K^+ ions, contributing to the maintenance of water and electrolyte homeostasis in the central nervous system (Harik, 1986).

In addition, it has been shown that the stimulation of adenylate cyclase at the capillary level is associated to an increase in pinocytotic vesicle number (Joo et al., 1975).

On the other hand, recent observations have revealed that in cultures of endothelial cells derived from brain vessels norepinephrine is able to activate glycolytic processes, suggesting a role of capillary adrenergic receptors also in the regulation of the general metabolism of the cells (Spatz et al., 1986).

Interestingly, beta-2 receptors have not been detected in microvessels isolated from cerebral tumors (Magnoni et al., 1988a). Although the factors leading to the loss of binding site expression are not known, this finding provides evidence that capillary function in neoplastic tissues of glial origin is regulated differently from that of normal cerebral tissue and, in particular, is devoid of neuronal control.

In pathological states, such as hypertension, brain microvessel function and morphology are altered. In fact, ultrastructural studies have shown increased diameter, endothelial degeneration and deposition of collagen in hypertensive monkeys, while physiological evidence indicates an increase in the permeability of the blood-brain barrier (Johansson et al., 1970; Garcia et al., 1981).

In microvessels isolated from hypertensive rats we have observed a diminished beta-receptor density (Magnoni et al., 1983), consistent with the results by Palmer (Palmer, 1978) showing in the same conditions a reduction in the responsiveness of capillary adenylate cyclase to adrenergic agonists. The impairment of receptor function may be the consequence of an altered pattern of the central adrenergic neurons which control brain microvasculature as well as of changes in circulating catecholamine level, and can be responsible, at least partially, for the BBB dysfunctions occurring in hypertension.

Diabetes is also reported to affect cerebral microvasculature, as indicated by the increased cerebrovascular permeability to albumin and the amplified transendothelial transport of pinocytotic vesicles (Schivers, 1979; Stauber et al., 1981). Experiments performed in our laboratories suggest that beta-adrenergic receptors are decreased in cerebral capillaries of diabetic rats (Magnoni et al, 1984) in correlation with the decreased activity of beta-adrenergic sensitive adenylate cyclase (Palmer et al., 1982). Recent studies on brain microcirculation in ischemic conditions have opened new perspectives on the existence of neuronal mechanisms controlling the BBB. Both neurotransmitter function and metabolism appear altered in various cerebral areas after ischemia. These events are accompained by changes in microvascular functions, such as increased permeability to water and decreased glucose uptake and oxygen consumption. In recent studies we have observed that experimental ischemia induced in the gerbil or in the rat by occlusion of one carotid artery causes a decrease in the density of capillary beta-receptors (Magnoni et al., 1985; Magnoni et al., 1986a). This event may reflect an adaptive response to both ischemia-related release of catecholamines and/or changes in vascular perfusion. Notably, as shown in table II, the reduction in receptor number occurs also in the hemisphere contralateral to the ligature and is always more pronounced and persistent in the left hemisphere, independently of the side of occlusion.

TABLE II

Bmax values of ICYP binding to rat cerebral microvessels after 48 hrs right or left carotid occlusion

	Right hemisphere	Left hemisphere
Sham operated	122 ± 11	116 ± 11
Right carotid occlusion	91 ± 8^{a}	$73 \pm 7^{a,b}$
Left carotid occlusion	96 ± 8^{a}	$78 \pm 6^{a,b}$

a = $p<0.05$ vs the nonligated control.
b = $p<0.05$ vs right hemisphere, carotid ligated.
Data were examined by analysis of variance followed by Tuckey's HSD test for multiple group comparison.
Kd values were not significantly modified, and ranged from 60 to 70 pM.

The mechanisms at the basis of the particular responsivity of the left hemisphere to ischemia are not known. Clinical data seem also to support the view of an asymmetric responsiveness of the two hemispheres to the ischemic insult. In fact, the susceptibility of catecholaminergic neurones and their functional recovery (reflected by the level of catecholamine metabolites in the cerebrospinal fluid of patients with cerebral infarction) seem to vary in dependence on the side of the injury. In particular, in patients with left-sided hemispheric infarction the increase in catecholamine metabolites is more pronounced than in patients with right-sided insult and correlates significantly with the clinical course of the disease (Ferrarese et al., 1986).

In order to investigate the role of neuronal pathways in the biochemical changes observed at capillary level we have followed two experimental approaches (Magnoni et al., 1986b): 1) measurement of capillary beta-receptors after righ or left carotid occlusion in split-brain rats; 2) measurement of capillary beta-receptors after carotid occlusion in rats in which the central noradrenergic system has been previously lesioned by intraventricular injection of 6-hydroxydopamine (6-OHDA).

The results in fig. 1a demostrate that corpus callosum section partially reverses the effects induced by carotid occlusion on the contralateral hemisphere. In addition, the lesion of central noradrenergic system nearly abolishes the effects of ischemia on ICYP binding to microvessels of both hemispheres (Bmax 134 ± 12 and 131 ± 11 fmol/mg prot. for right and left hemispheres of 6-OHDA treated rats, respectively;

117±9 and 119±9 fmol/mg prot. for right and left hemispheres of 6 OHDA-treated, right carotid-ligated rats, respectively).

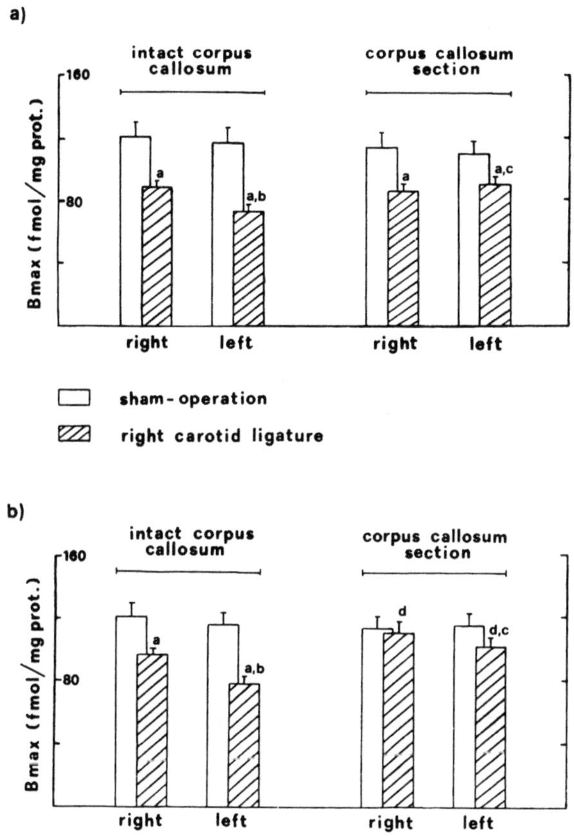

Fig. 1: Bmax values of ICYP binding to brain microvessels after right (a) or left (b) carotid occlusion in callosal-intact and callosal-sectioned rats. Data were examined by analysis of variance followed by Tuckey's HSD test.
a = $p<0.05$ vs the respective non ligated control.
b = $p<0.05$ vs right hemisphere, carotid ligated.
c = not significant vs right hemisphere, callosal-sectioned, carotid ligated.
d = not significant vs the respective non ligated control

The data provide further evidence that beta-receptors located on brain capillaries are, at least in part, regulated by neuronal pathways and suggest the relevance of transneuronal mechanisms in mediating the biochemical changes induced by ischemia in areas far from the insult. Surprisingly, in case of left carotid occlusion (fig. 1b) the section of the callosal connections not only abolishes the effects on the contralateral side, but also greatly attenuates those in the ipsilateral hemisphere.

In addition the results suggest a complex pattern of cerebral asymmetry showing a kind of "polarity" peculiar to the left side of the brain, the expression of which requires the integrity of the interhemispheric connections. On the other hand, the hypothesis of a lateralization in the neuronal mechanisms regulating microvessel function is in line with anatomical, biochemical and behavioral studies which have extensively documented the functional asymmetrical specialization of the cerebral hemispheres (Geschwind and Galaburda, 1985).

Along this line a particularly interesting condition is aging because of the age-related changes in neuronal activity and interhemispheric connections (Pradahan, 1980; Hunziker et al., 1979; Burns et al., 1979). In fact, a number of studies report changes in neurotransmitter function in various cerebral areas during senescence. Catecholaminergic transmission seems to be particularly affected by advanced age, as suggested by several observations on cerebral adrenergic receptors and catecholamine metabolism (Greenberg, Weiss, 1977; Greenberg et al., 1978).

We have studied the effect of ischemia induced by right or

left carotid artery occlusion on capillary beta-receptors in senescent rats (Magnoni et al., 1988b).

Although both human and animal studies report that there is no evidence for a relevant reduction in cerebral blood flow and metabolism in advanced age (DeLeon et al., 1983; Dastur, 1985), the reduced basal level in beta-receptor density (Kobayashi et al., 1982b) may indicate that the regulatory mechanisms of microcirculation may be vulnerable to particular demands of energy and metabolic supply.

The concomitant events of altered pattern of neuronal function and vascular regulation may be at the basis of the qualitative changes in brain capillary reactivity (measured in terms of beta-adrenergic receptor kinetic parameters) to ischemia during aging.

The results reported in fig. 2a and b are reminiscent of the previous data obtained with split-brain rats: in fact, right carotid occlusion was not able to induce any effect in the left hemisphere of aged rats, whereas the decrease in receptor density in the ipsilateral side of the brain was similar or even more pronounced as compared to adult controls. On the other hand, no effect was observed after left carotid ligature on both the hemispheres.

The data confirm the importance of a fully operative neuronal transhemispheric connections for both the expression and the asymmetry of the response of microvasculature to ischemia. However, the possibility that other factors, as changing in circulating catecholamines or blood pressure, may contribute to the observed modifications, cannot be ruled out.

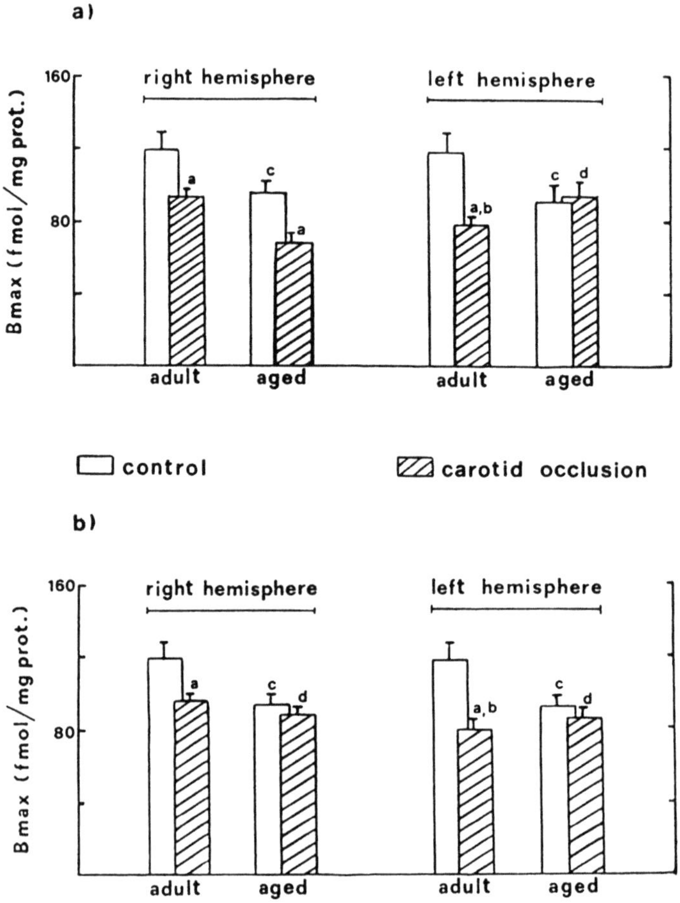

Fig. 2: Bmax (fmol/mg prot) values of ICYP binding to cerebral microvessels after right or left carotid occlusion in adult and aged rats.
a: $p<0.05$ versus the respective sham operated.
b: $p<0.05$ versus right hemisphere, carotid occlusion.
c: $p<0.05$ versus the respective adult sham operated.
d: not significant versus the respective aged, sham operated.
Kd values, ranging from 58-68 pM, were not significantly modified.
Data are means $\pm$ S.D. of four experiments and were examined using analysis of variance followed by Tuckey's HSD test for multiple group comparison.

The data are in line with emerging evidences indicating that neurochemical and behavioral asymmetries such as some psychological capacities and the control of sensory and pain threshold change during the life span (Malfara and Jones,1979; Klisz, 1978; Neri and Agazzani, 1984).

The study of beta-adrenergic receptor reactivity in pathological conditions as ischemia has provided an useful tool in support of the hypothesis of a direct neuronal influence on brain capillaries and has opened new perspectives on cerebral lateralization and its plastic responses to pharmacological manipulations and physiopathological conditions.

At the same time the data give origin to several questions: what is the functional role of beta adrenergic receptors in brain capillaries and the physiopathological meaning of their decrease in ischemic conditions ? Should this phenomenon be interpreted as a sign of damage or as a defense response to an altered local microenviroment ? What is the physiological significance of the asymmetric organization of the neuronal control of brain microvasculature?

REFERENCES

Bates D, Weinshilboum RM, Campbell RJ and Sundt MT Jr (1977) The effect of lesions of the locus coeruleus on the physiological responses of the cerebral blood vessels in cats. Brain Res 136:431-443

Burns EM, Kruckeberg TW, Comerford LE and Buschman MBT (1979) Thinning of capillary wall and declining number of endothelial mitochondria in the cerebral cortex of the aging primate, Macaca nemestrine. J Geront 34:642-650

Dastur DK (1985) Cerebral blood flow and metabolism in normal hunam aging, pathological aging and senile dementia. J Cereb Blood Flow Metab 5:1-9

DeLeon MJ, Ferris SH, George AE, Reisberg B, Chistman DR, Kricheff II, Wolf AP (1983) Computed tomography and positron emission tomography evaluation of normal aging and Alzheimer's disease. J Cereb Blood Flow Metab 3:391-394

Ferrarese C, Bassi S, Frattola L, Loactelli P, Piolti R, Trabucchi M (1986) Different patterns of CSF

neurotransmitter metabolism in patients with left or right hemispheric stroke. Acta Neurol Scand 73:581-585
Garcia JH, Ben-David E, Conger KA, Geer JL and Hollander WA (1981) Arterial hypertension injures brain capillaries. Stroke 12:410-414
Geschwind N, Galaburda AM (1985) Cerebral lateralization. Biological mechanisms, associations and pathology: I - A hypothesis and a program for research. Arch Neurol 42:428-459
Greenberg LH, Weiss B (1977) Beta-adrenergic receptors in aged rat brain: Reduced number and capacity of pineal gland to develop supersensitivity. Science 201:61-63
Greenberg LH, Dix RK, Weiss B (1978) Age-related changes in the binding of 3H-dihydroalprenolol in rat brain. Adv Exp Med Biol 97:245-249
Harik SI (1986) Blood-brain barrier sodium/potassium pump: modulation by central noradrenergic innervation. Proc Natl Acad Sci USA 83:4067-4070
Hartman BK, Zide D and Udenfriend S (1972) The use of dopamine beta-hydroxylase as a marker for the central noradrenergic nervous system in rat brain. Proc Natl Acad Sci USA 69:2722-2726
Hunziker O, Adbel'as S, Shulz V (1979) The aging human cerebral cortex: a stereological characterization of changes in the capillary network. J Gerontol 34:345-350
Johansson B, Li CL, Olsson Y and Klatzo I (1970) The effect of acute arterial hypertension on the blood-brain barrier to ptroein tracers. Acta Neuropath 16:117-127
Joo F et al (1975) cAMP-mediated regulation of the permeability in the brain capillaries. Experientia 32:582-583
Klisz D (1978) Neurophysiological evaluation in older persons In: Storendt M, Siegler IC, Elias MF (eds) The Clinical Physiology of Aging. Plenum Press New York pp 71-96.
Kobayashi H, Memo M, Spano PF and Trabucchi M (1981a) Identification of beta-adrenergic receptor binding sites in rat brain microvessels using 125I-iodohydroxybenzylpindolol. J Neurochem 36:1383-1388
Kobayashi H, Maoret T, Ferrante M, Spano PF and Trabucchi M (1981b) Subtypes of beta-adrenergic receptors in rat cerebral microvessels. Brain Res 220:194-197
Kobayashi H, Frattola L, Ferrarese C, Spano PF and Trabucchi M (1982a) Characterization of beta-adrenergic receptors on human cerebral microvessels. Neurology 32:1384-1387
Kobayashi H, Maoret T, Spano PF, Trabucchi M (1982b) Effect of age on beta-adrenergic receptors on cerebral microvessels. Brain Res 244:374-377
Kobayashi H, Wada A, Izumi F, Magnoni MS and Trabucchi M (1985) Alpha adrenergic receptors in cerebral microvessels of normotensive and spontaneously hypertensive rats. Circulation Res 56:402-409
Magnoni MS, Kobayashi H, Cazzaniga A, Izumi F, Spano PF and Trabucchi M (1983) Hypertension reduces the number of beta-adrenergic receptors in rat brain microvessels. Circulation Res 67:610-613

Magnoni MS, Kobayashi H, Trezzi E, Catapano A, Spano PF and Trabucchi M (1984) Beta-adrenergic receptors in brain microvessels of diabetic rats. Life Sci 34:1095-1100
Magnoni MS, Kobayashi H, Frattola L, Spano PF, Trabucchi M (1985) Effect of common carotid occlusion on beta adrenergic receptor function in cerebral microvessels. Stroke 16: 505-509
Magnoni MS, Frattola L, Pasinetti G, Govoni S, Battaini F, Trabucchi M (1986a) Time-related asymmetric changes of brain microvessel beta-adrenergic receptors in the two hemispheres after carotid occlusion. Stroke 17:966-96
Magnoni MS, Kobayashi H, Govoni S, Battaini F, Pasinetti G, Trabucchi M (1986b) Partial reversal of the asymmetry in microvessel neurochemical changes following ischemia by corpus callosum section. Circulation 73:784-788
Magnoni MS, Frattola L, Piolti R, Govoni S, Kobayashi H, Trabucchi M (1988a) Glial brain tumors lack microvascular adrenergic receptors. Eur Neurol 28:27-29
Magnoni MS, Govoni S, Kobayashi H, Battaini F and Trabucchi M (1988b) Aging modifies the asymmetry in brain microvascular regulation. Stroke 19:38-41
Malfara A, Jones B (1979) Hemispheric asymmetries in motor control of guided reaching with and without optic displacement. Neuropsychology 19:483-486
McKenzie ET, McCulloch J and O'Keane M (1976) Influence of endogenous norepinephrine on cerebral blood flow and metabolism. Am J Physiol 231:489-495
Neri A, Agazzani E (1984) Aging and right-left asymmetry in experimental pain measurement. Pain 19:43-48
Palmer GC (1978) Diminished adenyate cyclase responses in frontal cortex and cerebral capillaries of spontaneously hypertensive rats. Biochem Pharmac 28:2847-2848
Palmer GC, Wilson GL, Palmer SJ and Chronister B (1982) Streptoxotocin-induced diabetes influences receptor activation of adenylate cyclase in rat retina, microvessels, cerebrum and pia arachnoid. Fed Proc 41:1223
Pradahan SN (1980) Central neurotransmitters and aging. Life Sci 26:1643-1656
Raichle ME, Hartman BK, Eichlins JO and Sharpe LG (1975) Central noradrenergic regulation of cerebral blood flow and vascular permeability. Proc Natl Acad Sci USA 72:3726-3730
Rennels ML and Nelson E (1975) Capillary innervation in the mamalian central nervous system. An electron microscopic demonstration. Am J Anat 144:233-241
Schivers RR (1979) Effect of hyperglicemia on brain capillary permeability in the lizard anolis carolinensis. Brain Res 170:509-522
Spatz M, Marsulja BB, Wroblewska B, Merkel N, Bembry J (1986) Modulation of glycogen metabolism in cerebromicrovascular smooth muscle and endothelial cultures. Biochem Biophys Res Comm 134:484-491
Stauber WT, Ong SH and McCuskey RS (1981) Selective extravascular escape of albumin into the cerebral cortex of the diabetic rat. Diabetes 30:500-503

Swanson LW, Connelly MA and Hartman B.K (1977) Ultrastructural evidence for central monoaminergic innervation of blood vessels in the paraventricular nucleus of the hypothalamus. Brain Res 136:166-173

DISCUSSION

Chaired by Betz L. and Frattola L.

B.K. Siesjo:

I'm fascinated by the fact that the changes are also observed in the hemisphere contralateral to the ischemic insult. I was wondering how long do you keep the carotid artery occluded?

S. Govoni:

The occlusion is permanent. From literature data it appears that after single carotid occlusion the decrease in flow in both hemispheres is around 25%, at least at cortical level. Thus, from a vascular point of view the ligature causes equal flow reduction in the two sides of the brain, while at neuronal level we do observe an asymmetrical effect. It would be interesting to investigate the existence of functional asymmetries, for example in behavior (studies with sophisticated and subtle methods) and in other relevant biochemical parameters.

B.K. Siesjo:

Thus, this is actually a reaction to a sustained decrease in cerebral blood flow in the two hemispheres rather than a reaction to a short period of ischemia leading to long lasting changes. What about the sideness of the biochemical changes related to ischemia?

S. Govoni:

What I have shown is that in certain situations there is a major responsiveness of one hemisphere while in the other there is a complete loss of response. For example, after callosal section the effect of right carotid occlusion leads to the loss of the asymmetric response, whereas after lesion of the central noradrenérgic system by 6OH DA the effect is completely lost in both sides.

B.K. Siesjo:

Is there any asymmetry in the central noradrenergic system?

S. Govoni:

There are some data in literature, generally referred to parameters like the level and turnover of cathecholamines, probably reflecting an asymmetric projection to the cortex of the noradrenergic fibers arising from the locus ceruleus, but no data are so far known on an asymmetry in the vascular innervation.

B.K. Siesjo:

There are also differences described between the right and the left hemisphere in terms of lipid metabolism, as the content of diacylglycerol, and the kind and amount of free fatty acids released during ischemia.

S. Govoni:

So far as diacylglycerol and PKC are concerned, in our laboratory we have episodically observed an asymmetry in phorbol ester binding sites. Thus, also in other parameters there are demonstrations for the existence of asymmetry.
The important question is: are the neurochemical asymmetries changing during the process of aging as we have observed in our parameter in response to an ischemic insult?

J. Mc Culloch:

2 hrs after the occlusion you have described a 30% decrease in the number of beta receptors in the right hemisphere. This is a short time for such a large change. What do you think is the mechanism which underlies such a rapid reduction.

S. Govoni:

I cannot give a straight explanation for this finding. Receptors in other situations may do present quite rapid changes. For example, during ischemia catecholamine release occurs very soon, so that the beta-adrenergic reduction observed may represent a quick response to a massive release. Nevertheless, in this case one would rather expect a change in receptor affinity than in the number. This rapid changes in beta-adrenergic receptors, anyway, are not unique and not confined at capillary level. We are in fact performing a series of experiments on the effect of short term ischemia, induced using the same approach, on calcium antagonist binding sites. In hippocampus we have observed a very rapid change in both the affinity and the number of the binding sites, occurring already after 1 h of ischemia.
The mechanisms at the basis of such quick changes (internalization, modifications of conformation etc.) are still to be explored.

L. Betz:

Concerning the results on aging I have a comment.
The data on receptor binding or cAMP generation are expressed per mg of capillary protein. Thus, the assumption is that the mg of capillary proteins is the same under the various experimental conditions. However, during aging, at least in humans, there is an increase in the thickness of the basal membrane. Also in hypertension a deposition of collagen on the capillary basal membrane occurs. This may lead to a change in the microvascular protein content and result in the

lower receptor density values observed during aging or hypertension, so far attributed to chages in the number or properties of endothelial cells.

S. Govoni:

This is a point which may be of importance also in the interpretation of other neurochemical changes during aging: are they due to a modification in the number of neurons or to an alteration in the protein content? So far, we have no means to solve this question.

G.C. Palmer:

In 6OH-DA treated rats did you see any basal increase in beta-adrenergic receptor number as compared to controls?

S. Govoni:

We did not see any increase in the number of beta-adrenergic receptors after 6OH-DA treatment.
Not always after denervation we find receptor hypersensitivity; also in case of chronic treatment with beta-blockes we have not been able to detect receptor density upregulation. However, there is the possibility, still to be tested, of supersensitivity at the level of the cyclase response.

EXPERIMENTAL MODELS OF ISCHEMIA

S. Sampaolo
J. Cervós-Navarro
Institute of Neuropathology
Free University Berlin
Hindenburgdamm 30
1000 Berlin 45
West-Germany

Ischemia is an insult to brain tissue which commonly accompanies diseases of the central nervous system. Stroke, head injury, hydrocephalus intracranial hypertension and space occupying lesions may result in damage to the brain tissue, attributable to impairment of cerebral circulation.
Ischemic stroke caused by occlusions of single or multiple arteries is the most important cause of brain ischemia. Much of the current knowledge on the pathophysiology of brain ischemia is based upon information obtained in animal models of ischemic stroke. Animal experiments have often caused controversy, in the past as in modern times, due to limited extrapolability of the experimental results to the human specie, and because of ethical issues.
This review deals with experimental models which produce either global or focal ischemia. Valuable information derived from in vitro studies are also reported.

Animal models of ischemia:

The use of animal models in the investigation of cerebral

NATO ASI Series, Vol. H 33
Regulatory Mechanisms of Neuron to Vessel Communication in the Brain. Edited by F. Battaini et al.

vascular disease began in 1836 with the experiments of Cooper (Cooper, 1836) of extrinsic ligation of the carotid and vertebral arteries in dogs. In 1847 Flourens (1847) introduced arterial embolization by intravascular injection techniques. Almost all the subsequent animal experiments of brain ischemia have been variations based on these two methods.

Methods to induce ischemic stroke:

Experimental models of brain ischemia were comprehensively reviewed in the Princeton Conference on cerebral vascular disease in 1957 (Wright, Millikan 1957), during the Eight Princeton Conference in 1973 (McDowell, Brennan 1973) and again in 1979 (Price, Nelson 1979).
In 1976 and in 1977 two attempts of classification of the available stroke models have been published (Molinari 1976; Katzman, Clasen, Klatzo et al. 1977).
Garcia and Molinari reviewed the literature on experimental ischemia in 1984 (Garcia 1984) and in 1986 respectively (Molinari, 1986).
A summary of reproducible and reliable ischemic stroke models in animals are reported in table 1. References to original descriptions, validation techniques as well as advantages, disadvantages and specifics of each model are reported in the above mentioned review papers.

Global ischemia

Models of global ischemia are relevant in the study of clinical global ischemia and brain "death".

TABLE 1: ANIMAL MODELS OF CEREBRAL ISCHEMIA

GLOBAL ISCHEMIA

- CARDIAC ASYSTOLE BY KCl INJECTION
- CLAMPING OF INTRATHORACIC VESSELS SUPPLYING THE HEAD ± HYPOTENSION
- CERVICAL COMPRESSION WITH VERTEBRAL LIGATION

FOREBRAIN ISCHEMIA

- CLAMPING OF BOTH COMMON CAROTIDS (Gerbil)
  a) AND BOTH VERTEBRAL ARTERIES
  b) AND THE BASILAR ARTERY (Rat)
- OCCLUSION OF BOTH SUBCLAVIAN ARTERIES AND ONE COMMON CAROTID, AND PARTIAL OCCLUSION OF THE CONTRALATERAL COMMON CAROTID

HEMISPHERAL ISCHEMIA

- COMMON CAROTID LIGATION
  PLUS HYPOXEMIA
- INTRACAROTID PERFUSIONS:
  POLYVINIL ACETATE; ARACHIDONATE; ADP
- EMBOLISM:
  AIR; BLOOD CLOT; MICROSPHERES

FOCAL ISCHEMIA

- MIDDLE CEREBRAL ARTERY OCCLUSION:
  POINT: LIGATURE; CLIP; INFLATABLE CUFF; EMBOLISM WITH SINGLE MICROSPHERE;
  SEGMENTAL: SERIAL CLIPS; GRADED MICROSPHERE; MOLDED SILICONE CYLINDER; SEGMENTAL BLOOD CLOT; PHOTOCHEMICALLY INDUCED THROMBOSIS
- MICROVASCULAR PHOTOCHEMICALLY INDUCED THROMBOSIS

Most frequently global ischemia is caused by heart failure. Irreversible functional neurological injury occurs in 50% - 90% of survivors following resuscitation from cardiac arrest.

Both cardiac arrest induced by i.v. injection of potassium chloride (Myers and Yamaguchi 1977; White et al. 1983) and the model of systemic hypotension (caused by ganglionic blockers and/or bleeding) plus hypoxia (Yatsu et al. 1972) require minimal surgery and are therefore mechanically simple.

Conversely, clamping of the intrathoracic arterial trunks supplying the head (i.e. subclavian, brachiocephalic and mammary arteries) requires considerable dexterousness (Hossmann 1976; Zimmermann and Hossmann 1975).

Maintaining the intracranial pressure equal to or in excess of mean systemic arterial pressure induces impairment of cerebral blood perfusion (Hallenbeck and Bradley 1977). Intracranial pressure can be increased by either infusion of artificial CSF in the liquoral spaces (Ljunggren et al. 1974) or by means of an epidural inflatable balloon (Matakas et al. 1973).

Both this model and the model of cervical vessels occlusion caused by strangulation (Ames et al. 1968) are an example of clinical phenomena in which intracranial venous occlusion precedes arteriolar changes. Such sequence of events has only limited relevance to clinical strokes in humans, in whom the main finding is primary arterial occlusion.

Main disadvantage common to all the models of global ischemia is that they are pharmacologically quite complex. Resuscitation after cardiac arrest requires cardiac massage, ventilation with 100% O_2, rapid adjustment of arterial blood pressure and acid-base imbalance. Vasoactive drug administration, hemodynamic manipulations and control of the acid-base equilibrium are necessary in the procedures for global ischemia in order to keep the animals alive. All of these variables must be taken into account and controlled for by the investigators in their experiments.

Methods producing global ischemia permit one to investigate vulnerability and destruction of various neuroanatomical areas within the brain.

In experimental global ischemia the restoration of cerebral blood flow (CBF) results in two readly distinguishable groups of animals: one with signs of electrophysiological recovery, and the other in which such recovery is absent. The most obvious difference is normal or slightly elevated CBF on recirculation in the group with recovery, and reduced CBF in the animals without recovery (Zimmermann and Hossmann 1975; Hossmann and Zimmermann 1974). Minimal blood flow during ischemia is associated with poor or no recovery, perhaps because of subsequently impaired reperfusion or because of accumulation of acidic metabolites. In all animals intracranial pressure rises immediately after recirculation begins, but remains elevated only in the not recoverable animals. Morphologically, swelling of perivascular elements and of endothelia is demonstrated at the end of the ischemic period. This becomes more pronounced on recirculation, particularly in animals without electrophysiological recovery. Morphologically characteristic of brain death in baboon and cats are autolytic changes in the brain. Ischaemic neuronal changes, midbrain haemorrhages, focal necroses in the brain-stem, demarcation of the C1/C2 cord segment, and displacement of cerebellar tissue are present. Ultrastructural examination reveals extreme brain edema, autolytic changes, complete obstruction of capillaries by astrocytic and endothelial swelling and intravascular blebs (Matakas et al. 1973). These facts indicate that brain death develops in several stages. The circulatory arrest is initially caused by venous compression but becomes irreversible when vascular obstruction develops.

Forebrain ischemia

Forebrain ischemia models may be regarded as a particular case of global ischemia. Indeed cerebellum and brain stem are infrequently damaged (Pulsinelli and Brierley, 1979). The gerbil preparation of bilateral common carotid occlusion has been used extensively in stroke research (Crockard et al., 1980; Kelly and Halsey, 1976; Levine and Payan, 1966). Bilateral common carotid occlusion in this model resulting in a diffuse no flow situation yielded a reliable demonstration of the principle "no flow - no edema" (Kelly and Halsey, 1976; Crockard et al., 1980). The cervical four-vessel occlusion model by Pulsinelli and Brierly (1979) offers several advantages over previously published models (Pulsinelli et al., 1982a-b; Todd et al. 1986a-b). Craniotomy is not required, carotid arteries are ligated in conscious animals, seizures occur rarely. However, the technic and the criteria proposed by the author for evaluating the extent of damage do not yield reproducible. Modifications of the model have been proposed. Coupling of hemorrhagic hypotension to four-vessel occlusion allows the gradation of the level of ischemia (Smith et al., 1984).

Four-vessel occlusion in the rat lowers the forebrain blood flow to less than 10% of control values. Reperfusion results in hyperemia followed by hypoperfusion (Todd et al., 1986 a-b; Pulsinelli et al., 1983).

The three-vessel occlusion model of Kameyama et al. (Kameyama et al., 1985) substitutes basilar artery occlusion and transection for cauterization of the vertebral arteries; this procedure ensures quite constant ischemic damage of brain stem and cerebellum.

These models remain somewhat limited in scope as transient occlusion of the carotids cannot be manipulated to produce different levels of ischemia. A model of forebrain ischemia permitting the control of the levels of ischemia has been recently described (Boehme et al., 1988). In this model the vertebral arteries and one carotid are occluded. The input of blood to the brain depends solely on one carotid artery. Perfusion pressure is controlled by partial compression of this artery with a balloon.

Brief periods of ischemia by four vessels occlusion are quite well tolerated. After 20 minutes of forebrain ischemia the brain shows only scattered neuronal damage at postmortem examination (Plum, 1983). After 30 minutes of ischemia the brains disclose selective neuronal necrosis limited to specific areas of the forebrain and involving, especially in the hippocampus, zones CA_1, CA_3 and to a less extent the paramedian area, layers 3,5 and 6 of the cortex. Damage to small to medium-sized neurons in the dorsolateral crescent of the striatum is a common finding (Plum, 1983).

The phenomenon of delayed postischemic neuronal damage has been described in forebrain ischemia (Pulsinelli et al., 1982a; Kirino, 1982; Brown et al. 1979; Ito et al., 1975).

In different parts of the forebrain, cells showing with ischemic neuronal damage become visible at different times after the ischemic injury. Proliferation and increased number of astrocytes in proportion to the total numbers of neurons and glia cells have been observed in area of ischemic neuronal damage after forebrain ischemia (Petito and Babiak, 1982). Swelling of astrocytes perikarya and foot processes precedes widening of the extracellular space. Increase of brain water content by 1% at 15, 30 and 60 minutes seems to correlate positively to increased pinocytotic activity of the vascular endothelium (Plum, 1983). Furthermore, 15 minutes reperfusion after 30 minutes of 4-vessel-occlusion in the rat results in increased blood brain barrier permeability to small molecules (Sage et al., 1984).

Hemispheral ischemia

The historical evolution of small mammalian models for cerebrovascular ischemia begins with the combined insult method of Levine (Levine, 1960). Unilateral common carotid ligation together with exposure to nitrogen or nitrous oxide produced hemispheric lesions in the rat, the extent and severity of which are determined largely by the depth and duration of the anoxic period. An attractive feature of this and similar methods is their technical simplicity (Brown and Brierley, 1968; Salford et al., 1973). Unfortunately, the need to include a systemic derangement, be it anoxia or hemorrhagic hypotension, complicates the interpretation of ischemic injury. The simplicity of the gerbil model is attractive because of the absence of posterior communicating arteries in many animals which makes these species susceptible of developing hemispheric ischemia after unilateral carotid ligation (Levine and Sohn, 1969; Kahn, 1972). However the attractiveness is partially offset by genetic propensity of the Mongolian Gerbil to seizures (Loskata and Lomax, 1975) and by the inability to identify prospectively the animals with concomitant absence of the anterior communicating arteries (Levy and Brierley, 1974). Furthermore, unilateral occlusion of the gerbil which carotid artery results in a rather wide spectrum of postocclusive CBFs (Crockard et al., 1980) and neurologic deficits (Ito et al., 1975).

High variability of postocclusive CBF is a disadvantage common to all the models utilizing intravascular embolization or injection of substances activating intravascular thrombosis. Consequently size and location of the tissue damage are unpredictable, and an appreciable sample size is essential for statistical significance.

Focal ischemia

Experimental methods to induce both transient and permanent focal brain ischemia have been developed in monkeys, cats, dogs and various other species (Spetzler et al., 1980; Morawetz et al., 1978; Little, 1977; Kamijyo and Garcia, 1975; O'Brien and Waltz, 1973; Hudgins and Garcia, 1970; Sundt and Waltz, 1966; Harvey and Rasmussen, 1951). The method adapted to anesthesized or unanesthesized animals utilize generally the occlusion of a middle cerebral artery (MCA).

The transorbital surgical approach to occlude the proximal segment of the middle cerebral artery in subhuman primates was first described in 1970 (Hudgins and Garcia, 1970).

The results of pathophysiological and clinical observations performed in the MCA-occlusion model in subhuman primates show that this method is the best mimicking human brain ischemia. Nowadays, due to ethical issues, and because of almost prohibitive expenses in obtaining and maintaining subhuman primates their use for laboratory investigations has been drastically reduced.

The rat model of focal cerebral ischemia through the occlusion of a MCA has found wide popularity among laboratory investigators (Hatashita et al., 1988; Young et al., 1987; Rappaport et al., 1987; Bederson et al., 1986; Sako et al., 1985; Gotoh et al., 1985; Shigeno et al., 1985; Teasdale et al., 1983; Tamura et al. 1981; Albanese et al., 1980).

The post-operative mortality rate of about 1% and the predictability of distribution of the brain lesion make this model suitable for both acute and chronic studies of incomplete focal ischemia. The need for anesthesia and craniotomy as well as technical difficulties in achieving reperfusion of the ischemic territory are major disadvantages of this method.

A method to achieve both transient and permanent occlusion of a MCA by <u>in situ</u> platelet thrombosis induced by a photochemical technique in the rat has been described recently (Nakayama et al., 1988; Dietrich et al., 1986a-b; Watson et al., 1985). Although this method requires craniectomy, dura opening and direct management of the artery are avoided. This model may offer unique opportunities to study intravascular clotting events and therapeutic fibrinolysis.

Various methods to induce embolic occlusion of an intracranial artery at predictable sites have been published (Molinari and Laurent, 1976; Molinari et al., 1974; Hill et al., 1955). Despite the resemblance of several of these methods to events occurring in clinical embolic stroke, hese techniques have not enjoyed wide acceptance. Basic disadvantage common to all of the models of embolization is that the methods preclude reperfusion (Molinari, 1979).
A rat model to produce MCA occlusion and reperfusion by the insertion and removal of a silicone coated nylon thread has been described recently (Koizumi J et al., 1986).
The data on CBF changes following a single artery occlusion, are apparently inconsistent. Heterogeneity of CBF changes and structural damage are typical of single artery occlusion (Garcia, 1984). After MCA-occlusion, circulation persists in many of the areas normally supplied by the occluded artery. In the central areas of the ischemic territory, CBF may remain as low as 50% of normal (Sundt and Waltz, 1971; Waltz et al., 1966) Peripheral to the area of ischemia, local loss of autoregulation can be demonstrated about one hour after the MCA-occlusion (Symon, 1975; Shibata, 1974). Animal experiments with MCA occlusion permit to relate certain CBF values to partial or complete, reversible or irreversible disturbances of function (Bell et al., 1985; Heiss, 1983; Astrup et al., 1981). Evoked potentials begin to decrease at blood flow values below 20 ml/100 m/min and are completely suppressed below 15. Neurons can survive for some time in a sublethal state of injury, as reperfusion can restore evoked potential, as well as normalize extracellular K+, and pH. Extracellular potassium rises when CBF drops below 8-10 ml/100 g/min. Disturbance water and electrolyte content becomes evident at flow values below 10 ml/100 m/min. These observations support the concept of an "ischemic penumbra" during which neurons may remain structurally viable, but are functionally inactive (Strong et al., 1983a-b). While function is impaired immediately when flow drops below a certain level, the development of morphological

changes is time dependent (Heiss, 1983).
Focal incomplete ischemia induces typical ischemic neuronal changes (Fig. 1-2).
Evidence for defective electrical function without either membrane or energy failure in the ischemic penumbra has lead to the concept of selective failure of synaptic transmission. The fact that transient amelioration of neurologic deficits resulting from ischemia can be accomplished by neurotransmitter drugs (such as amphetamine) suggest that viable non communicating neurons are responsible for the dysfunction (Glick and Zimmerberg, 1978).
We have developed in our laboratories a model of chronic hypoxic hypoxia in which cats are exposed to stepwise decreasing inspiratory fractions of oxygen (15%, 12%, 10%, 8%, 7%, 6%, 5%) during 7 months. Purkinje cells in Golgi preparations of the hypoxic brains showed marked loss of dendrites and dendritic spines. Cruz-Sanchez et al. (1986), suggesting elective vulnerability of dendrites and synapsis to hypoxia.
Brain water and sodium content increase during the initial three hours of incomplete focal ischemia (Hatashita, 1988; Garcia, 1984). In the rat model of MCA occlusion increase of brain water content has been observed as early as 30 minutes after the occlusion of the artery (Shigeno, 1985). During this time interval fluid accumulates in response to an increase in tissue osmolarity, however the blood brain barrier permeability to proteins and sodium remains normal (Lo et al., 1987; O'Brien, 1979). While part of the increase in tissue osmolarity is due to breakdown of cellular constituents (Hossmann, 1976), change of the blood-brain barrier permeability to small osmotically active molecules may contribute to the development of ischemic edema. We have observed extravasation of sodium fluorescein (MW 376) and ionic lanthanum (MW 138.9) at 15 and 30 minutes respectively, after occlusion of a MCA in the rat (Sampaolo et al., 1988) (Fig. 3).
The "no-reflow" phenomenon reported by Ames et al. in a model

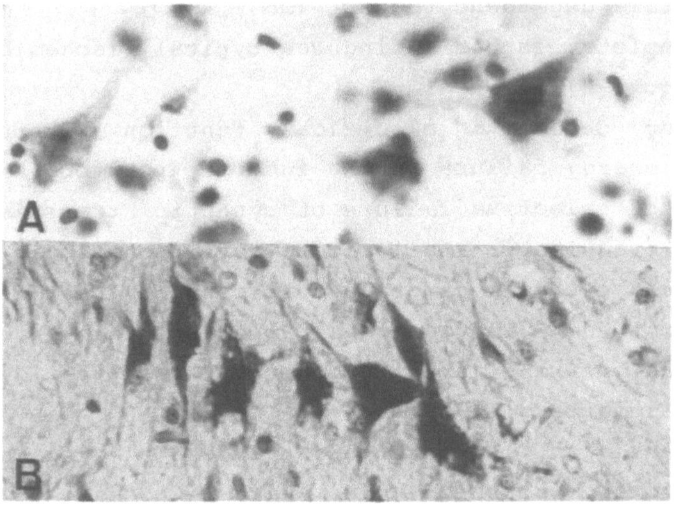

Fig. 1. Ischemic neuron, rat. (A) one hour occlusion of a MCA. A microvacuolated pyramidal cell on the left. On the right the darkly stained neuron with triangular nucleus displays a later stage of ischemic change. (B) : three hours MCA occlusion. Ischemic cell change with calcium incrustation. x 160.

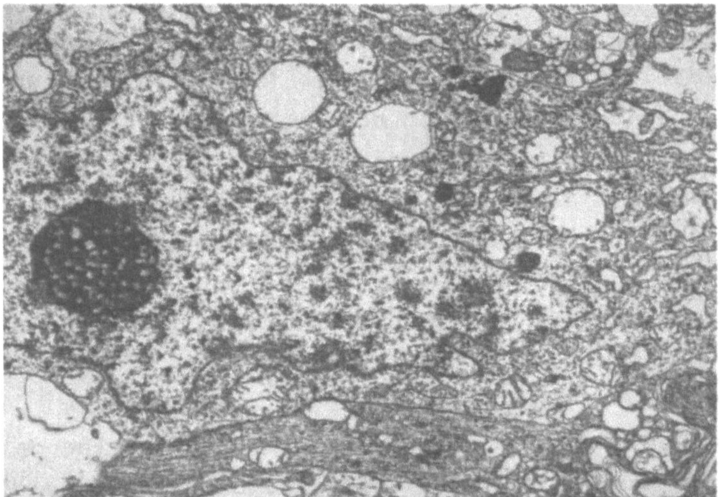

Fig. 2. Rat preparation as in fig. 1-A. Electronmicrograph of a cortical pyramidal cell. Microvacuolation. Swollen mitochondria with disorganized cristae lie within the cytoplasm. A few microvacuoles correspond to dilated cisternae of the endoplasmic reticulum. x 12500.

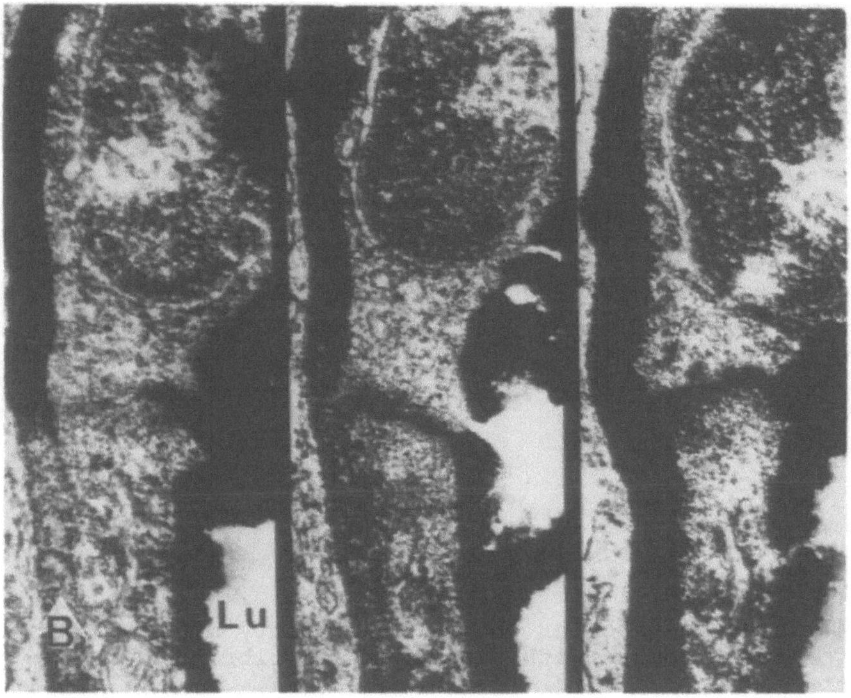

Fig. 3. Ischemic vascular change. Rat, 30 minutes MCA occlusion. In this series of electronmicrographs ionic lanthanum (black) is seen crossing the interendothelial cleft of a brain capillary, and accumulating in the basal membrane. B:basal membrane. Lu:capillary lumen. x 13800.

of strangulation (Ames et al., 1968) suggested that microcirculatory occlusions may play a relevant role in the production of an infarction. Furthermore, this concept indicates that vasculature may have a lower threshold to ischemic injury than neurons and glia. The no-reflow phenomenon seems to depend on the method of ischemia and pressure of reperfusion (Levy et al., 1975). However, post-ischemic obstruction with absence of circulation through small brain vessels (microcirculation) has been demonstrated in several models of ischemia (Sampaolo et al., 1987). Formation of microthrombi begins as early as 10 minutes after the occlusion of a MCA in the rat and their number increases during the first 7 days after the stroke (Sampaolo et al., 1987) (Fig. 4).

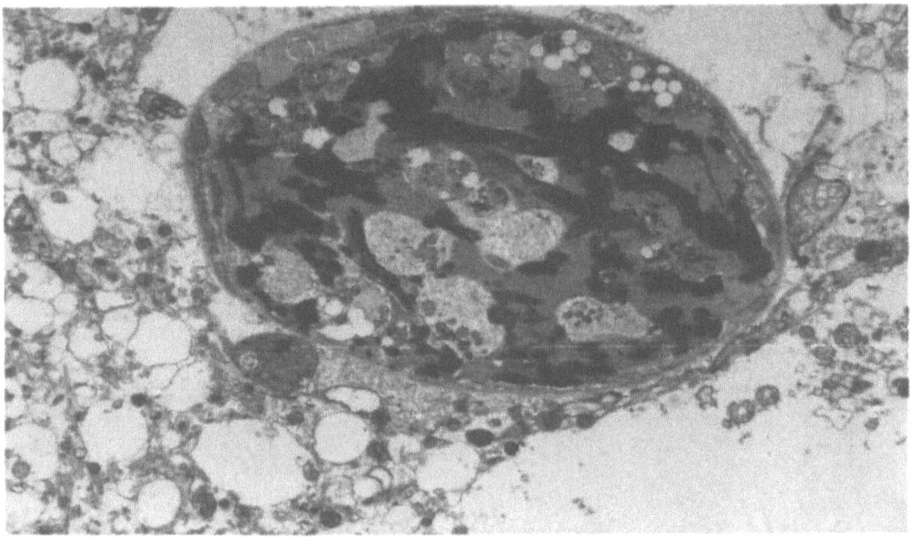

Fig. 4. Ischemic vascular change. Rat, forty min. MCA occlusion. Microthrombus with platelets and fibrin. x 1370.

In vitro models

The use of in vitro models to study stroke has both advantages and disadvantages. In vitro tissue is isolated from both cerebral circulation and outside electrical activity. Although this is disadvantageous in relation to total function, the model allows evaluation of specific biochemical reactions uninfluenced by factors such as changes in blood flow or alterations in electrical activity. In vitro tissue also metabolizes at a slower rate than tissues in vivo. As long as these disadvantages are taken into account valuable information can be obtained from in vitro technics.

Brain slices and dissociated co-cultures (astrocytes-neurons) have also been used in studies of cerebral ischemia in vitro.

In vitro ischemia has been produced by completely removing oxygen and glucose from the culture medium and surrounding atmosphere (Goldberg et al. 1987b, Whittingham et al. 1984). Necrosis of the tissue was observed. The neuronal population was most affected. The astrocytes in postischemic cultures closely resembled reactive astrocytes observed <u>in vivo</u> following ischemia (Goldberg et al. 1987a).

The <u>in vitro</u> preparation which most closely resembles the <u>in vivo</u> brain is the cerebral slice. If the slice is between 0.10 and 0.35 mm thick, oxygen can diffuse in freely and substrates can be incorporated so that stable metabolic conditions are achieved. Stable electrophysiologic conditions are also maintained. Electrical activity can be recorded from cerebral slices (McIlwain and Bachelard 1971). This in vitro preparation has been employed to study the mechanisms of neuronal excitability and epileptogenesis (Cohen et al. 1983). There are, however, both structural and chemical differences between tissue prepared in this fashion and the intact brain.

Usefulness of brain slices in studies of the ischemic penumbra has been suggested. Schiff and Somjen (1987) observed the disappearance of synaptic transmission and increases of extracellular potassium in hippocampal slices exposed to 30 minutes of moderate or mild hypoxia. The extracellular pH of slices decreased in response to both moderate and mild hypoxia. Extracellular calcium did not change.

Newman et al. (1988) studying 2-deoxyglucose metabolism in hypothalamic brain slices, varying in thickness from 400 to 1000 found evidence that the metabolic profiles depend on the depth of the tested fraction in the slices. Increase in lactate and anaerobiosis appear in slice thicker than 600 nm.

Brain slices provide a useful model for studying the neurochemistry and neuropharmacology of the ischemic penumbra.

References

Albanese V, Tommasino C, Spadaro A, Tomasello F (1980) A transbasisphenoidal approach for selective occlusion of the middle cerebral artery in rats. Experientia 36:1302-1304

Ames A, Wright RJ, Kowader M (1968) Cerebral ischemia II. The no reflow phenomenon. Am J Path 52:437-443

Astrup J, Siesjö BK, Symon L (1981) Thresholds in cerebral ischemia - The ischemic penumbra. Stroke 12:723-725

Bederson JB, Pitts LH, Tsuji M, Nishimura MC, Davis RL, Bartkowski H (1986) Rat middle cerebral artery occlusion: Evaluation of the model and development of a neurologic examination. Stroke 17:472-476

Bell BA, Symon L, Branston NM (1985) CBF and time thresholds for the formation of ischemic cerebral edema, and effect of reperfusion in baboons. J Neurosurg 62:31-41

Boehme RJ, Conger KA, Anderson ML (1988) Computer-regulated constant pressure ischemia in the rat: The animal model. J Cereb Blood Flow Metab 8:236-243

Brown AW, Levy DE, Kublik M (1979) Selective chromatolysis of neurons in gerbil brain: A possible consequence of "epileptic" activity produced by common carotid artery occlusion. Ann Neurol 5:127-138

Brown AW, Brierley JB (1968) The nature, distribution, and earliest stages of anoxic-ischemic nerve cell damage in the rat brain as defined by the optical microscope. Br J Exp Pathol 49:87-106

Cohen MM, Bluhm R, Gloneck T, Kopp S, Patel-Mandlik K, and Pettegrew J (1983) In vitro models of stroke. In: Stefanovich V (ed). Stroke animal models. Adv Biosciences, vol. 43, Pergamon Press, New York, pp. 1-12

Cooper A (1836) Some experiments and observations on tying the carotid and vertebral arteries. Guys Hosp Rep 1:458

Crockard A, Iannotti F, Hunstock AT, Smith RD, Harris RJ, Symon L (1980) Cerebral blood flow and edema following carotid occlusion in the gerbil. Stroke 11:494-498

Cruz-Sanchez F, Cervós-Navarro J, Artigas J, Aruffo C, Sampaolo S (1986) Morphologic and morphometric study of the Purkinje cells in chronic hypoxia. Clin Neuropathol 5:106-107

Dietrich WD, Ginsberg MD, Busto R, Watson BD (1986a) Photochemically induced cortical infarction in the rat. 1. Time course of hemodynamic consequences. J Cereb Blood Flow Metab 6:184-194

Dietrich WD, Ginsberg MD, Busto R, Watson BD (1986b) Photochemically induced cortical infarction in the rat. 2. Acute and subacute alterations in local glucose utilization. J Cereb Blood Flow Metab 6:195-202

Flourens JPM (1847) Note touchant l'action de divers substances injecteés dans les artèries. Acad Sci 24:905

Garcia JH (1984) Experimental ischemic stroke: A review. Stroke 15:5-14

Glick SD, Zimmerberg B (1978) Pharmacological modification of brain lesion syndromes. In: Finger S (ed) Recovery from brain damage, research and theory. Plenum Press, New York, pp. 190-201

Goldberg WJ, Connor JR, Bernstein JJ (1987a) Glial fibrillary acid protein, immunohistochemistry of spinal cord, astrocytes after induction of ischemia or anoxia in culture. J Neurosci Res 17:168-175
Goldberg WJ, Kadingo RM, Barrett JN (1987b) Effects of ischemia-like conditions on cultured neurons: protection by law Na, low Ca solutions. J Neurosci 6:3144-3151
Gotoh O, Asana T, Koide T, Takakura K (1985) Ischemic brain edema following occlusion of the middle cerebral artery in the rat. I: The time courses of the brain water, sodium and potassium contents and blood-brain barrier permeability to 125I-albumin. Stroke 16:101-109
Hallenbeck JM, Bradley M (1977) Experimental model for systematic study of impaired microvascular reperfusion. Stroke 8:238-243
Hatashita S, Hoff JT, Salamat SM (1988) Ischemic brain edema and the osmotic gradient between blood and brain. J Cerebr Blood Flow Metab 8:552-559
Heiss WD (1983) Flow thresholds of functional and morphological damage of brain tissue. Stroke 14:329-331
Hill ND, Millikan CH, Wakin KG (1955) Studies in cerebrovascular disease: Experimental production of cerebral infarction by intracarotid injection of homologous blood clot. Mayo Clin Proc 30:625-629
Hossmann KA (1976) Development and resolution of ischemic brin swelling. In: Pappius HM, Feindel W (eds). Dynamics of Brain Edema. Springer-Verlag, Heidelberg, pp. 219-227
Hossmann KA, Zimmermann V (1974) Resuscitation of the monkey brain after 1 hour complete ischemia. Physiological and morphological observations. Brain Res 81:59-74
Hudgins WR, Garcia JH (1970) Transorbital approach to the middle cerebral artery of the squirrel monkey: A technique for experimental cerebral infarction applicable to ultrastructural studies. Stroke 1:107-111
Ito U, Spatz M, Walker JT Jr (1975) Experimental cerebral ischemia in Mongolian gerbils. I. Light microscopic observations. Acta Neuropathol 32:209-223
Kahn KJ (1972) The natural course of experimental cerebral infarction in the gerbil. Neurology 22:510-515
Kameyama M, Suzuki J, Shirane R, Ogawa A (1985) A new model of bilateral hemispheric ischemia in the rat - three vessel occlusion model. Stroke 16:489-493
Kamijyo Y, Garcia JH (1975): Carotid arterial supply of the feline brain. Applications to the study of regional cerebral ischemia. Stroke 6:361-369
Katzman R, Clasen R, Klatzo I et al. (1977) Report of joint committee for stroke resources. IV. Brain edema in stroke. Stroke 8:512-540
Kelly RA, Halsey HJ Jr (1976) Comparison of local blood flow and oxygen availability at the same locus in the ischemic gerbil brain. Stroke 7:274-278
Kirino T (1982) Delayed neuronal death in the gerbil hippocampus following ischemia. Brain Res 239:57-69
Koizumi J, Yoshida Y, Nakazawa T, Ooneda G (1986) Experimental studies of ischemic brain edema I. A new experimental model of cerebral embolism in rats in which recirculation can be introduced in the ischemic area. Jpn J Stroke 8:1-7

Levine S, Sohn D (1969) Cerebral ischemia in infant and adult gerbils. Relation to incomplete circle of Willis. Arch Pathol 87:315-317
Levine S, Payan H (1966) Effects of ischemia and other procedures on the brain and retina of the gerbil. Exp Neurol 16:255-263
Levine S (1960) Anoxic-ischemic encephalopathy in rats. Am J Pathol 36:1-12
Levy DE, Brierley JB, Plum F (1975) Ischemic brain damage in the gerbil in the absence of "no-reflow". J Neurol Neurosurg Psychiatry 38:1197-
Levy DE, Brierley JB (1974) Communications between vertebro-basilar and carotid arteril circulation in the gerbil. Exp Neurol 45:503-508
Little JR (1977) Implanted device for middle cerebral artery occlusion in conscious cats. Stroke 8:258-260
Ljundgren B, Norberg K, Siesjö BK (1974) Influence of tissue acidosis upon restitution of brain energy metabolism following total ischemia. Brain Res 77:173-186
Lo WD, Betz AL, Schielke GP, and Hoff JT (1987) Transport of sodium from blood to brain in ischemic brain edema. Stroke 18:150-157
Loskota W, Lomax P (1975) The mongolian gerbil as a model for the study of epilepsies. Electroencephalogr Clin Neurophysiol 38:597-604
Matakas F, Cervos-Navarro J, Schneider H (1973) Experimental brain death. 1. Morphologic and fine structure of the brain. J Neurol Neurosurg Psychiatr 36:497-508
McDowell F, Brenan W (eds) (1973) Cerebral Vascular Diseases, 8th conference. Grune and Stratton, New York
McIllwain H, Bachelard HS (1971) Biochemistry and the central nervous system. Churchill Livingstone, Edinburgh
Molinari G (1986) Experimental Models of ischemic stroke. In: Barnett HJM, Mohr JP, Stein BM, Yatsu FM (eds). Stroke pathophysiology, diagnosis and management. Churchill Livingstone, Edinburgh, pp. 57-74
Molinari GF (1979) Clinical relevance of experimental stroke models. In: Cerebrovascular Diseases, eleventh Princeton Conference, TR Price, E Nelson (eds), Raven Press, New York, pp. 19-28
Molinari GF, Laurent JP (1976) A classification of experimental models of brain ischemia. Stroke 7:14-17
Molinari GF, Moseley JI, Laurent JP (1974) Segmental middle cerebral artery occlusion in primates: an exerimental method requiring minimal surgery and anestesia. Stroke 5:334-339
Morawetz RB, DeGirolami U, Ojemann RG, et al. (1978) Cerebral blood flow determined by hydrogen clearance during middle cerebral artery occlusion in unanesthetized monkeys. Stroke 9:143-149
Myers RE, Yamaguchi S (1977) Nervous system effects of cardiac arrest in monkeys. Arch Neurol 34:65-73
Nakayama H, Dietrich WD, Watson BD, Busto R, Ginsberg MD (1988) Photothrombotic occlusion of rat middle cerebral artery: histopathological and hemodynamic sequelae of acute recanalization. J Cerebr Blood Flow Metab 8:357-366
O'Brien MD (1979) Ischemic cerebral edema. A review. Stroke 10:623-628

O'Brien MD, Waltz AG (1973) Transorbital approach for occluding the middle artery without craniectomy. Stroke 4:201-206
Petito CK, Babiak T (1982) Early proliferative changes in astrocytes in postischemic noninfarcted rat brain. Ann Neurol 11:510-518
Plum F (1983) What causes infarction in ischemic brain?: The Robert Wartenberg Lecture. Neurology (NY) 33:222-233
Price TR, Nelson E (eds) (1979) Cerebrovascular Diseases, 11th Princeton conference. Raven Press New York
Pulsinelli WA, Levy DE, Duffy TE (1983) Cerebral blood flow in the four-vessel occlusion rat model. Stroke 14:832-833
Pulsinelli WA, Brierley JB, Plum F (1982a) Temporal profile of neuronal damage in a model of transient forebrain ischemia. Ann Neurol 11:491-498
Pulsinelli WA, Levy DE, Duffy TE (1982b) Regional cerebral blood flow and glucose metabolism following transient forebrain ischemia. Ann Neurol 11:499-509
Pulsinelli WA, Brierley JB (1979) A new model of bilateral hemispheric ischemia in the unanesthetized rat. Stroke 10:267-272
Rappaport ZH, Young W, Flamm ES (1987) Regional brain calcium changes in the rat middle cerebral artery occlusion model of ischemia. Stroke 18:760-764
Sage JI, Van Uitert RL, Duffy TE (1984) Early changes in blood brain barrier permeability to small molecules after transient cerebral ischemia. Stroke 15:46-50
Sako K, Kobatake K, Yamamoto L, Diksic M (1985) Correlation of local cerebral blood flow, glucose utiliztation, and tissue pH following a middle cerebral artery occlusion in the rat. Stroke 16:828-834
Salford LG, Plum F, Siesjö BK (1973) Graded hypoxia-oligemia in rat brain. I. Biochemical alterations and their implications. Arch Neuro 29:227-233
Sampaolo S, Heye N, Cervós-Navarro J (1988) Blood-brain barrier permeability to micromolecules after focal cerebral ischemia. In: Hartmann A, Kuschinsky W (eds) Cerebral ischemia and calcium metabolism, Springer, Berlin, Heidelberg, New York, (in press).
Sampaolo S, Cervós-Navarro J, Djouchadar D, and Figols J (1987) Clinical and experimental evidence of microthrombosis in cerebral ischemia. In: Hartmann A, Kuschinsky W (eds) Cerebral ischemia and hemorheology. Springer, Berlin Heidelberg New York, pp. 386-393
Shibata S, Hodge CP, Pappius HM (1974) Effect of experimental ischemia in cerebral water and electrolytes. J Neurosurg 41:146-159
Shigeno T, Teasdale GM, McCulloch J, Graham DI (1985) Recirculation model following MCA occlusion in rats. Cerebral blood flow, cerebrovascular permeability, and brain edema. J Neurosurg 63:272-277
Smith ML, Auer RN, Siesjö BK (1984) The density and distribution of ischemic brain injury in the rat following 2-10 min of forebrain ischemia. Acta Neuropathol (Berl) 64:319-332
Spetzler RF, Selman WR, Weinstein P, et al. (1980) Chronic reversible cerebral ischemia: Evaluation of a new baboon model. Neurosurgery 7:257-261

Strong AJ, Venables GS, Gibson G (1983) The cortical ischaemic penumbra associated with occlusion of the middle cerebral artery in the cat: 1. Topography of changes in blood flow, potassium ion activity, and EEG. J Cereb Blood Flow Metab 3:86-96

Strong AJ, Tomlinson BE, Venables GS, Gibson G, Hardy JA (1983) J Cereb Blood Flow Metab 3:97-108

Sundt TM, Waltz AG (1971) Cerebral ischemia and reactive hyperemia. Studies of cortical blood flow and microcirculation before, during, and after temporary occlusion of middle-cerebral-artery of squirrel monkeys. Cir Res 28:426-433

Sundt TM Jr, Waltz AG (1966) Experimental cerebral infarction: retroorbital extradural approoach for occluding the middle cerebral artery. Mayo Clin Proc 41:159-168

Symon L (1975) Experimental model of stroke in the baboon. In: Meldrum BS, Marsden CD (eds) Primate Models of Neurological Disorders. Adv Neurol, vol. 10, Raven Press, New York, pp. 199-212

Tamura A, Graham DI, McCulloch J, Teasdale GM (1981) Focal cerebral ischemia in the rat: I. Description of technique and early neuropathological consequences following MCA occlusion. J Cerebr Blood Flow Metab 1:53-60

Teasdale G, Tyson G, Tamura A, Graham DI, McCulloch J (1983) In: Stefanovich V (ed) Stroke: Animal Models. Adv Biosciences, vol. 43, Pergamon Press, pp 83-97

Todd NV, Picozzi P, Crockard HA, Russell RR (1986) Reperfusion after cerebral ischemia: Influence of duration of ischemia. Stroke 17:460-466

Todd NV, Picozzi P, Crockard HA, Russell RR (1986) Duration of ischemia influences the development and resolution of ischemic brain edema. Stroke 17:466-471

Waltz AG, Sundt TM, Owen CA (1966) Effect of middle cerebral artery occlusion on cortical blood flow in animals. Neurology 16:1185-1190

Watson BD, Dietrich WD, Busto R, Wachtel MS, Ginsberg MD (1985) Induction of reproducible brain infarction by photochemically initiated thrombosis. Ann Neurol 17:497-504

White BC, Winegar CD, Jackson RE, Joyce KM, Vigor DN, Hoehner TJ, Wilson (1983) Cerebral cortical perfusion during and following resuscitation from cardiac arrest in dogs. Am J Emerg Med 1:128-138

Whittingham TS, Lust WD, Passonneau JV (1984) An in vitro model of ischemia: metabolic and electrical alterations in the hippocampal slice. J Neurosci 4:793-802

Wright IS, Millikan CH (eds) (1957) Cerebral vascular diseases, 2nd conference. Grune and Stratton New York

Yatsu FM, Diamond L, Graziano C (1972) Experimental brain ischemia: protection froom irreversible damage with a rapid-acting barbiturate (methotrexital). Stroke 3:726-734

Young W, Rappaport ZH, Chalif DJ, Flamm ES (1987) Regional brain sodium, potassium, and water changes in the rat middle cerebral artery occlusion model of ischemia. Stroke 18:751-759

Zimmermann V, Hossmann KA (1975) Resusitacion of the monkey brain after one hour's complete ischemia. II. Brain wter and electrolytes. Brain Res 85:1-12

DISCUSSION

Chaired by Pumain R.

B.K. Siesjo:

Two types of edema have been classically defined: cytogenic (or cellular) and vasogenic. Cellular edema involves the accumulation of sodium and chloride ions in the cytoplasm, leading to cell swelling. This implies an ionic gradient between plasma and extracellular fluid. Vasogenic edema follows a compromission of the BBB, with extravasation of plasma proteins. Which are your auguments against this classical view or definition of edema?

J. Cervós Navarro:

The shift of water from the neurons (which shrink) to the glial cells (which swell) occurring during ischemia can not yet be considered as edema. In my opinion, edema is defined as a quantitative net increase of water in the brain, coming from the vasculature. For example, in an edema caused by proteins an increased number of vesicles in the vessels can be found.

B.K. Siesjo:

I do not agree. I think it is possible to have edema without water coming from the plasma. Cellular edema is characterized by increase in cellular volume at the expenses of the extracellular fluid and does not necessarily involve a passage of water from the vascular bed.

J. Cervós Navarro:

In my opinion, water redistribution between the intracellular and extracellular compartiments can be better defined as cell swelling than edema.

B.K. Siesjo:

Cellular swelling does reflect an increase in the content of water, which comes from the extracellular space: this can be indeed defined as edema, which may develop without any damage to the vasculature. This is my point. I think that vasogenic edema, which reflects a compromission of the BBB, is a separate and additional factor contributing to glia swelling. This occurs when the integrity of endothelial cells is impaired to the extent that there is an opening of pores for macromolecules which normally do not penetrate the brain parenchyma.

F. Joo:

I agree with the view of Dr. Cervos-Navarro about the origin of water in cytotoxic edema. Recent leterature shows a clearcut dissociation between water and protein transport through the endothelium. Even in the early phase of ischemia, cerebral capillaries show a sensitivity to the insult comparable to the neurons. For example, at the very beginning of ischemia there are changes in the activity of xantine oxydase, leading to peroxyde formation and damage of the endothelium function which ultimately can cause increase in water influx.

B.K. Siesjo:

I do not agree with this view. Water can be transported only if there is an osmotic gradient, which during ischemia is created either by Na+ and Cl- leak or by idiogenic osmosis. By restricting the extracellular concentration of Na+ and Cl- to 25% of the normal value we create a diffusion ion gradient which is coupled to the water transport. It is impossible to open endothelial pores only for water: its permeability is strictly linked to the osmotic gradients on the basis of the thermodynamic laws. In case of damage of the endothelial well we cannot observe water flow unless the osmotic gradient increases at the other side of the lumen.

F. Joo:

I do not argue this. What we are emphasizing is the origin of water in edema formation. Certainly the extraction of water from the BBB is coupled to ions as Na+ and Ca2+, but not to proteins, at least in the early phase of the post ischemic period.

MECHANISMS OF ISCHEMIC DAMAGE TO NEURONS, GLIAL CELLS AND VASCULAR TISSUE.

By Bo K. Siesjö and Maj-Lis Smith
Laboratory for Experimental Brain Research
University of Lund
Lund, Sweden

During recent years, new information has amassed on mechanisms of brain damage during ischemia, hypoglycemia, and status epilepticus. The results obtained, which have been described in a large number of primary publications and review articles, form the basis of some promising and challenging therapeutic principles. It is the objective of this brief article to summarize current information on mechanisms of brain damage in ischemia. Since there are several recent review articles from our own laboratory (Siesjö 1984, 1988 a and b, 1989; Siesjö and Wieloch 1985, Wieloch 1985) and from those of others (Hossmann 1985, Rothman and Olney 1986, Kaplan et al. 1987, Choi 1988), we will confine original references to a minimum. The interested reader is referred to the articles quoted for further literature.

A. The past: pathophysiology of ischemia.

Several decades ago, the prevailing dogma was that brain cells tolerate anoxia of only 3-4 min duration. This concept was based both on clinical experience in resuscitating patients following cardiac arrest, and on several experimental studies in which the circulation to the brain was temporarily arrested. However, experimental work conducted in the period 1950-1960 by Hirsch, Schneider and their collaborators made it likely that such short revival times (defined as the longest ischemic periods which could be tolerated without functional or structural brain damage) reflected the susceptibility of the heart to anoxia, rather than that of the brain (see Hirsch et al.

NATO ASI Series, Vol. H 33
Regulatory Mechanisms of Neuron to Vessel Communication in the Brain. Edited by F. Battaini et al.

1957, Hirsch and Müller 1962). Thus, when measures were taken to protect the heart against anoxic damage revival times were found to be 8-10 min at temperatures of 37°C. It was also established that resuscitation of the brain was optimal only if the perfusion pressure of the brain could be promptly restored at the termination of the ischemia. This finding seemed to provide an explanation for the short revival periods following cardiac arrest, a condition in which post-ischemic blood pressure is often suboptimal. Quite logically, the findings inspired both clinical and experimental workers to raise postischemic perfusion pressure by administration of pressure agents.

Subsequent work centered on the reasons for why brain cells tolerated anoxia better *in vitro* than *in vivo*. Thus, even with a seemingly adequate restoration of perfusion pressure, brain cells seemed inappropriately sensitive to anoxia. The hypothesis was advanced that resuscitation of the brain was limited by vascular factors in the sense that, following ischemia of a certain duration, adequate reflow was hindered by increased viscosity of stagnating blood and by swelling of both parenchymal and endothelial cells, intravascular coagulation playing a contributory role ("the no-reflow phenomenon" of Ames et al. 1968). It was established that ischemia, even when complete, leads to the production of endogenous osmoles. The hyper-osmolarity thus created caused rapid swelling of the tissue, with an increase in the intracranial pressure, once perfusion was restored (see Hossmann 1982, 1985). Because of stagnation of blood, vascular narrowing, edema formation, and the risk of coagulapathy, efforts were made to raise postischemic perfusion pressure above normal, to decrease cell swelling by hyperosmotic solutions, and to combat intravascular coagulation. The results were striking since Hossmann, Kleihues and collaborators achieved remarkable short-term recovery of neurophysiological and metabolic brain functions after complete ischemia of up to 60 min duration (Hossmann and Kleihues 1973, Hossmann 1982, 1985). Using similar principles, Safar and colleagues could achieve good long-term recovery after 12 min of cardiac arrest (Safar et al. 1976, Safar 1986).

It soon became clear that short-term recovery after brain ischemia was not synonymous with long-term recovery. Thus, following the pioneering observation of Ito et al. 1975, who found that cell death after ischemia requires a "maturation" period, it was demonstrated that following transient ischemia, death of some neuronal populations may be delayed by 1-2 days, or even longer (Kirino 1982, Pulsinelli et al. 1982). Clearly, an impressively complete short-term recovery does not preclude the subsequent development of extensive brain damage. These findings suggested that pharmacological intervention may be possible in the free interval between the primary insult and the secondary damage.

Obviously, the scientific community was still confronted with the question: why are brain cells so sensitive to anoxia? However, it had been known since the days of Spielmayer, and of the Vogts, that some neurons are selectively sensitive to insults. It may be more to the point, therefore, to ask why some cells are so sensitive to anoxia, and others quite resistant. The question is more complicated than so since there is, in reality, two types of ischemic brain damage: selective neuronal necrosis and pan-necrosis, i.e. infarction. It could be deduced from a number of studies that one factor which can convert selective neuronal vulnerability to infarction is the duration of ischemia, longer ischemic periods being prone to cause destruction of neurons, glial cells, and vascular tissue and, thus, to infarction.

However, studies performed more than 10 years ago in our own and in Myers´ laboratories showed that feeding or glucose infusion aggravated the ischemic damage, converting selective vulnerability to pan-necrosis. It was speculated that exaggerated acidosis was responsible (Myers 1979 a,b, Siesjö 1981, 1984, 1985, Plum 1983, Rehncrona 1985). Accordingly, efforts are now made in many centers to correct or prevent hyperglycemia in cases with ongoing or impending ischemia.

In regional ischemia caused by obstruction of the middle cerebral artery (MCA), i.e. in stroke, both time and acidosis contribute to cause infarction. Since stroke usually allows some remaining circulation, even in the focus, and since flow in perifocal areas ("the penumbra") may be only moderately reduced, a contributing factor to the developing edema is influx of osmotic equivalents from blood. It is accepted that ischemic edema is of two types, cytotoxic (or cellular) and vasogenic (see Katzman et al. 1977, Baethmann 1978, Chan and Fishman 1985, Klatzo 1985). Cellular edema is characterized by accumulation of Na^+ and Cl^- (with loss of K^+) while vasogenic edema is caused by extravasation of proteins, considered to be due to structural damage to vessels. During ischemia, and in the early recirculation period following an ischemic insult, edema formation is held to be mainly of a cytotoxic nature and no or little extravasation of large molecules, as Evan´s Blue, has been shown in this period (Petito et al. 1982, Shigeno et al. 1985). Transient openings of the blood-brain barrier for small molecules have been demonstrated in the early recirculation phase after ischemia (Sage et al. 1984), but the formation of vasogenic edema seem closely related to inflow of large molecules, as serum proteins, into the extracellular space (Kuroiwa et al. 1985). However, the latter type of extravasation seem to be at hand mainly in very severe situations, encompassing infarctions and cerebral bleedings (Petito et al. 1982, Avery et al. 1984), and is mostly seen at longer recirculation times.

B. Mechanisms of ischemic damage

Intense attempts are now being made to define the cellular and molecular mechanisms of ischemic damage, and to find pharmacological means of ameliorating the damage. Three mediators of cell damage are in the focus of interest: calcium overload, acidosis, and free radical formation.

a. Calcium-mediated damage. We and others have assumed that ischemia, like hypoglycemia and status epilepticus, is accompanied by influx of calcium into cells, with molecular damage occurring by uncontrolled proteolysis, lipolysis, and protein phosphorylation, or by mitochondrial overload (Siesjö 1981, Meldrum 1983, Raichle 1983; for further literature, see Siesjö 1988a). Originally, calcium was assumed to enter neurons by voltage-sensitive channels (VSCCs), inspiring hope in the therapeutic efficacy of calcium antagonists. Later, it became clear that a pathogenetically important part of the calcium influx occurs via agonist-operated calcium channels (AOCCs). The impetus for this development came from Olney's excitotoxic hypothesis (see Olney 1978, Rothman and Olney 1986), which predicts that excitatory amino acids (EAAs), particularly glutamate and its structural analoges, are toxic to cells containing receptors for the EAAs whenever these acids are released in increased amounts, or their uptake is blocked. The major receptor types are that activated by kainate (K)/quisqualate (Q), which gates an ion channel allowing Na^+ to enter and K^+ to leave, and the one activated by N-methyl-D-aspartate (NMDA), which gates a channel permeable to calcium. It was originally believed that enhanced influx of Na^+ (with Cl^-) caused osmolytic damage to postsynaptic structures. It now seems established, though, that EAAs exert their major toxicity in vitro and in vivo by allowing (uncontrolled) calcium influx (Choi 1985, 1987, Rothman and Olney 1986).

The calcium hypothesis of ischemic brain damage has led to attempts to ameliorate damage by blocking VSCCs and AOCCs by dihydropyridine and other blockers of VSCCs and by NMDA antagonists, respectively. Blockers of VSCCs seem to have a marginal effect only, unless given before experimental stroke, or in postischemic situations in which neuronal hyperactivity and increased calcium cycling are present. More promising results have been obtained with competitive and non-competitive NMDA-antagonists, the latter including the dissociative anaesthetics ketamine and phencyclidine, and MK 801 (see Choi 1988). However, results with NMDA antagonists are disappointing in the sense that they seem to have only marginal effects in global or forebrain ischemia of the type seen during cardiac arrest. Much more promising results have been obtained with regional ischemia (induced by experimental stroke). This is probably because calcium antagonists act by reducing calcium influx and depolarization in incomplete ischemia (or in hypoxia), i.e. in situations in which cellular oxygenation has been critically reduced, but has not ceased altogether (Siesjö 1988b, Siesjö and Bengtsson 1989).

b. Acidosis and free radical formation. Exaggerated tissue acidosis has been linked to postischemic seizures, to edema formation, to destruction of glial cells, and to infarct development (for recent literature, see Siesjö 1985, 1989, Kraig et al. 1987, Norenberg et al. 1987, Siesjö et al. 1987,, Petito et al. 1987). It is of interest that if acidosis during ischemia is exaggerated by preischemic hyperglycemia, seizures and edema appear after a free interval which may be as long as 18-24 hrs (Warner et al. 1987).

Acidosis probably does not act by enhancing the rise in Ca^{2+}_i. Although the molecular mechanisms have not been defined, it seems more likely that acidosis acts primarily by accelerating the formation of harmful free radicals, thereby contributing to postischemic membrane damage ("reperfusion damage") (Siesjö et al. 1985, Rehncrona et al. 1988). Part of this damage seems to affect the vasculature, which have the potential of forming superoxide anions (O_2^-) upon reoxygenation (see Kontos 1985). Thus, promising results have been published with superoxide dismutase (SOD), administered in conjugated form, or trapped in liposomes (Chan et al. 1987, Snelling et al. 1987, Michelson et al. 1988). Probably, though, free radicals are formed also in parenchymal cells. It is now believed by many that acidosis wrecks the machinery by releasing pro-oxidant iron from proteins such as transferrin and ferritin, and that iron (chelated to low molecular weight compounds) triggers the formation of toxic free radicals, such as •OH. In support, a novel group of free radical scavengers (21-aminosteroids or "lazaroids") seem to markedly ameliorate ischemic damage, at least that incurred by reversible focal ischemia (Hall et al. 1988, Young et al. 1988).

c. Hypothermia and ischemia brain damage. It was reported already by Hirsch and Müller (1962) that brain temperature differences of only 1-2°C could alter revival times, i.e. the longest ischemic periods tolerated without neurological or structural damage. Novel results have reconfirmed the protective effect of hypothermia, and shown that a lowering of brain temperature by only 2-4°C can have pronounced effects on selective neuronal vulnerability (Busto et al. 1987, Minamisawa et al. in preparation). Even more interestingly, these results have suggested that amelioration of ischemic damage seems possible with post-ischemic brain cooling. One can foresee that, in the future, pharmacological treatment during brain resuscitation will include a combination of at least two types of calcium antagonists (directed against VSCCs and AOCCs), and of free radical scavengers/iron chelators, perhaps supplemented with moderate hypothermia.

REFERENCES

Ames A III, Wright RL, Kowada M, Thurston JM, Majno G (1968) Cerebral ischemia. II. The no-reflow phenomenon. Am J Pathol 52:437:453

Avery S, Crockard HA, Russell RR (1984) Evolution and resolution of oedema following severe temporary cerebral ischaemia in the gerbil. J Neurol, Neurosurg and Psychiatry 47:604-610

Baethmann A (1978) Patophysiological and pathochemical aspects of cerebral edema. Neurosurg Rev 1:85-100

Busto R, Dietrich WD, Globus M Y-T, Waldés I, Scheinberg P, Ginsberg MD (1987) Small differences in intraischemic brain temperature critically determine the extent of ischemic neuronal injury. J Cereb Blood Flow Metabol 7:729-738

Chan PH, Fishman RA (1985) Brain edema. In: Handbook of Neurochemistry (A Lajtha, ed) New York, Plenum Publishing, pp 153-174

Chan PH, Longar S, Fishman RA (1987) Protective effects of liposome-entrapped superoxide dismutase on posttraumatic brain edema. Ann Neurol 21:540-547

Choi DW (1985) Glutamamte neurotoxicity in cortical cell culture is calcium dependent. Neurosci Lett 58:293

Choi DW (1987) Ionic dependence of glutamate neurotoxicity. J Neurosci 7:369-379

Hall ED, Pazara KE, Braughler JM (1988) 21-aminosteroid lipid peroxidation inhibitor U74006F protects against cerebral ischemia in gerbils. Stroke 19:997-1002

Hirsch H, Euler KH, Schneider M (1957) Ueber die erholung und Wiederbelebung des Gehirns nach Ischämie bei Normothermie. Pflügers Arch 265:281-313

Hirsch H, Müller HA (1962) Funktionelle und histologische Veränderungen des Kaninchengehirns nach kompletter Gehirnischämie. Pflügers Arch 275:277-291

Hossmann K-A (1982) Treatment of experimental cerebral ischemia. J Cereb Blood Flow Metabol 2:275-297

Hossmann K-A (1985) Post-ischemic resuscitation of the brain: selective vulnerability versus global resistance. In: Progress in Brain Research, Vol 63 (K Kogure, K-A Hossmann, BK Siesjö and Welsh FA, eds) Elsevier, Amsterdam, pp 3-17

Hossmann K-A, Kleihues P (1973) Reversibility of ischemic brain damage. Arch Neurol (Chic.) 29:375-382

Ito U, Spatz M, Walker JT, Klatzo I (1975) Experimental cerebral ischemia in mongolian gerbils. I. Light microscopic observations. Acta Neuropathol (Berl.) 32:209-223

Kaplan J, Dimlich RVW, Biros MH, Herris J (1987) Mechanisms of ischemic cerebral injury. Resuscitation 15:149-169

Katzman R, Clasen R, Klatzo I, Meyer JS, Pappius HM, Waltz AG (1977) IV. BrainEdema in Stroke: Study group on brain edema in stroke. Stroke 8:512-540

Kirino T, Sano K (1984) Selective vulnerability in the gerbil hippocampus following transient ischemia. Acta Neuropathol (Berl.) 62:201-208

Klatzo I (1985) Brain oedema following brain ischaemia and the influence of therapy. Br J Anaesth 57:18-22

Kontos HA (1985) Oxygen radicals in cerebral vascular injury. Circ Res 57:508-516

Kraig RP, Petito CK, Plum F, Pulsinelli WA (1987) Hydrogen ions kill brain at concentrations reached in ischemia. J Cereb Blood Flow Metabol 7:379-386

Kuroiwa T, Cahn R, Juhler M, Goping G, Campbell G, Klatzo I (1985) Role of extracellular proteins in the dynamics ov vasogenic brain edema. Acta Neuropathol (Berl.) 66:3-11

Meldrum BS (1983) Metabolic factors during prolonged seizures and their relation to nerve cell death. Adv Neurol 34:261-275

Michelson AM, Jadot G, Puget K (1988) Tratment of brain trauma with liposomal superoxide dismutase. Free Rad Res Comms 4:209-224

Myers RE (1979a) A unitary theory of causation of anoxic and hypoxic brain pathology. In: Cerebral hypoxia and its consequences (S Fahn, JN Davis, LP Rowland, eds) Raven Press, New York, pp. 195-213

Myers RE (1979b) Lactic acid accumulation as a cause of brain edema and cerebral necrosis resulting from oxygen deprivation. In: Advances in Perinatal Neurology (R Korobkin, G Guilleminault, eds) New York, Spectrum, pp. 85-114

Norenberg MD, Mozes LW, Gregorios JB, Norenberg L-O B (1987) Effects of lactic acid on astrocytes in primary culture. J Neuropath Exp Neurol 46(2):154-166

Olney JW (1978) Neurotoxicity of excitatory amino acids. In: Kainic acid as a tool in neurobiology (EG McGeer, JW Olney and PL McGeer, eds) Raven Press, New York, pp. 95-121

Petito CK, Pulsinelli WA, Jacobson G, Plum F (1982) Edema and vascular permeability in cerebral ischemia: Comparison between ischemic neuronal damage and infarction. J Neuropath Exp Neurol 41(4):423-436

Petito CK, Kraig RP, Pulsinelli WA (1987) Light and electron microscopic evaluation of hydrogen ion-induced brain necrosis. J Cereb Blood Flow Metabol 7:625-632

Plum F (1983) What causes infarction in ischemic brain?: The Robert Wartenberg lecture. Neurology (NY) 33:222-233

Pulsinelli WA, Brierley IB, Plum F (1982) Temporal profile of neuronal damage in a model of transient forebrain ischemia. Ann Neurol 11:491-498

Raichle ME (1983) The patophysiology of brain ischemia. Ann Neurol 13:2-10

Rehncrona S, Rosén I, Smith M-L (1985) Effect of different degrees of brain ischemia and tissue lactic acidosis on the short-term recovery of neurophysiologic and metabolic variables. Experimental Neurology 87:458-473

Rehncrona S, Nielsen Hauge H, Siesjö B K (1988) Enhancement of iron-catalyzed free radical formation by acidosis in brain homogenates: differences in effect by lactic acid and CO_2. J Cereb Blood Flow Metabol. In press

Rothman SM, Olney JW (1986) Glutamate and the pathophysiology of hypoxic-ischemic brain damage. Ann Neurol 19:105-111

Safar P (1986) Cerebral resuscitation after cardiac arrest. A review. Circulation 74:138-153

Safar P, Stezoski W, Nemeto EM (1976) Amelioration of brain damage after 12 minutes cardiac arrest in dogs. Arch Neurol 33:91-95

Sage JI, van Uitert RL, Duffy TE (1984) Early changes in blood brain barrier permeability to small molecules after transient cerebral ischemia. Stroke 15:46-50

Shigeno R, Teasdale GM, McCulloch J, Graham DI (1985) Recirculation model following MCA occlusion in rats. J Neurosurg 63:272-277

Siesjö BK (1981) Cell damage in the brain: A speculative synthesis. J Cereb Blood Flow Metabol 1:155-185

Siesjö BK (1984) Cerebral circulation and metabolism. J Neurosurg 60:883-908

Siesjö BK (1985) Acid-base homeostasis in the brain: Physiology, chemistry, and neurochemical pathology. In: Progress in Brain Research (K Kogure, K-A Hossmann, BK Siesjö, FA Welsh, eds) Elsevier, Amsterdam 63:121-154

Siesjö BK (1988a) Historical overview: Calcium, ischemia and death of brain cells. Ann NY Acad Sci 522:638-661

Siesjö BK (1988b) Mechanisms of ischemic brain damage. Crit Care Med 16:954-963

Siesjö BK (1988c) Calcium and cell death. Europ Neurol, in press

Siesjö BK (1989) Transmembrane ionic fluxes in the brain in energy-depleted cells. Neurochem Pathol, in press

Siesjö BK, Wieloch T (1985) Cerebral metabolism in ischaemia: neurochemical basis for therapy. Br J Anaesth 57:47-62

Siesjö BK, Bengtsson F (1989) Calcium fluxes, calcium antagonists, and calcium-related pahtology in brain ischemia, hypoglycemia and spreading depression: a unifying hypothesis. J Cereb Blood Flow Metabol, in press

Siesjö BK, Bendek G, Koide T, Westerberg E, Wieloch T (1985) Influence of acidosis on lipid peroxidation in brain tissues in vitro. J Cereb Blood Flow Metabol 5:253-258

Siesjö BK, Smith M-L, Warner DS (1987) Acidosis and ischemic brain damage. In: Cerebrovascular Diseases (M E Raichle and WJ Powers, eds) Raven Press, New York, pp. 83-95

Snelling LK, Ackerman AD, Dean JM, North MC, Traystman RJ (1987) The effects of superoxide dismutase on neurologic injury (cerebral blood flow and evoked potentials) following global cerebral ischemia. Anestesiology 67(3A):A153

Warner DS, Smith M-L, Siesjö BK (1987) Ischemia in normo- and hyperglycemic rats: effects on brain water and electrolytes. Stroke 18:464-471

Wieloch T (1985) Neurochemical correlates to selective neuronal vulnerability. In: Progress in Brain Research, Vol 63 (K Kogure, K-A Hossmann, BK Siesjö, FA Welsh, eds) Elsevier Science Publishers B.V., pp. 69-85

Young W, Wojak JC, DeCrescito V (1988) 21-aminosteroid reduces ion shifts and edema in the rat middle cerebral artery occlusion model of regional ischemia. Stroke 19:1013-1019

DISCUSSION

G.C. Palmer:

Have you ever studied the hyperglycemic damage in the gerbils? In our experiments this species does not seem to be very sensitive to such insult.

B.K. Siesjo:

The hyperglycemic damage was originally described by Meyers in the monkey and in the cat. The fact that fasting gerbils do not show attenuation of the ischemic damage by hypoglycemia may be due to a very high level of lactate, leading to acidosis, even in fasted animals. I think that you should measure the lactate content in the gerbils at the end of ischemia. It is possible that starting from a lower degree of lactic acidosis you might see the effect of hyperglycemia.

R. Pumain:

Why are glutammate and aspartate preferentially released during ischemia?

B.K. Siesjo:

Since K^+ is gradually released in the extracellular fluid in ischemic conditions, the increasing concentration of this ion causes depolarization of the presynaptic terminals and release of the neurotransmitters. Furthermore, literature data show that the iontophoretic release of glutammate is associated with a decrease in the extracellular calcium concentration, particularly in the layers II on V of the cortex. The terminals of these layers are thought to be glutamatergic. However, the reason of the particular vulnerability to hypoxia of glutamatergic neurons is not known.

L. Hertz:

The glutamate uptake into the astrocytes is severely impaired during hypoxia, and this could contribute to the major role of this neurotransmitter in ischemic damage. I have also a question: astrocytes are reported to have voltage dependent calcium channels. In our experiments on cell cultures we have observed dihydropyridine-inhibited potassium-induced Ca2+ uptake; however, other authors have not been able to detect the presence of these channels in astrocytes derived from hippocampus slices. Could that mean that in this area of the brain Ca2+ does not enter the astrocytes through voltage dependent channels?

B.K. Siesjo:

I do not know. There is evidence from electrophysiological

studies that a calcium electrode implanted into the cortex and the hippocampus records the same degree of ion influx into the cells; however, this does not discriminate the type of cell which preferentially receives calcium.
I do not know how to interpret Ca influx into glial cells. The presence of calcium precipitates has been demonstrated in both neurons and astrocytes after ischemia: however, the astrocytes, in contrast to the neurons, survive after the insult. This may suggest that Ca2+ influx is greater in neurons than in glial cells; it could be of interest to verify this hypothesis by measuring the intracellular calcium concentration in primary cultures of astrocytes.
Concerning the first comment, it is important to underline that the excytotoxic hypothesis is based on the assumption that the accumulation of glutamate at the receptors is due either to an excessive release or to a deficient uptake. In conditions of hyperglicemia or ischemia, with very low levels of ATP, the uptake mechanism is likely to be compromised.

L. Betz:

Your presentation on the unifying hypothesis of calcium role in ischemic damage was very interesting and provocative, and I'm happy to see that you have taken in consideration also the role of other agents, such as free radicals in the mediation of the ischemic effects; in fact, we have recently been working with free radical scavengers and allupurinol, a inhibitor of the xantine oxydase (rich in capillaries): both drugs significantly reduce the size of the infraction in rats after middle cerebral artery occlusion.

B.K. Siesjo:

I dont' think that the ischemic damage by free radicals interferes with the unifying hypothesis of Ca-mediated insult. I have tried to unify the discordant results in the calcium field but I do believe that other factors, such as free radicals, may contribute to the genesis and progression of the ischemic injury.

F. Joo:

Calcium may enter the cell via the different subtypes of voltage-dependent calcium channels and the NMDA-gated channels. Are there studies in literature which have investigated the effect of the combination of NMDA receptor blockers plus dihydropiridines on the consequences of ischemia?

B.K. Siesjo:

This is a point of great theoretical and clinical interest. I can answer that we are starting this investigation now.

R.Pumain:

Do your think that one of the mechanisms of acidification-induced damage could be the fact that during acidosis the intracellular binding of Calcium is reduced and, as a consequence, the cytoplasmatic free Calcium concentration is higher?

B.K. Siesjo:

That is certainly a possibility; the only reason why I think this may be not a major mechanism are the results by two groups of researchers showing that the acidification of the cytoplasm causes the release of Ca2+, but the effect is not very dramatic (Ca2+ increases from 100 nM to 500 nM for 1 degree decrease in pH). Furthermore, low pH is known to block Ca influx through voltage-dependent channels in many cell and to influence Na/Ca exchange in many tissues.

L. Sokoloff:

What about the inhibitors of xantine oxydase in the protection of ischemic injury. Do the inhibitors of superoxide dismutase enhance the effects of ischemia?

B.K. Siesjo:

Some experiments have been performed using tungsten administration in the diet, as inhibitor of the xantine oxydase, and resulted in reduction of ischemic damage (i.e. edema) and increase in survival.

L. Betz:

Brain capillaries have a notably low content in superoxide dismutase.

B.K. Siesjo:

My impression, from literature data, particularly on reperfusion injury, is that free radicals cause damage especially to the vessels, in a massive and rapid manner. My question: is xantine oxydase localized to the microvascular fraction or is it also present in the brain parenchyma?

L. Betz:

Some lines of evidence suggests that the enzyme is probably esclusively localized at vascular level.

B.K. Siesjo:

This would indicate that during reperfusion there is a high production of oxyl and peroxyl-radicals at endothelial level, leading to leukocyte and platelet aggregation.

L. Sokoloff:

Are there data in literature on the reperfusion recovery in a low oxygen athmosphere?

B.K. Siesjo:

Recent data show that in a gerbil model of reversible ischemia the production of ATP and the neurological behavior are more impaired in the group of animal exposed to oxygen as compared to normal air. Other Authors have observed the same effect by reducing the flow to the tissues in the reperfusion phase.
However, in our model of forebrain ischemia we did not observe any effect. Another example is MK-801, which is uneffective in our experimental conditions. Probably this type of ischemia is very "dense", but the recirculation is optimal, whereas in the gerbil model the recirculation is lousy.
Also clinical data reveal that many patients have blood pressure of 60-65 mm Hg in the recovery phase: this means that the possibility of endothelial and platelet interactions with the endothelium are less.

DYNAMIC VARIATIONS OF THE EXTRACELLULAR SPACE IN THE CENTRAL NERVOUS SYSTEM

R. Pumain*, J. Louvel and I. Kurcewicz
Unité de Recherches sur l'Epilepsie, INSERM U 97
2ter rue d'Alésia
75014 Paris
FRANCE.

*To whom correspondence should be addressed
Tel: (1) 45 89 89 07

SUMMARY

During neuronal hyperactivity, including repetitive stimulations, applications of excitatory amino acids, epilepsy, spreading depressions and anoxia, large changes in the concentration of extracellular ion occur. Such changes are accompanied by changes in the extracellular space (ES) size, which can shrink by up to 60%. The changes in ES size can be rapid, localized and also rapidly reversible. Several mechanisms can account for the ES changes: KCl uptake in glia, the glial spatial buffering, NaCl uptake into neurons, metabolic increase of the intracellular osmolarity. The observed ES changes may have a number a consequences, among which the maintenance of neuronal functioning appears to be the most important.

KEYWORDS

Extracellular space, Cerebral cortex, Ion selective electrode, Glutamate, N-methyl-D-aspartate, Kainate, Epilepsy.

NATO ASI Series, Vol. H 33
Regulatory Mechanisms of Neuron to Vessel Communication in the Brain. Edited by F. Battaini et al.

INTRODUCTION

It has been a common opinion that the extracellular space (ES) of the central nervous system of vertebrates, particularly of mammals, provides for the neurons a stable and very well regulated milieu, necessary for the healthy functioning of the nerve cells. This standing belief applies especially to the ionic composition of the neuronal microenvironment, since the neurons are exquisitely sensitive to changes of the concentration of several ionic species in the external medium. However, it has been increasingly apparent, with the development of ion-selective microelectrodes, that the ionic composition of the ES can indeed undergo substantial changes in a variety of conditions. If such changes appear to be relatively modest in physiological situations, they may be rather dramatic in pathological conditions such as epilepsy, anoxia or spreading depressions. In fact, since the neuronal activity is underlain by ion transfer across the plasmatic membranes, any sustained activity should modify the ionic content of both the intra- and extra-cellular milieu. The changes in the extracellular milieu depend on the intensity of the ion fluxes, the volume of dispersion of the ions and the characteristic of ion migration in the ES, together with the involvement of regulatory mechanisms. The redistribution of ions between the various compartments should also produce osmotic unbalance and therefore water fluxes, inducing further changes in the relative volume fraction of the intra- and extra-cellular spaces. In this chapter, we shall describe succinctly the methods which have been used for measuring possible ionic and ES changes, then the characteristic features of the extracellular space. We shall describe further the changes that have been observed in various conditions, but essentially during applications of excitatory amino acids, and finally discuss some of the possible consequences of these changes. Since the changes that we are reporting have been observed both in vivo and in brain slices, we will not consider the possible contribution of intravascular material.

1. METHODOLOGY

1.1 Ion selective electrodes

The methodology of ion selective electrodes has recently been extensively presented in a number of several excellent reviews (cf Ammann, 1986) and will not be detailed here at length. In brief, the microelectrodes have usually two channels, which can be coupled or separated: the first is filled with the liquid sensor at its tip, surmounted by a column of aqueous solution containing the ion to be measured; the second one is filled with an electrolyte-containing solution , usually sodium chloride for measurements in the ES, and is referred to as the reference electrode. The selective channel, like a selective membrane, develops a potential difference according to the Nerst equation according to the concentration of the ion in the electrode and in the external medium. However, the selective channel, which can have a very high electrical resistance (up to several Gigaohms), is also sensitive to potential changes. Therefore, this potential signal, which is also measured through the reference channel, has to be subtracted from the signal recorded through the selective channel in order to obtain the ion signal in isolation. This substraction is most often performed using an electronic device.

The reliability of the recordings depends critically on the selectivity of the sensors. Fortunately, in the recent past years, several neutral sensors with a high selectivity have been developed, owing essentially to the efforts of Prof. Simon and his colleagues (Simon et al, 1984). Sensors for measuring sodium, calcium, potassium, pH, and chloride ions are presently available. Sensors for measuring magnesium ions are being developed, but till now with a poor selectivity vis-à-vis the other cations.

Another problem is that such electrodes indeed measure the activity of the ions, not their concentration. However, since the activity of ions is difficult to determine in multi-ionic solutions and since for the calibration solutions, it is the ion concentrations that are really known, the results are usually expressed in terms of concentration.

The tip of the electrode when the two channels are coupled, which is usually the case for ES measurements, has usually an external diameter of 2 µm. Such electrode, when inserted in the nervous tissue, can create a microcavity. Although the diffusion of ions in such microcavity may slow down the speediness of measurements, the use of electrodes with very rapid response time has shown that equilibration time with the ES is short, in the order of a few tens of ms (Pumain et al, 1983). Nevertheless, the intrinsic response time of conventional microelectrodes is longer, due to the high electric resistance, in the order of 200 ms for a calcium selective electrode.

1.2 Techniques used for measuring ES changes

The first techniques used to determine the volume fraction of the extracellular space in the brain involve determination of the clearance rate of extracellular markers, impedance measurements and morphometric studies of electron micrographs. Indeed, the use of the last technique yielded values which were very small, in the order of 5% or even less (Horstmann and Meves, 1959). However, it has been recognized that the fixation procedures induced cell swelling and a resulting shrinkage of the ES (Torack, 1965; Van Harreveld et al, 1965), and indeed, measurements obtained from tissue fixed through rapid freeze techniques (Van Harreveld, 1972; Cragg, 1979), yielded much larger values for the volume fraction of the ES.

Clearance methods, for instance using inulin (Levin et al, 1970) or radioactive molecules as extracellular markers, yielded also larger values. However, a proportion of the perfused inulin is cleared away in the ventricles and such method therefore tends to underestimate the volume fraction value.

Impedance measurements can be performed by inserting a current source in a given structure and measuring the resulting signal at a known distance (Freygang and Landau, 1955; Adey et al, 1962; Van Harreveld et al, 1963) . However, the precision of the measurements may be decreased because the distance between the electrodes are sometimes difficult to determine precisely, because of the geometry of the structure, of the capacity of cell membranes when using a sinusoidal signal, of the estimation of fraction of current that flows through cells (Gardner-Medwin, 1983; Gardner-Medwin and Nicholson, 1983).

A more recent approach to measure the volume fraction of the ES has been the use of ion selective electrodes: an extracellular marker is introduced into the nervous tissue, either from a point source or by superfusion. The resulting concentration change is monitored continuously through one or several ion selective microelectrodes sensitive for the marker, placed at various distances from the source. Using the appropriate diffusion equation, it is then possible to calculate the relevant parameters (Nicholson et al, 1979; Nicholson and Phillips, 1981). The extracellular marker should diffuse exclusively and freely in the ES, not to have physiologic or toxic properties at the concentrations achieved during the experiment, and of course, a sensor selectively sensitive to such substance must be available. Both anions and cations have been tested (Nicholson and Phillips, 1981), but the most widely used ion has been tetramethylammonium (TMA): indeed, a potassium sensor developed by Corning (477315) is sensitive to quaternary ammonium compounds in the micromolar range, and at that concentration, TMA appears to be devoid of effect, at least during the duration of the experiment (Hille, 1967).

The TMA method may also be used to monitor continuously possible changes in the ES size: TMA can be applied through a micropipette by constant ionophoresis pulses and the resulting TMA concentration changes recorded using a nearby ion selective microelectrode. The amplitude of the TMA signals is constant if the volume of repartition of TMA, i.e. the ES volume is constant. If this volume increases, the amplitude of the TMA signals will decrease, and conversely, the amplitude of the TMA signals will increase in case of reduction of the ES size.

2. CHARACTERISTICS OF THE ES.

2.1 Volume fraction of the total volume taken by the ES.

The volume fraction of the ES has been determined using the various methods described above. A typical value of 23% has emerged, ranging from 0.18 to 0.28 in cerebella of various animal species (Nicholson and Rice, 1986). Similar values has been obtained in the neocortex (Lehmenkuhler et al, 1985). The volume fraction of the ES appears to be usually about constant in the brain. However, its value may vary substantially from place to place, even in a given structure, owing to differences in cell packing density. For instance, in the CA1 field of the hippocampus, the volume fraction of the ES has the usual value of 20-22% in the stratum radium, but only around 5% in the Stratum Pyramidale, where the cell bodies of the pyramidal neurons are very densely packed (Heinemann, in prep).

Further, it appears that the ES size can change with development, the ES becoming slightly smaller with age.

2.2 Tortuosity

Another important feature is that migration of substances appears to be hindered in the ES . For instance, when a substance which diffuses freely in water is applied to brain structures, the kinetics of the concentration changes within the nervous tissue are slower than in water, or than in an agar gel. This is due to the fact that the diffusion pathway in the nervous tissue is shortened, due to the presence of impermeable cellular elements. In fact, the average spacing between cells may be between 20 and 50 nm, while the average free path for an ion is, at the ionic strength of the ES is only about 0.01 nm. The ions therefore behaves most of the time in the ES as if they were in an infinite medium and rarely "bounce" on a cell membrane. Although then the microscopic behavior of ions follows the laws of diffusion in a continuum, the presence of the cell membranes may be viewed as boundary conditions, the migration of a small fraction of the ions being restricted. The resolution of the corresponding equations is not straightforward but, making simplifying assumptions, the boundary conditions are replaced by a new parameter known as the tortuosity factor (Nicholson, 1980), since the substances migrate following tortuous pathways in the ES. For more details about the calculations, the reader is referred to the paper by Nicholson and Phillips (1981). The diffusion coefficient in water D is replaced by a factor called the apparent diffusion coefficient, the value of which being D/λ^2.

If the substance, in most cases TMA, is released from a micropipette by ionophoresis, and at equilibrium, the concentration achieved at a distance r from the pipette will be $In\lambda^2/4\pi FDra$, I being the ionophoretic current, n the transport number of the pipette (Purves, 1979), F the Faraday constant, a the volume fraction of the ES. The values obtained for the tortuosity factor ranged from 1.49 to 1.84 in the cerebella of various species (Nicholson and Rice, 1986), the typical value being 1.6.

It is not clear at present whether there is a relationship between the volume fraction and the tortuosity. However, this problem should be borne in mind, especially when measuring extracellular space changes. Preliminary data indicate that they are related by a power law (Nicholson and Rice, 1986)

2.3 Diffusion in the ES

It has been a matter of debate whether cations, especially divalent cations diffuse freely in the ES, due to possible interactions with the fixed surface negative charges of the cellular membranes (Mc Laughlin et al, 1971), many of which being located on the long chains of glycosaminoglycan that protrude in ES. A report by Morris and Krnjevic (1980) indeed indicated that the migration of calcium was slowed down in the ES. However, subsequent studies did not confirm such view, and the migration of calcium ions appears to be quite similar to that of other ions (Nicholson and Rice, 1986, 1987). This finding implies that, at the ionic strength of the extracellular milieu, there is little interaction between divalent cations and the molecules of glycosaminoglycan.

Besides, the diffusion of substances in the ES may appear to be hindered when uptake processes for this substance are present.

3. IONIC AND EXTRACELLULAR SPACE CHANGES.

Possible ES changes appear to be linked with water fluxes between the intra- and extra- cellular compartments, themselves due to changes in osmotic pressure. These changes appear to be mainly the result of alterations of ion fluxes across cellular membranes, either through opening of membrane channels and/or through modifications of the efficacy of metabolic pumps. An increase in the intracellular metabolic activity, by augmenting the number of osmotically active particles may also play a role. we shall first describe briefly the ionic changes in the ES observed in a number of situations, and afterwards the ES changes that have been observed.

3.1 Ionic changes in the ES.

Changes in the concentration of extracellular potassium ($[K^+]_o$) have been described in a number of situations. The base-line level of $[K^+]_o$ is between 2.8 and 3.2 mM in mammals. When neurons are activated, there is an afflux of potassium ions and an accumulation in the ES (Houndsgaard and Nicholson, 1983). Thus, an increase in $[K^+]_o$ of a few mM can already be observed during exercise in the spinal cord (Heinemann U, Schaible HG and Schmidt RF, in prep.), while larger increases have been observed during repetitive stimulations in various structures (Heinemann and Lux, 1977; Nicholson et al, 1978; Prince et al, 1973; Somjen, 1980; Sykova et al, 1976; Urbanics et al, 1978), during application of excitatory amino acids (Heinemann and Pumain, 1980; Pumain and Heinemann, 1985; Pumain et al, 1987), during epileptiform episodes (Lothmann and Somjen, 1976; Heinemann and al, 1977; Pumain et al, 1985; Heinemann et al, 1986), during spreading depression and anoxia (Vyskocil et al, 1972; Hansen and Zeuthen, 1981) and hypoglycemia (Wieloch et al, 1984). However, during repetitive stimulation (Heinemann and Lux, 1977), or during epileptiform episodes (Heinemann et al, 1977; Pumain et al, 1985; Heinemann et al, 1986), the increase in $[K^+]_o$ never goes beyond 10 mM (the "ceiling level"), while during spreading depression, it can exceed 40 or even 50 mM (Kraig and Nicholson, 1978; Hansen and Zeuthen, 1981). During applications of excitatory amino acids, the rises in $[K^+]_o$ are generally modest, being maximally of about 12mM for the largest doses (Heinemann and Pumain, 1980; Pumain and Heinemann, 1985; Pumain et al, 1987) At cessation of the various stimulations, there is often an undershoot of $[K^+]_o$ below the base line level (Heinemann and Lux, 1975; Pumain and Heinemann, 1985), undershoot attributed to the persistence of activity of metabolic pumps.

Changes in the concentration of extracellular sodium ($[Na^+]_o$) have also been observed in a variety of conditions, most usually decreases. The base line level is about between 140 and 150 mM in the ES of mammals (Dietzel et al, 1980; Pumain et al, 1987). During electrical stimulations, falls of at most 10 mM have been measured (Dietzel et al, 1982), but if the stimulation was prolonged, the signal returned steadily to the base line value. During excitatory amino acid applications, decreases of up to 50 mM have been observed in the cortex (Pumain et al, 1987). Large decreases in $[Na^+]_o$ have also been measured during seizures (Pumain et al, in prep) and spreading depressions (up to 60 mM) (Kraig and Nicholson, 1978; Hansen and Zeuthen, 1981).

Concerning the concentration of extracellular chloride ions ($[Cl^-]_o$), rises of 2 to 6 mM were detected during repetitive electrical stimulation in the cortex of cats, from a base line level of 135 to 140 mM (Dietzel et al, 1981). In contrast, excitatory amino acid applications induced decreases in $[Cl^-]_o$ by up to 15 mM (Pumain et al, in prep). During generalized seizures, $[Cl^-]_o$ may initially rise by a few mM, and subsequently may fall by about 10 mM under the base line value, while during spreading depressions, decreases by up to 90 mM (Kraig and Nicholson, 1978).

Calcium ions, although less significant for osmotic equilibrium due to its smaller concentration, is very important for synaptic transmission and several metabolic processes (cf Pumain, 1988). The base line level of the concentration of extracellular calcium ($[Ca^{2+}]_o$] is between 1.2 and 1.3 mM in mammals. $[Ca^{2+}]_o$ falls by maximally 0.4-0.5 mM during electrical stimulations in the cerebellar (Nicholson et al, 1978) and the cerebral cortex (Heinemann and Pumain, 1980; Somjen, 1980; Pumain and Heinemann, 1985). Local applications of the excitatory amino acid glutamate or of glutamate agonists that activate the N-methyl-D-aspartate (NMDA) subtype of glutamate receptors type produces large decreases in $[Ca^{2+}]_o$ at all depths in the cortex. In contrast, the glutamate agonists that activate the other receptor subtypes, such as quisqualate (Quis) or kainate (KA), induce increases in $[Ca^{2+}]_o$ in the superficial cortical layers or small decreases at deep layers (Heinemann and Pumain, 1980; Pumain and Heinemann, 1985; Pumain et al, 1987). These experiment show that the NMDA-operated channels are permeable to calcium ions in addition to sodium and potassium ions (Mac Dermott et al, 1986; Pumain et al, 1987). Large decreases in $[Ca^{2+}]_o$ have been observed to accompany epileptiform activity (Heinemann et al, 1977; Lehmann et al, 1986), down to maximally 0.15 mM during seizures (Pumain et al, 1985) or to 0.6 mM during single interictal spikes (Pumain et al, 1983). The corresponding intracellular calcium accumulation may be toxic to the neurons (Siesjö et al, 1986). Spreading depressions are also accompanied by very large decreases in $[Ca^{2+}]_o$, maximally to a level of 0.1 mM (Kraig and Nicholson, 1978; Hansen and Zeuthen, 1981).

3.2 Extracellular space changes

It is clear from the above-mentioned data, that such very large extracellular ionic changes must affect the osmotic equilibrium between the extra- and intra-cellular compartments. When looking at numbers, it is apparent that osmotically-active particles disappear from the ES when nerve cells are synchronously activated. This must result in an increase of the osmolarity of the intracellular compartment and a concurrent water influx. The resulting shrinkage of the ES should in turn modify the extracellular concentration of the ions that do not cross readily the cellular membranes.

Indeed, shrinkage of the ES have been demonstrated during electrical stimulations (Dietzel et al, 1980; Sykova, 1987)), application of excitatory amino acids (Pumain et al, 1987), epileptiform episodes (Dietzel and Heinemann, 1986; Lux et al, 1986), spreading depression and anoxia (Hansen and Olsen, 1980) and changes in PCO_2 (Lemenkühler et al, 1985). Thus, decreases of the size on the ES by up to 60% have been described in the most extreme conditions. In order to study in more details the spatial variations and the time course of such changes, we measured, in the rat cortex, the ES changes induced through applications of excitatory amino acids.

3.3. Changes in the ES size induced through applications of excitatory amino acids.

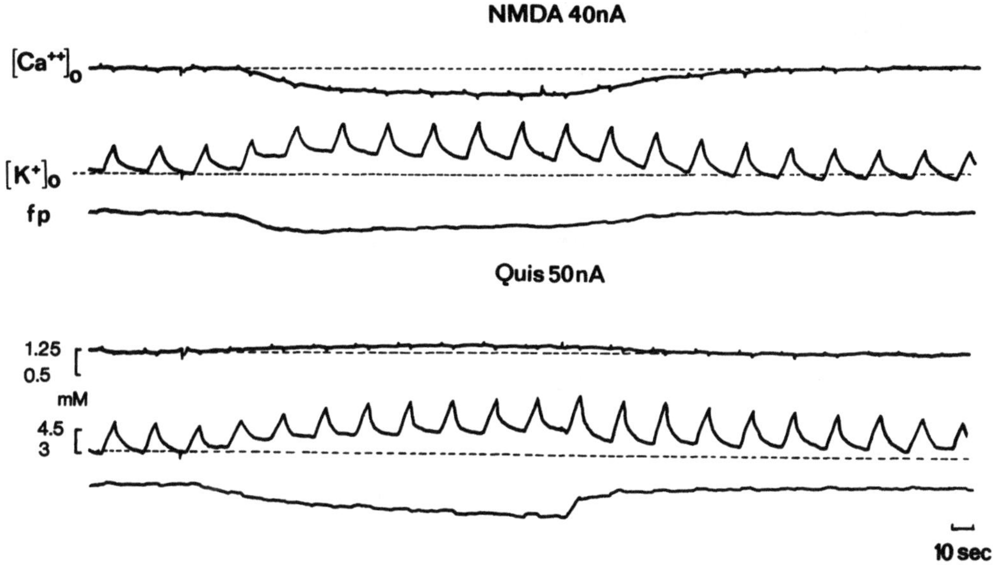

Fig. 1: Simultaneous recordings of changes in the extracellular calcium and potassium concentrations, using ion selective microelectrodes, during local ionophoretic applications of the excitatory amino acids N-methyl-D-aspartate (NMDA) and Quisqualate (Quis). Both recordings were performed at the same place, in the cerebral cortex of a rat in vivo, at a depth of 300 µm under the pial surface. The triangular signals that ripple the potassium trace are due to constant ionophoretic pulse applications of tetramethylammonium (TMA), substance to which the potassium electrode is also sensitive. The progressive increase of the amplitude of TMA signals indicate the development of the reduction of the extracellular (ES) size (see text). After the applications, the recovery is rather rapid, especially for NMDA.

The experiments were performed in the rat cortex in vivo. An ion selective microelectrode sensitive to both potassium and TMA ions (see above), was glued to an ionophoresis microelectrode through which TMA could be applied locally, such that their tips were 10-20 µm distant. The other channels of the six-channel ionophoresis electrode contained several excitatory amino acids, namely glutamate (Glu), DL-homocysteate (DLH), Quis, KA, or NMDA, and NaCl for current balancing. With such technique, it is thus possible to apply various substances very locally, thus allowing a laminar study of ES changes within a given structure. In several experiments, another ion selective electrode was attached

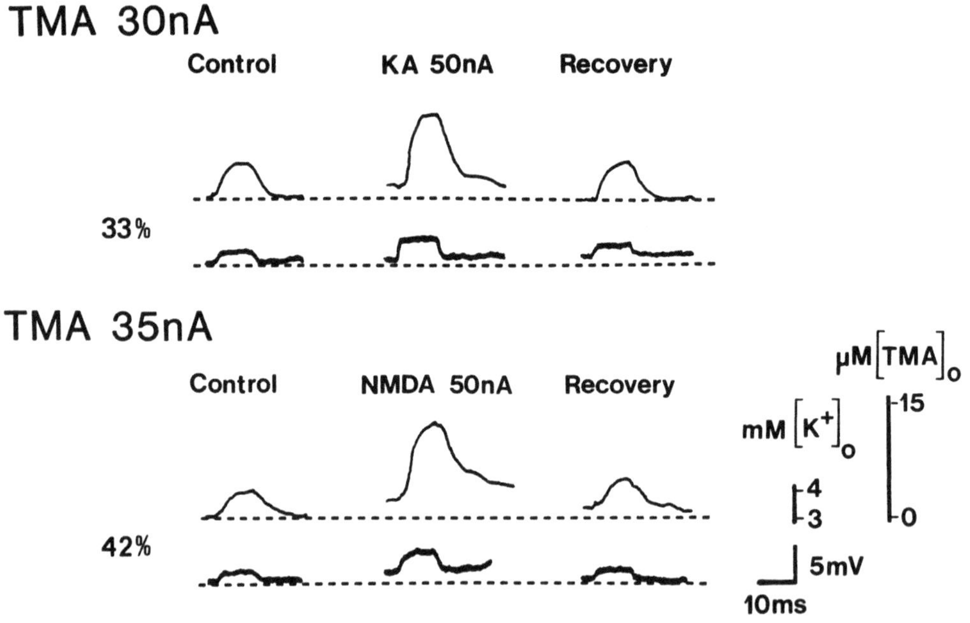

Fig. 2: Amplitude of TMA signals before, during and after applications of the excitatory amino acids kainate (KA) and N-methyl-D-aspartate (NMDA). Recording depth 350 μm. Otherwise as in Fig. 1. The corresponding percentage of the ES shrinkage are indicated on the left.

to the electrode assembly, in order to measure the concomitant changes of another ion, in most cases calcium ions. On Fig. 1, NMDA and Quis were successively applied in the cortex at a depth of 300 μm under the pial surface, while TMA was being released in pulses through another channel. During NMDA applications, there was a negative shift on the field potential (fp) recording, an increase in $[K^+]_o$ from 3 mM to 4.5 mM and a decrease in $[Ca^{2+}]_o$ from 1.25 mM to 0.5 mM. The triangular signals that ripple the potassium trace represent the TMA concentration changes induced by the constant ionophoretic pulses. It is clear that the amplitude of these signals increases during the NMDA application, thus showing that there was a concomitant decrease in the ES size. During the Quis application, there was also a negative shift on the fp recording, indicating that the nearby neurons were depolarized, and an increase in $[K^+]_o$ (from 3 mM to 5 mM),

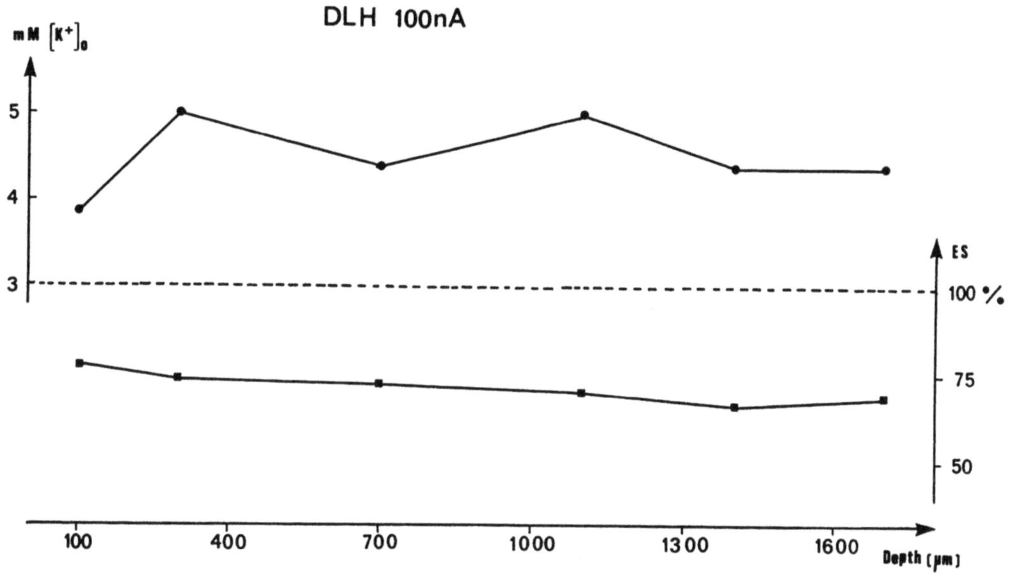

Fig. 3: Laminar profiles of the increase of the concentration of extracellular potassium and of the decrease of the ES induced through constant applications of DL-Homocysteate at different depths in the cerebral cortex of a rat.

but in contrast, a slight increase in $[Ca^{2+}]_o$ occurred. Meanwhile, one could observed an increase in the amplitude of the TMA signals, showing a reduction in the ES size. It is interesting to note that the changes in ES size paralleled the ionic changes and that their recovery after the application was rather rapid, in a few tens of sec.

The same phenomenon is shown on Fig. 2, in which KA and NMDA were applied, in another experiment, at a depth of 350 µm under the pial surface. In this case, the resulting changes in ES size were calculated from the relative amplitude of the TMA signals and indicated as percentages. It is apparent that with moderate doses, the reductions (33% during KA and 42% during NMDA) are already rather large.

In Fig. 3, a laminar profile of the $[K^+]_o$ increases and of the concomitant ES reduction induced through constant applications of DLH at various depths in the cortex, from surface

Depth 200µm	ΔV %	Recovery time
GLU 60	12	~ 30 sec
NMDA 50	32	~ 130 sec
QUIS 50	35	~ 150 sec
KA 60	31	> 300 sec

Table 1: ES changes and recovery times of these changes induced through ionophoretic applications of the following excitatory amino acids: glutamate (Glu: 60 nA), N-methyl-d-aspartate (NMDA: 50 nA), Quisqualate (Quis: 50 nA) and Kainate (KA: 60 nA). The recording site was in a rat cortex, at a depth of 200 µm below the pial surface.

to deep layers: the reductions of the ES size appear to be identical throughout the cortex, indicating that there were no regional differences of such responses. In the underlying white matter, there were very little, if any, ionic and ES changes.

The time course of the ES changes are indicated on table 1. We determined the amplitude of the changes and the recovery time of the ES after cessation of application of four different excitatory amino acids, at a recording site located at 200 µm below the cortical surface. It appears that the recovery times depend of the excitatory amino acid applied, the shorter times being observed for the agent for which powerful uptake mechanisms exist (glu) and the longer times for the agents for which it is weak or absent.

These experiments show that the activation of nerve cells induce both ionic and ES changes, which seem to parallel each other. The changes in ES size can be rather rapid and localized, but readily reversible. In the cerebral cortex, the ES changes in response to excitatory amino acid applications are about identical at all layers.

3.4 Changes in the ES during development

Ransom and coworkers (1985) studied the amount of ES change induced by electrical stimulation in the optic nerve of rats at different stages of their development. In neonate rats, stimulus-dependent ES shrinkage was absent, showing a gradual increase in the next two weeks. Such capability is correlated in time with that of maintaining a ceiling level (about 12 mM in the adult) of maximal increases in $[K^+]_o$ during stimulation. It is also roughly correlated with the proliferation and maturation of glial cells after birth.

4. MECHANISMS OF ES CHANGES.

The observed ES changes can arise from several mechanisms, among which KCl uptake, spatial glial potassium buffer mechanism, NaCl entry into neurons and a metabolic increase in the intracellular osmolarity.

KCl uptake occurs when an there is an increase in $[K^+]_o$ (Boyle and Conway, 1941; Gardner-Medwin, 1983) and operates even with uniform changes, without concentration gradient. Indeed when $[K^+]_o$ is elevated there is a potassium uptake in the glial cells (Walz and Hertz, 1983; Ballanyi et al, 1987) which can produce a cellular swelling (Walz, 1987). The KCl uptake rate is substantially higher in glial cells (astrocytes) than in neurons (Hertz, 1986). The glial potassium uptake may be passive or due to different active systems, the Na^+,K^+ -ATPase, and a furosemide-sensitive uptake (Hertz, 1986). Although these mechanisms have been demonstrated in vitro, there is some evidence that they operate also in vivo.

The spatial potassium buffering requires inhomogeneous levels of potassium in the ES. When $[K^+]_o$ is locally elevated, potassium is dispersed by current flow through glial cells, due to their high resting permeability to this ion (Orkand et al, 1966; Somjen, 1975; Newman, 1988). Such dispersion can be

extended since the glial cells appear to be connected through gap junctions. The resulting unbalance in the extracellular electroneutrality is compensated by a current carried in the ES by the majority ions, Na^+ and Cl^-, which migrate respectively towards and away from the site where the maximum K^+ accumulation occurs. Hence, there is a net loss of extracellular osmotic particles at this site and a water influx
into glia accompanied by a shrinkage of the ES, while there is a widening of the ES at remote sites.

Besides, the current thus generated in the ES gives rise to DC potential shifts Krnjevic and Morris, 1972), negative at the site of maximum K^+ accumulation and positive at remote sites.

Dietzel and Heinemann (1986) developed a model to determine the relative efficacy of the various systems in the decrease of the ES size. This model was applied to the changes measured in the cat cortex during repetitive direct electrical stimulation, and to the changes observed in the retina of the honeybee drone during light stimulation (Orkand et al, 1984, Coles et al, 1986). They had to postulate, in addition to the above-mentioned mechanisms, an supplementary increase in the intracellular osmolarity, possibly of metabolic origin, to explain the increases in Cl^- observed during the stimulations.

5. CONSEQUENCES OF ES SPACE CHANGES.

It is difficult to separate the effects of the changes in ES size from those of the ES ion concentration changes. However, it appears that the ES changes are due for a good part to mechanisms aimed at maintaining the level of $[K^+]_o$ below the "ceiling" level during neuronal activity, and in which the glial cells have a large part. Thus, it can be calculated that without such mechanisms, the level of $[K^+]_o$ would rise to over 30 mM during intense stimulation.

First, the most obvious consequence is to maintain the ionic concentrations to a level at which neuronal functioning is still possible, for instance preventing too large decreases in $[Na^+]_o$, or in 'Ca^{2+}§$_o$, which is necessary for synaptic transmission (Dingledine and Somjen, 1981). For instance, the decreases in 'Ca^{2+}§$_o$ recorded during NMDA applications would have been much larger without the concomitant ES decrease. Besides, the increases in $[Ca^{2+}]_o$ observed during applications of Quis or KA in the cortex can be explained by the ES space change, postulating that the ionic channels operated by these agents are not or poorly permeable to calcium.

A second consequence is that the resistivity of the ES is increased, thus favoring possible ephaptic interactions between nerve cells. This factor may be important for maintenance of epileptic activities. Although it may be not very significant in the cerebral cortex, it may be so in the hippocampus, where ephaptic synchronizing influences have been demonstrated to take place (Jefferys, 1981; Taylor and Dudek, 1982; Yim et al, 1986).

A third consequence is that during periods during which the ES size is diminished, the diffusion processes may be impaired. To what extent this phenomenon is implicated during epileptiform or anoxic episodes remains to be determined. Further, Local changes of the ES may also influence the microcirculation in a given zone, thus producing metabolic impairment. The possible role of this factor remains also to be clarified.

REFERENCES

Adey WR, Kado RT, Didio J (1962) Impedance measurements in brain tissue of animals using microvolt signals. Exp Neurol 5:47-66.

Ammann D (1986) Ion-selective micro-electrodes: priciples, design and application. Springer-Verlag Berlin Heidelberg New-York Tokyo pp. 346.

Ballanyi K, Grafe P, ten Bruggencate G (1987) Ion activities and potassium uptake mechanisms of glial cells in guinea-pig olfactory cortex slices. J Physiol London 382:159-174.

Boyle PJ, Conway EJ (1941) Potassium accumulation in muscle and associated changes. J Physiol London 100:1-63.

Coles JA, Orkand RK, Yamate CL, Tsacopoulos M (1986) Free concentration of Na, K, and Cl in the retina of the honeybee drone: stimulus-induced redistribution and homeostasis. Ann NY Acad Sci 481:303-316.

Cragg B (1979) Brain extracellular space fixed for electron microscopy. Neurosci Lett 321:301-306.

Dietzel I, Heinemann U (1986) Dynamic variations of the brain cell microenvironment in relation to neuronal hyperactivity. Ann NY Acad Sci 481:72-86.

Dietzel I, Heinemann U, Hofmeier G, Lux HD (1980) Transient changes in the size of the extracellular space in the sensorimotor cortex of cats in relation to stimulus-induced changes in potassium concentration. Exp Brain Res 40:432-439.

Dietzel I, Heinemann U, Hofmeier G, Lux HD (1982) Stimulus induced changes in extracellular Na^+ and Cl^- concentration in relation to changes in the size of the extracellular space. Exp Brain Res 46:73-84.

Dingledine R, Somjen GG (1981) Calcium dependence and synaptic transmission in the hippocampal slice. Brain Res 207:218-222.

Freygang jr WH, Landau WM (1955) Some relations between resistivity and electrical activity in cerebral cortex of cat. J cell and comp Physiol 45:377-392.

Gardner-Medwin AR (1983) A study of the mechanisms by which potassium moves through brain tissue in the rat. J Physiol London 335:353-374.

Gardner-Medwin AR, Nicholson C (1983) Changes of extracellular potassium activity induced by electric current through brain tissue in the rat. J Physiol London 335:375-392.

Hansen AJ, Olsen CE (1980) Brain extracellular space during spreading depression and ischemia. Acta Physiol Scand 108: 355-365.

Hansen AJ, Zeuthen T (1981) Extracellular ion concentrations during spreading depression and ischemia in the rat brain cortex. Acta Physiol Scand 113:437-445.

Heinemann U, Konnerth A, Pumain R, Wadman WJ (1986) Extracellular calcium and potassium concentration changes in chronic epileptic brain tissue. In: Advances in Neurology 44, Delgado Escueta AV, Ward jr AA, Woodbury DM, Porter RJ (eds). Raven Press New-York pp. 641-661.

Heinemann U, Lux HD (1975) Undershoots following stimulus induced rises of extracellular potassium concentration in cerebral cortex of cat. Brain Res 93:63-76.

Heinemann U, Lux HD (1977) Ceiling of stimulus induced rises in extracellular potassium concentration in the cerebral cortex of cats. Brain Res 120:231-249.

Heinemann U, Lux HD, Gutnick MJ (1977) Extracellular free calcium and potassium during paroxysmal activity in cerebral cortex of the cats. Exp Brain Res 27:237-243.

Heinemann U, Pumain R (1980) Extracellular calcium activity changes in cat sensorimotor cortex induced by iontophoretic application of aminoacids. Exp Brain Res 40:247-250.

Hertz L (1986) Potassium transport in astrocytes and neurons in primary cultures. Ann NY Acad Sci 481:318-330.

Hille B (1967) The selective inhibition of delayed potassium currents in nerve by tetraethylammonium ion. J Gen Physiol 50: 287-302.

Hortsmann E, Meves H (1959 Die feinstructur des molecularen rindengrues und irhe physiologische bedeutung. Z Zellforsch 49:569-604

Hounsgaard J, Nicholson C (1983) Potassium accumulation around individual Purkinje cells in cerebellar slices from the guinea-pig. J Physiol London 340:359-388.

Jefferys JGR (1981) Influence of electric fields on the excitability of granule cells in guinea-pig hippocampal slices. J Physiol 319:143-152.

Kraig RP, Nicholson C. (1978) Extracellular ionic variations during spreading depression. Neuroscience 3:1045-1059.

Krnjevic K, and Morris ME (1972) Extracellular K+ activity and slow potential changes in spinal cord and medulla. Can J Physiol Pharmacol 50:1214-1217.

Lehmann A, Hagberg H, Lazarewicz JW, Jacobson I, Hamberger A (1986) Alterations in extracellular amino acids and Ca^{2+} following excitotoxin administration and during status epilepticus. Adv Exp Med Biol 203:363-373.

Lemenkühler A, Caspers H, Kersting U (1985) Relations between DC potentials, extracellular ion activities, and extracellular volume fraction in the cerebral cortex with changes in PCO_2. In: Ion measurements in biology and Medecine. Kessler M, Harrison DK, Hoper J (eds). Springer-verlag Berlin pp 199-205.

Levin VA, Fenstermacher JD, Patlak CS (1970) Sucrose and inulin space measurements of cerebral cortex in four mammalian species. Am J Physiol 219:1528-1533.

Lothman EW, Somjen GG (1976) Function of primary afferents and responses of extracellular K+ during spinal epileptiform seizures. Electroenceph Clin Neurophysiol 41:253-267.

Louvel J, Heinemann U (1980) Diminution de la concentration extracellulaire des ions calcium lors des crises épileptiques focales induites par l' oenanthotoxine dans le cortex du chat. C R Acad Sci Paris 291:997-1000.

Lux HD, Heinemann U, Dietzel I (1986) Ionic changes and alterations in the size of the extracellular space during epileptic activity. In: Advances in Neurology. Delgado Escueta AV, Ward jr AA, Woodbury DM, Porter RJ (eds). Raven Press New-York 44:619-639.

Mac Dermott AB, Mayer ML, Westbrook GL, Smith SJ, Barker JL (1986) NMDA-receptor activation increases cytoplasmic calcium concentration in cultured spinal cord neurones. Nature 321: 519-522.

Mc Laughlin SGA, Szabo G, Eisenman G (1971) Divalent ions and the surface potential of charged phospholipid membranes. J Gen Physiol 58:667-668.

Morris ME, Krnjevic K (1981) Slow diffusion of Ca^{2+} in the rat's hippocampus. Can J Physiol Pharmacol 59:1022-1025.

Newman EA (1988) Potassium conductance in Müller cells of fish. Glia 1:275-281.

Nicholson C (1980) Dynamics of the brain cell microenvironment. Neurosci Res Prog Bull 18:1-113.

Nicholson C, ten Bruggencate G, Stöckle H, Steinberg R (1978) Calcium and potassium changes in extracellular microenvironment of cat cerebellar cortex. J Neurophysiol 41:1026-1039.

Nicholson C, Phillips JM (1981) Ion diffusion modified by tortuosity and volume fraction in the extracellular microenvironment of the rat cerebellum. J Physiol London 321: 225-257.

Nicholson C, Phillips JM, Gardner-Medwin AR (1979) Diffusion from an iontophoretic point source in the brain: role of tortuosity and volume fraction. Brain Res 169:580-584.

Nicholson C, Rice ME (1986) The migration of substances in the neuronal micorenvironment. Ann NY Acad Sci 481:55-66.

Nicholson C, Rice ME (1987) Calcium diffusion in the brain cell microenvironment. Can J Physiol Pharmacol 65:1086-1091.

Orkand RK, Dietzel I, Coles JA (1984) Light-induced changes in extracellular volume in the retina of the drone, apis mellifera. Neurosci lett 45:273-278.

Orkand RK, Nicholls JG, Kuffler SW (1966) Effect of nerve impulses on the membrane potential of glial cells in the central nervous system of amphibia. J Neurophysiol 29:788-806.

Prince DA, Lux HD, Neher E (1973) Measurement of extracellular potassium activity in cat cortex. Brain Res 50:489-495.

Pumain R (1988) Calcium ions. In: Neuromethods: The neuronal microenvionment, Boulton AA, Baker BA, Walz W (eds). The Humana Press Clifton pp. 589-650.

Pumain R, Heinemann U (1985) Stimulus- and amino-acid induced calcium and potassium changes in rat neocortex. J Neurophysiol 53:1-16.

Pumain R, Kurcewicz I, Louvel J (1983) Fast extracellular calcium transients: Involvement in epileptic processes. Science 222: 177-179.

Pumain R, Kurcewicz I, Louvel J (1987) Ionic changes induced by excitatory amino acids in the rat cerebral cortex. Can J Physiol Pharmacol 65:1067-1077.

Pumain R, Menini C, Heinemann U, Louvel J, Silva Barrat C (1985) Chemical synaptic transmission is not necessary for epileptic seizures to persist in the baboon Papio papio. Exp Neurol 89: 250-258.

Purves RD (1979) The physics of iontophoretic pipettes. J Neurosci Meth P1:165-178.

Ransom BR, Yamamate CL, Connors BW (1985) Activity dependent shrinkage of the extracellular space in rat optic nerve: a developmental study. J Neurosci 5:532-535.

Siesjö BK, Wieloch T (1986) Epileptic brain damage: pathophysiology and neurochemical pathology. In: Adv Neurol Delgado-Escueta AV, Ward Jr AA, Woodbury DM, Porter RJ (eds). Raven Press New York 44: 813-847.

Simon W, Ammann D, Anker P, Oesch U, Band DM (1984) Ion-selective electrodes and their clinical application in the continuous ion monitoring. Ann NY Acad Sci 428:279-299.

Somjen GG (1975) Electrophysiology of neuroglia. Ann Rev Physiol 37:163-190.

Somjen GG (1980) Stimulus-evoked and seizure-related responses of extracellular calcium activity in spinal cord compared to those in cerebral cortex. J Neurophysiol 44:617-632.

Sykova E (1987) Modulation of spinal cord transmission by

changes in extracellular K^+ activity and extracellular volume. Can J Physiol Pharmacol 65:1058-1066.

Sykova E, Shirayev B, Kriz N, Vyklycky L (1976) Accumulation of extracellular potassium in the spinal cord of frog. Brain Res. 106:413-417.

Taylor CP, Dudek FE (1982) Excitation of hippocampal pyramidal cells by an electrical field effect. J Neurophysiol 52:126-142.

Torack RM (1965) The extracellular space of rat brain following perfusion with glutaraldehyde and hydroxyadipaldehyde. Z Zellforsch Mikroskop Anat 66:352-364.

Urbanics RE, Leninger-Follert E, Lübbers DW (1978) Time course of changes of extracellular H^+ and K^+ during and after direct electrical stimulation of the brain cortex. Pflg Arch 378:47-53.

Van Harreveld A (1972) The extracellular space in the vertebrate central nervous system. In: The structure and function of the nervous tissue, Vol 4. Bourne GH (ed). Academic Press New York pp 447-511.

Van Harreveld A, Crowell J, Malhotra SK (1965) A study of extracellular space in central nervous tissue by freeze substitution. J Cell Biol 25:117-137.

Van Harreveld A, Murphy T, Nobel KW (1963) Specific impedance of rabbits's cortical tissue. Am J Physiol 205:203-207.

Vyskocil F, Kriz N, Bures J (1972) Potassium-selective microelectrodes used for measuring the extracellular brain potassium during spreading depression and anoxic depolarization in rats. Brain Res 39:255-259.

Walz W (1987) Swelling and potassium uptake in cultured astrocytes. Can J Physiol Pharmacol 65:1051-1057.

Waltz W, Hertz L (1983) Intracellular ion changes of astrocytes in response to extracellular potassium. J Neurocsi Res 10:411-423.

Wieloch T, Harris RJ, Symon L, Siesjö BK (1984) Influence on severe hypoglycemia on brain extracellular calcium and potassium activities, energy and phospholipid metabolism. J Neurochem 43: 160-168.

Yim CC, Krnjevic K, Dalkara T (1986) Ephaptically generated potentials in CA1 neurons of rat's hippocampus in situ. J Neurophysiol 56:99-122.

DISCUSSION

Chaired by Lenzi G.L. and Siesjo B.K.

B.K. Siesjo:

You have discussed the uptake of Ca^{2+}, Na^{+} and Cl^{-} elicited by glutamate, quisqualate and kainate. In particular, the entry of Na Cl into the post synaptic neuronal elements may be related, in certain situations, to dendritic swelling. Thus, it appears that neurons, more than glia, are responsible for the process of swelling. Is it also your conclusion?

R. Pumain:

I think that in the cortex the swelling is due both to the neurons and the to glial cells. In fact, ultrastructural studies on edema show that the first elements to swell are the dendrites.

R.K. Siesjo:

There are also classical observations showing, by electromicroscopy technique, that in spreading depression the structures which are swelling are the apical dendrites. Furthermore, histochemical studies have demonstrated the entry of Cl-ions into the swollen elements.
Another comment concerns the increase in K+ concentration induced by electrical stimulation of the cortex or by eliciting speading depression: what is the difference between the electrical stimulation which only causes increase in K+ values and the one which leads to spreading depression?

R. Pumain:

That is not clear. What we know is that spreading depression was observed to be more frequently elicited when the tissue is a little dry: this may mean that in case of extracellular space shrinkage, this phenomenon may be more easily triggered.

B.K. Siesjo:

Do you observe spreading depression when you put the pipette in contact with the tissue and measure glutamate release?

R. Pumain:

This may occur sometimes with Kainate, very rarely with glutammate. Anyway, this phenomenon is very unfrequent and I do not think that it is due to the trauma caused by the micropipette.

W. Meier-Ruge:

In case of ischemia we observe shrinkage of the brain, as

indicated also by the increase in the capillary diameter and the reduction of the intercapillary distance. Since the early stage of hypoxia is accompained by shifts in the ion composition of the extracellular space,without normally observing swelling of glial cells or terminal elements, my question is: is it possible that the ischemia-induced decrease in Na+ and Ca2+ in the extracellular space is due to a loss of these ions into the cerebrospinal fluid or through the microvessels into the blood and not to sequestration in the synapses or astrocytes?

R. Pumain:

This is a possibility that we had to take into account, especially when we started the experiments in vivo. However, in vitro evidence with brain slices, which are devoid of blood supply, resulted in the same ionic changes, suggesting that the electrolyte sequestration takes places into the cells. Furthermore, NMDA receptors are located on neurons and their activation appears to be linked to the ionic changes, which are abolished by application of selective NMDA antagonists.

B.K. Siesjo:

What do you mean, precisely, with shrinkage of the brain ?

W. Meier-Ruge:

Morphometric data indicate a decrease in the inter-capillary distance under ischemic conditions. This could also explain the fact that tissue reperfusion leads to hyperemia.

B.K. Siesjo:

In this case brain water content should decrease. Is it not possible that what you measure (capillary distance) is an artifact?

W. Meier-Ruge:

We work on fresh tissue, without fixation procedures which could intoduce artifacts. We do not know, however, whether the morphometric changes are accompained by water movements.

B.K. Siesjo:

Why do you think that glutamate induces changes in the extracellular space smaller than NMDA, quisqualate or kainate?

R. Pumain:

I think that one explanation is the possibility of uptake mechanisms for glutammate, so that we have to increase the dose 2-3 times in order to have more or less the same magnitude of effect as observed with the other amino acids.

ROLE OF GLIAL CELLS IN THE CONTROL OF THE NEURONAL MICROENVIRONMENT

Richard K. Orkand
Institute of Neurobiology
University of Puerto Rico Medical Sciences Campus
201 Blvd. del Valle
San Juan PR 00901 USA

There is increasing evidence that neuroglia have a functional role in the homeostasis of the fluid bathing central neurons with regard to ions and small molecules. The microenvironment of neurons in the central nervous system is the system of narrow extracellular clefts, 20-30 nm wide, separating neurons from each other and from glial cells. The glial membranes constitute an immense surface area for exchange. The clefts are in diffusion equilibrium with the bulk of the cerebrospinal fluid as well as the space around the capillary endothelial cells. In most species the ring of endothelial cells held together by tight junctions constitutes the blood- brain-barrier (Cserr, 1986). A striking feature of this anatomical arrangement is that with rare exception, for example the nerve terminals of neurosecretory cells, neurons themselves are always separated from the CSF and extracapillary space by neuroglia. It was this observation which led Golgi in 1885 to suggest that glial cells controlled the flow of materials to and from the neurons. Subsequently, Lugaro, in 1907, postulated that glial cells exchanged substances with the extracellular fluid and thereby regulated the neuronal microenvironment (see reviews by Kuffler and Nicholls, 1966; Somjen, 1988).

This is a review of relatively recent work on glial cells of the optic nerve of amphibia carried out at the University of Pennsylvania and at the University of Puerto Rico. It does not contain original data and therefore the appropriate cited material should be used for reference purposes. The studies support the early hypotheses of the Italian anatomists that proposed an active role for glial cells in the homeostasis of the extracellular fluid of the brain.

For a more comprehensive recent summary of neuroglial

NATO ASI Series, Vol. H 33
Regulatory Mechanisms of Neuron to Vessel Communication in the Brain. Edited by F. Battaini et al.

electrophysiology the review by Ransom and Carlini (1987) should be consulted.

The large glial membrane area forming the boundary of the narrow extracellular clefts provides the anatomical basis for the movement of ions and small molecules across glial membranes to determine the composition of the extracellular space. Since the glial cell volume is much greater than that of the extracellular space, transfers of substances across the glial membrane produce large changes in extracellular concentrations while changes in glial concentrations are relatively small. For example, if the ratio of glial to extracellular volumes is 5:1, an increase in $[K^+]o$ to 300% of normal (2 to 6 mM) can be absorbed by the glial cells with an increase in $[K^+]i$ of less than 1% (from 100 to 100.8 mM). In fact, with the addition of spatial buffer currents (see below) the increase in glial K^+ will be even less!

Substances move across glial membranes as a result either of space-independent net transport processes or space-dependent spatial buffering (Gardner-Medwin, 1980).

Space-Independent Net Transport by Neuroglia

Fig. 1 is a cartoon illustrating some of the net transport systems for which there is some evidence in glial cells in the amphibian optic nerve. They include the electrogenic Na^+/K^+ ATPase (Tang et al. 1980), an electrogenic Na^+/glutamate symport (Tang and Orkand, 1986) and an electrogenic Na^+/HCO_3^- cotransporter (Astion and Orkand 1988). There are preliminary indications of other transport systems, e.g. an amelioride sensitive Na^+/H^+ exchanger (Astion and Orkand, unpublished) and a passive K^+Cl^- uptake (Coles and Orkand, 1984), but they have not been studied in sufficient detail to warrant description here. Space-independent transport systems can operate when there is a uniform increase in the concentration of a substrate around the glial cell. The energy for the process is derived either directly from the hydrolysis of ATP or from electrochemical ion gradients which were established by the hydrolysis of ATP.

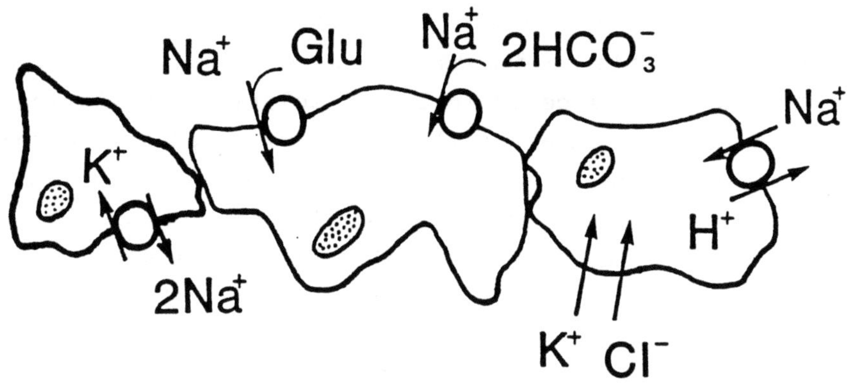

Fig. 1. Space-independent net transport systems in amphibian neuroglia. Three glial cells coupled by gap junctions are depicted. Note that only the K^+Cl^- uptake and Na^+/H^+ exchange are electroneutral. The others are electrogenic because more net charge is transported in one direction. The actual stoichiometry of the transport is not definitely known. All these processes can operate with uniform concentrations of the ions or amino acids in the extracellular fluid.

Electrogenic Na^+/K^+ Pump

Under physiological conditions the glial membrane potential behaves as expected for a K^+ electrode (Kuffler et. al 1966). That is, varying K^+ or temperature produces potential changes across the membrane precisely as expected from the Nernst equation. There is no indication that a significant portion of the membrane potential is due to electrogenic pump activity either at rest or when the cleft K^+ is elevated as a result of nerve impulses. This conclusion in strengthened by the observation that the addition of strophanthidin, an inhibitor of the Na^+/K^+ ATPase, does not have a marked depolarizing effect under normal conditions.

To demonstrate the electrogenic Na^+/K^+ ATPase, Tang et. al (1980) isolated the glial cells by removing the retina and waiting for two months while all the axons in the optic nerve degenerated. This eliminated the possibility that changes in glial membrane potential were secondary to extracellular K^+

changes caused by ion fluxes in axons. In the isolated glial cells it is possible to demonstrate an electrogenic pump after first Na^+-loading the cells in low K^+ solutions. Following Na^+-loading an increase in extracellular K^+ produced a much smaller depolarization than predicted by the Nernst relation. The reduced depolarization is expected if the reduction in K^+ equilibrium potential, which normally determines glial membrane potential, is partially antagonized by the activation of a hyperpolarizing electrogenic pump. After the addition of strophanthidin the depolarizing effect of K^+ was much greater. In addition, strophanthidin by itself had a marked depolarizing effect in Na^+-loaded glia bathed in normal K^+ but not in low K^+-Ringer. The electrogenic Na^+/K^+ ATPase is involved in the maintenance of ionic gradients across glial membranes. Its role in K^+ homeostasis of the extracellular clefts is less certain.

Electrogenic Na^+/glutamate Symport

The Schwann cell at the lobster neuromuscular junction was found to be the principal site of uptake of GABA, and this observation led to the hypothesis that glial cells were involved in transmitter uptake and inactivation (P. Orkand and Kravitz, 1971). About ten years later we found that glutamate depolarized glial cells in the optic nerve of Necturus (Tang and Orkand 1982; Tang and Orkand 1986). The Na^+-dependence, increase in membrane conductance and the blocking effect of strophanthidin on the afterhyperpolarization all suggest that Na^+ entry is primarily responsible for the depolarization (Fig. 2). However, the results do not distinguish clearly between two possible mechanisms, an electrogenic Na^+/Glu symport (Johnstone, 1979) or a transmitter type increase in Na^+ conductance (Fonnum, 1984). Nevertheless, it is known that glial cells take up amino acids via a Na^+-dependent transport system. The remaining question is whether they have neuronal-type glutamate receptors and, if so, what the function of these receptors might be. There is some evidence that neuro-transmitters may be used in signalling between neurons and glial cells (Ransom and Carlini, 1987).

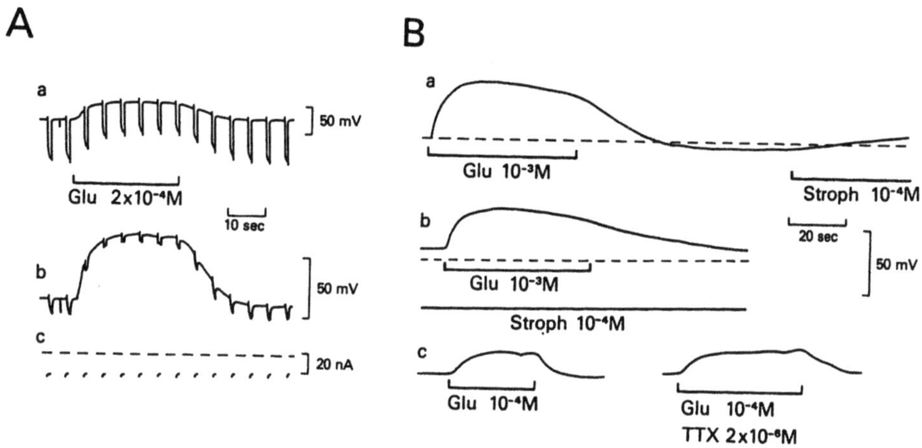

Fig. 2. Properties of glial cell depolarization by L-Glutamate. A: Increase in membrane conductance. Two intracellular, 3M KCl, electrodes, separated by about 300 µm, recorded the membrane potentials of glial cells during Glu superfusion. A current pulse, lowest trace, of 20 nA for 1s every 5s was passed via a bridge circuit through the electrode and the resulting electrotonic potential was recorded by that electrode, upper trace, and the distant second electrode (middle trace, higher gain). During Glu superfusion the cells are depolarized and the amplitudes of the electrotonic potentials decrease indicating an increase in membrane conductance. B, a,b: Effect of strophanthidin. The trace in b is a continuation of the trace in a. Ba: After washing out the L-Glu the membrane hyperpolarizes and the hyperpolarization is blocked by strophanthidin. Bb: In the presence of strophanthidin the repolarization is slowed and there is no hyperpolarization. Bc: The addition of TTX has no effect on the Glu depolarization. (from Tang and Orkand, 1986)

Electrogenic Na^+/HCO_3^- Cotransport

When the dominant K^+ conductance of the glial membrane is blocked by the addition of barium the glial membrane potential is about -40 mV. Under these conditions, Astion et. al (1987) observed that the addition of HCO_3^- caused a large (up to 40 mV) transient Cl^--insensitive hyperpolarization. Close examination in the absence of barium revealed the same phenomenon only at much reduced amplitude (2-4 mV). Subsequent study (Astion and Orkand 1988) has suggested that the glial cell membrane contains an electrogenic Na^+/HCO_3^- cotransporter similar to that described in a variety of epithelial cells (e.g. Lopes et al., 1987). The transporter is electrogenic because the ratio of HCO_3^- to Na^+ being transported is greater than 1. That is, the transporter causes a net negative charge to cross the membrane. The presence of the transporter is inferred from the following observations illustrated in Fig. 4: The HCO_3^--induced hyperpolarization is Na^+-dependent and blocked by SITS. In addition, the depolarization produced by Na^+ withdrawal depends on the presence of HCO_3^-. This transporter is present in glia but apparently not in neurons. Thus, it is tempting to suggest that it might contribute not only to pH control of glial cytoplasm but also to the ability of glial cells to buffer the extracellular pH. According to this scheme, an increase in extracellular acid, resulting from neuronal activity, would result in a neutralizing secretion of HCO_3^- by the glial cells. If the glial cell buffering power were greater than that of the smaller volume of extracellular space the acid could be neutralized with relatively little insult to glial cell pH.

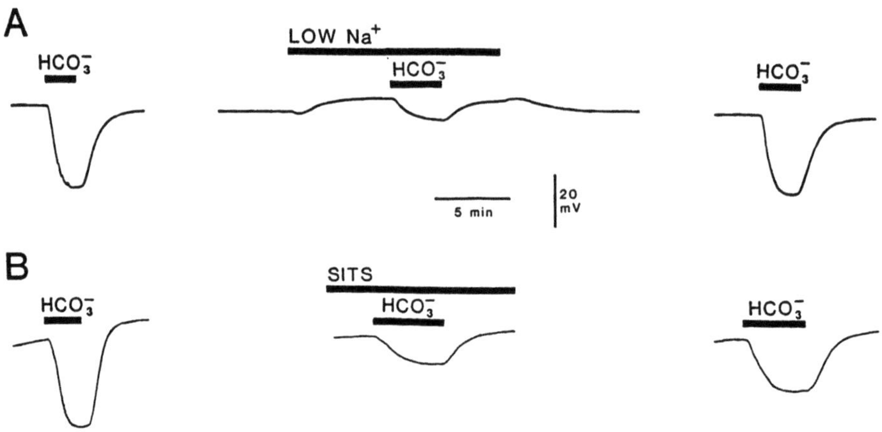

Fig. 3. The hyperpolarizing effect of HCO_3^- requires Na^+ and is SITS sensitive. All solutions contained 2 mM barium which depolarized the cell and blocked the K^+ conductance. A:left: control glial membrane potential response to the addition of 10 mM HCO_3^-. middle: after bathing the preparation in 10% Na. Na^+ replaced by N-methyl D-glucamine. right: control after Na^+ was restored. 30 minutes between records. B: Effect of SITS, 1 mM,

on the response to 10 mM HCO_3^-. SITS had little effect on the resting potential in the absence of HCO_3^-. Right control taken 60 minutes after removal of SITS (from Astion and Orkand, 1988).

Space-Dependent Transport

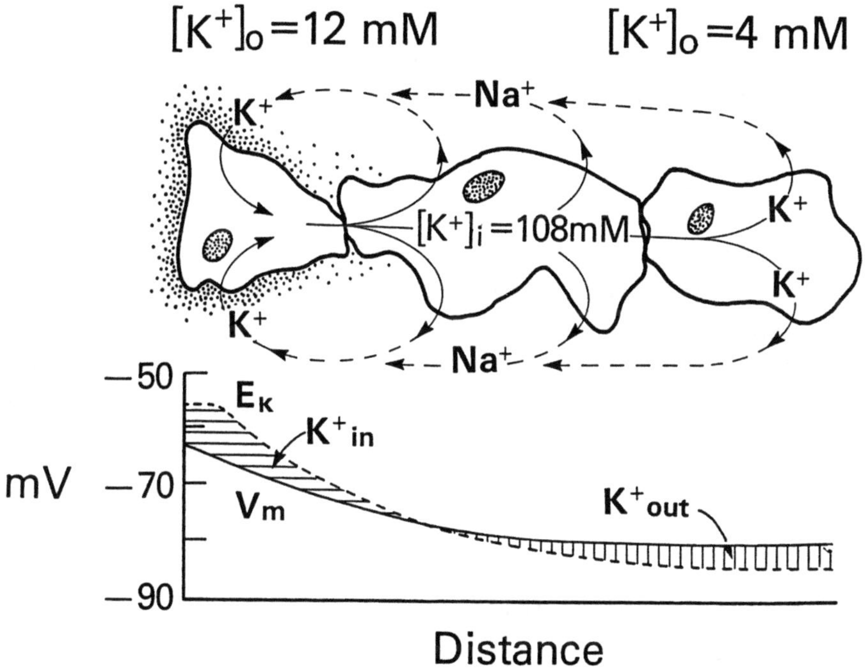

Fig. 4. Space-dependent spatial buffering. Activity in a group of neurons has produced a local increase in K^+ from 3 to 12 mM (shaded area left side). This increase produces a glial depolarization that spreads passively to the coupled glial cells. The difference between Vm and E_K drives K^+ inward at the region where it is raised and outward at distant regions. As a result of the selectivity of the membrane, all the transmembrane current is carried by K^+. The extracellular current is carried by Na^+ (and Cl^-, not shown). The result is a net flux away from the region where it has accumulated. The average $[K^+]_i$ is not affected. The graph shows the distribution of E_K^+ and Vm as a function of distance along the glial syncytium. During recovery the neurons take up K^+ and all the processes reverse. In the intact nervous system the extracellular volume is only a fraction of the glial volume (from Orkand, 1986).

The best studied space-dependent transport system is that for the redistribution of K^+ following its release from active axons and accumulation in the extracellular clefts. This process, termed spatial buffering (Orkand, Nicholls and Kuffler, 1966), or siphoning (Newman, 1984,1986), results from a non-uniform distribution of both membrane potential and K^+. The concentration gradient for potassium provides a difference of membrane potential which leads to a flux of potassium through the glial syncytium; this tends to reduce the gradient. It is, like many homeostatic mechanisms, a negative feedback process and is illustrated in Fig. 4. Spatial buffering produces a passive flux of K^+ from regions where it is high to those of lower K^+ via the glial cell cytoplasm. The high relative permeability of glial cells for K^+ and their electrical coupling make them well suited to redistribute K^+ via this passive process. As Newman (1986) suggests, an uneven distribution of K^+ channels along the glial membrane can further add to the efficiency of this mechanism for K^+ homeostasis in the extracellular space of the central nervous system.

Optical recording of glial membrane potential.

Since the distribution of glial membrane potential determines the efficacy of spatial buffering it is of interest to record this parameter in the intact nervous system. Recently, voltage-sensitive dyes have been used in an attempt to record changes in glial membrane potential (Konnerth, Obaid and Salzberg, 1987; Konnerth and Orkand, 1986; Konnerth, Orkand and Orkand, 1988; Lev-Ram and Grinvald, 1986). The method is based on the observation that certain dyes bind to cell membranes and change their absorption of light or fluorescence linearly with changes in membrane potential. Fig 5 is two electron micrographs of cross sections of the frog optic nerve. The periphery of the nerve is covered by a glia limitans composed of the processes of astrocytes separated occasionally by intercellular clefts of about 20 nM that usually appear patent.

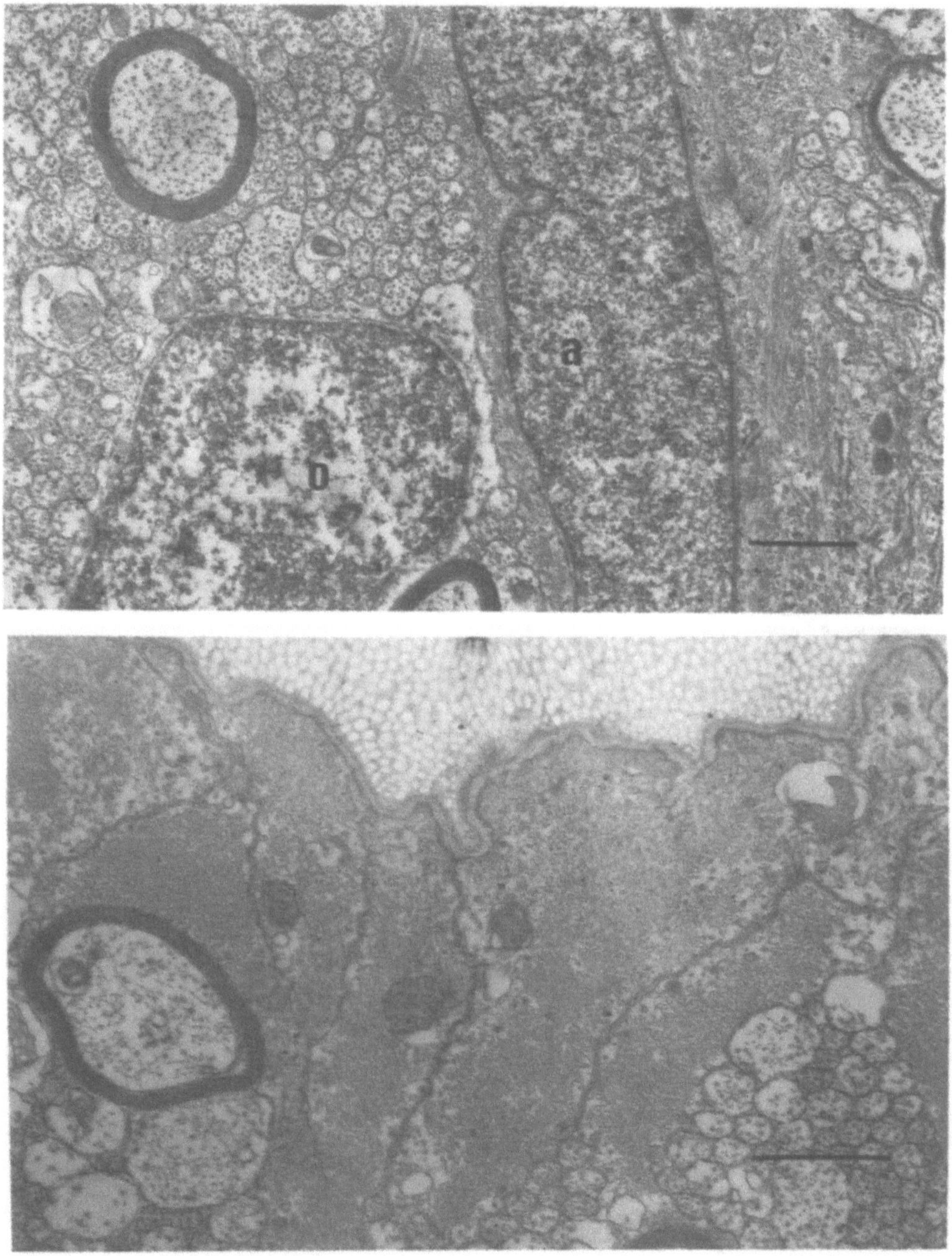

Fig.5. Electron micrographs of cross section of frog optic nerve. Top: The interior of the nerve contains myelinated (m) axons and groups of unmyelinated axons. In this selected micrograph an astrocyte (a) with its filamentous cytoplasm and an oligodendrocyte (o) are shown. Bottom: The periphery of the

nerve with its surrounding glia limitans composed of astrocyte processes. Bar=1 μm (from Konnerth et al., 1988).

In some areas gap junctions and desmosomes between glial cells are observed. In contrast to the optic nerve of Necturus which contains only astrocytes and unmyelinated axons (Orkand, et al., 1981) that of the frog contains unmyelinated axons and astrocytes as well as myelinated axons and their associated oligodendrocytes (Konnerth et al., 1988). Fig 6 illustrates optical records during stimulation of the frog optic nerve after it has been stained with voltage sensitive dyes.

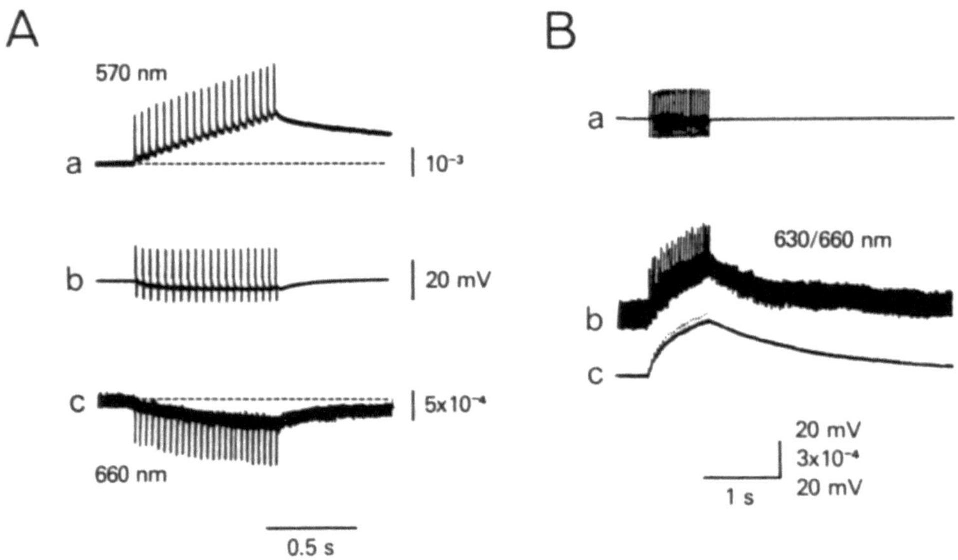

Fig. 6. Optical recordings from frog optic nerve stained with voltage-sensitive dyes. The optic nerve was stimulated with trains of stimuli (30 Hz for 0.8 sec) by a suction electrode at one end and the field potential was recorded by a suction electrode at the other end (A,b and B,a). A: the nerve was stained with dye NK2367, and changes in light absorption were recorded at the indicated wavelengths. The reversal of the optical signal indicates that the change in absorption is due to a dye response and not simply a change in the optical properties of the nerve. The relative amplitude of the spikes and slow potentials are the same at both wavelengths. B: This nerve was stained with dye WW781, and the change in fluorescence measured with an exciting filter of 630 nm and a barrier filter of 660 nm (trace Bb). Trace Bc is an

intracellular recording from a glial cell obtained simultaneously with the trace in b. The time-course of the glial depolarization is similar to that of the slow optical depolarization (from Konnerth and Orkand, 1986).

It can be seen in Fig.6 that the optical record appears to be a mixture of the membrane potential changes in both the axons and the glial cells. Although the relative amplitude of the signals from the axons and the glial cells varies with the type of dye used (Konnerth et al., 1987), dyes specific for either neurons or glial cells have not yet been identified.

Astion and colleagues (Astion et al., 1989) were able to record voltage-sensitive dye signals solely from glial cells by staining the optic nerve of Necturus and producing a selective depolarization of the glial cells by passing current through an intracellular electrode. Fig 7 illustrates the passive spread of an electrotonic potential through the glial network as measured optically with a photodiode array. In Fig. 7B the effect of increasing the glial membrane resistance by blocking the K^+ channels with barium is shown. Following the addition of barium the potential decayed less rapidly with distance and the amplitude and rise-time of the potential change increased. The results were as expected for a syncytium in which barium increased the membrane resistance more than the resistance across the gap junctions connecting the glial cells (Kettenmann and Ransom 1988).

Voltage-gated Channels in Glial Cells

The discovery of the presence of voltage-gated ion channels in glial membranes (for review see Ransom and Carlini, 1987) raises the possibility that ion fluxes through these channels serve to further modify the composition of the extracellular space. Using the loose patch-clamp technique, Ortiz et al (1988) recorded voltage-dependent currents from the surface of the frog optic nerve. Fig 8 illustrates these currents and demonstrates their sensitivity to the voltage-dependent Na^+ channel blocker TTX and the K^+ channel blocker 4-aminopyridine.

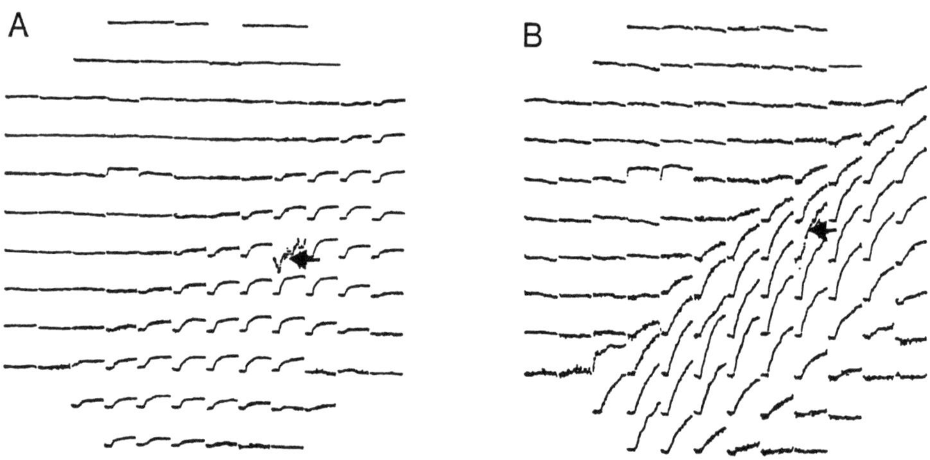

Fig. 7. Optical recording of the passive spread of an electrotonic potential through the glial network of Necturus optic nerve. The preparation was stained with dye RH155. The nerve was imaged onto a photodiode array so that each of the 144 photodiodes detected changes in light absorption from a 50 X 50 micron region. The nerve lies at a 45° angle in the lower right potion of the optical field. A: Control, normal Ringer's solution; depolarizing current injection 50 nA, 200 msec duration. The photodiode element indicated by the arrow was situated over the area of the nerve where the current electrode was located. The signal displayed in each element of the array represents an average of 16 sweeps. B: Effect of barium on the spread of potential. Record obtained 10 min after the addition of 2 mM barium. Same impalement and arrangement as in A. In A and B the decay of the electrotonic potential was cut off; the duration for individual sweeps was 175 msec (from Astion et al., 1989).

Recordings made under a variety of conditions support the conclusion that these voltage-gated currents arise from the glia limitans at the surface of the nerve.For example, when all the Na^+ is removed from the bath and the loose patch pipette is placed at the surface no inward currents are recorded. If only the patch electrode is filled with normal Ringer, so that the Na^+ is only high at the glia limitans, inward currents are recorded. In addition, when the K^+ is raised 10-fold the axons become inexcitable. But, after the surface is polarized inward currents are again obtained. Furthermore, such currents are still recorded from the nerve surface four months after retinal

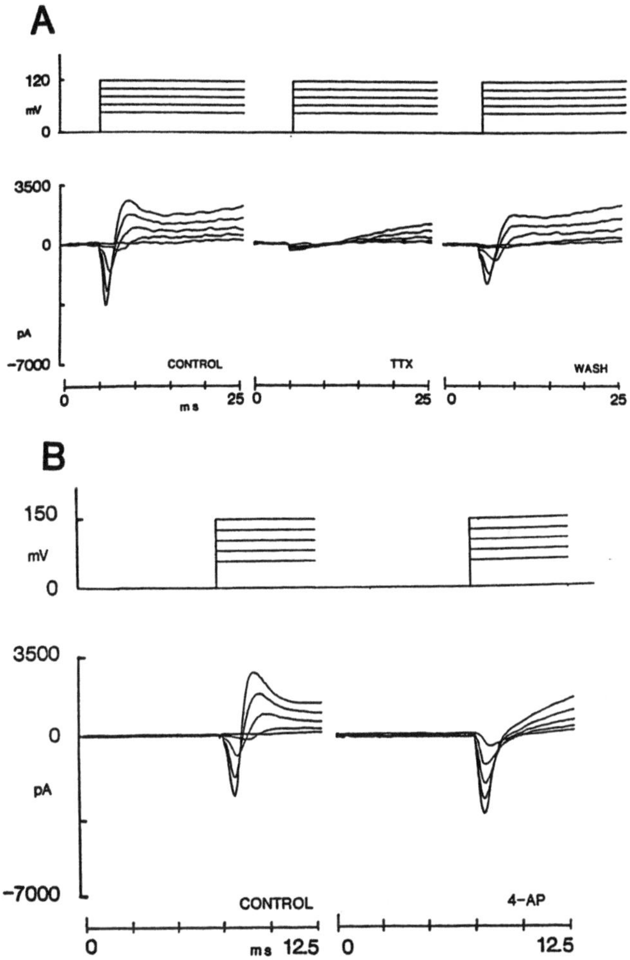

Fig. 8. Voltage-dependent currents recorded from the surface of the frog optic nerve using the loose patch-clamp technique. A Top trace: The voltage was stepped in six 20 mV increments. The voltage and currents shown (lower trace) are from steps 2-6. The records are the sum of the currents from symmetrical hyperpolarizing and depolarizing steps recorded with an electrode of 27 um. Left to right: Control, after TTX 400 nM, after washing. B. Effect of 4-aminopyridine (0.5 mM). Same protocol as in A only the steps are each 25 mV and electrode diameter 30 microns. Bath temperature 10° C throughout.

ablation when electron microscopy shows that almost all the axons have degenerated. The presence of these voltage-gated currents is puzzling because glial cells are not normally depolarized by the 40-50 mV necessary to open the channel gates.

There is obviously a great deal more to learn about the chemical and voltage control of ion movements across glial membranes. Moreover, there is still a need to define all the glial membrane transport processes. It is also necessary to sort out 1) which of these processes contribute to the role glial cells play in the control of the composition of the extracellular space and 2) which processes are solely concerned with the maintenance of the glial cell itself.

Acknowledgement

I thank Dr. Paula M. Orkand for helpful suggestions during the preparation of the manuscript. Work in our laboratory has been made possible by grants from the National Institutes of Health (NS-07464, NS-12253) and the National Science Foundation (RII 8705802).

REFERENCES

Astion ML., Obaid AL, Orkand RK (1989) Effects of barium and bicarbonate on glial cells of Necturus optic nerve: Studies with microelectrodes and voltage sensitive dyes. J gen Physiol In Press

Astion ML, Orkand RK (1988) Evidence for electrogenic Na/HCO_3 cotransport in glial cells in Necturus optic nerve. Glia 1: In Press

Bracho H, Orkand PM, Orkand RK (1975) A further study of the fine structure and membrane properties of neuroglia in the optic nerve of Necturus. J Neurobiol 6:395-410

Coles JA, Orkand RK (1984) Chloride enters receptors and glia in response to light or raised external potassium in the retina of the bee drone, Apis Mellifera. J Physiol Lond 357:11P

Cserr HF (1986) (ed) The neuronal microenvironment. Ann NY Acad Sci 481:1-391

Fonnum F (1984) Glutamate: a neurotransmitter in mammalian brain. J Neurochem 42:1-11

Gardner-Medwin AR (1980) Membrane transport and solute migration affecting the brain cell microenvironment. Neurosci Res Prog Bull 18:208-226

Johnstone RM (1979) Electrogenic amino acid transport. Can J Physiol Pharmacol 57:1-15

Kettenmann H, Ransom BR (1988) Electrical coupling between astrocytes and between oligodendrocytes studied in mammalian cell cultures. Glia 1:64-73.

Konnerth A, Obaid AL, Salzberg BM (1987) Optical recording of electrical activity from parallel fibres and other cell types in skate cerebellar slice in vitro. J Physiol Lond 393:681-702

Konnerth A, Orkand PM, Orkand RK (1988) Optical recording of electrical activity from axons and glia of frog optic nerve: Potentiometric dyes responses and morphometrics. Glia 1:225-232

Konnerth A, Orkand RK (1986) Voltage sensitive dyes measure potential changes in neurons and glia of the frog optic nerve. Neurosci Lett 66:49-54

Kuffler SW, Nicholls JG (1966) The physiology of neuroglial cells. Ergeb Physiol 57:1-90

Kuffler SW, Nicholls JG, Orkand RK (1966) Physiological properties of glial cells in the central nervous system of amphibia. J Neurophysiol 29:768-787

Lev-Ram V, Grinvald A (1986) Ca and K dependent communication between central nervous system myelinated axons and oligodendrocytes revealed by voltage-sensitive dyes. Proc Nat Acad Sci USA 83:6651-6655

Lopes AG, Siebens AW, Giebisch G, Boron WF (1987) Electrogenic Na/HCO_3 cotransport across basolateral membrane of isolated perfused Necturus proximal tubule. Am J Physiol 253:F340-F350

Newman EA (1984) Regional specialization of retinal glial cell membrane. Nature 309:155-157

Newman EA (1986) High potassium conductance in astrocyte end feet. Science 233:453-454.

Orkand PM, Kravitz E (1971) Localization of the sites of τ-aminobutyric acid (GABA) uptake in lobster nerve-muscle preparations. J Cell Biol 49:75-89

Orkand RK (1986) Glial-interstitial fluid exchange. Ann N Y Acad Sci 481 The Neuronal Microenvironment. Ed HF Cserr pp 269-272

Orkand RK, Nicholls JG, Kuffler SW (1966) The effect of nerve impulses on the membrane potential of the glial cells in the central nervous system of amphibia. J Neurophysiol 29:788-806

Ortiz S, Rodriguez O, Orkand PM, Orkand RK, Marrero H (1988) Voltage-gated currents recorded from the surface of the frog optic nerve. Puerto Rico Health Sci J 7:141-143

Ransom BR, Carlini WG (1987) Electrophysiological properties of astrocytes. In: Astrocytes Vol 3 Federoff (ed) Academic Press Orlando pp 1-49

Somjen G (1988) Nervenkitt: Notes on the history of the concept of neuroglia. Glia 1:2-9

Tang CM, Cohen MW, Orkand RK (1980) Electrogenic pumps in axons and neuroglia and extracellular potassium homeostasis. Brain Res 194:283-286

Tang CM, Orkand RK (1982) Glutamate directly depolarizes glial cells. Neuroscience 7:S208

Tang CM, Orkand RK (1986) Glutamate depolarization of glial cells in the Necturus optic nerve. Neurosci Lett 63:300-304

DISCUSSION

Chaired by Joo F. and Carruba M.O.

B.K. Siesjo:

Several years ago two scientists described, in invertebrate preparations, an hydrogen ion secreting mechanism, involving essentially the exchange between Na+ and bicarbonate ions which enter the cell, with H+ and Cl-ions which exit to the extracellular medium. This movement is driven energetically by sodium gradient.
Apparently this mechanisms, which is Cl- dependent and non electrogenic, differs from the one that you have now described and shown being non Cl- dependent and electrogenic. Can you further comment on the differences between these two systems?

R. Orkand:

Essentially, both mechanisms are Na^+ dependent and sensitive to stilbenes; the differences, as you have mentioned, are Cl- dependency and electrogenicity. In the most simple case all the exchange processes are determinated the Na+ and bicarbonate gradient on Vm. Now we are trying to find out how the process is set in resting conditions. For this purpose, we need to know the internal Na+ and bicarbonate concentrations.

B.K. Siesjo:

How could the mechanism that you have described be by itself a regulator of pH in any of the compartments, if we consider the constraints that this electrogenic system will impose on the ion fluxes?

R. Orkand:

We have shown the results in conditions in which K^+ fluxes have been cut off so that the system looks more electrogenic than it really is.

B.K. Siesjo:

If the mechanism works close to electroneutrality, it cannot be reversibile because than it cannot be posed close to the equilibrium. The Na+ gradient is much larger than the bicarbonate one and therefore it would always be posed out for transporting bicarbonate in if it did not have a large electrogenic component.

R. Orkand:

That depends on the stoichiometry (2:1 or 3:1) and on where the equilibrium potential for the reaction exactly is. At present we still have to find out the equilibrium potential for the process and to define all the gradients.

R. Pumain:

Concerning the action potential in glial cells, is there any evidence that the processes that you have described exist also in mammalian glial cells?

R. Orkand:

We are picking up the action potential currents between the patch electrode and the ground. With the patch electrode we are sitting on the membrane, and not inside the cells.
By recording with an intracellular electrode we can detect an external spike when we are outside the glial cell, and when we are inside we can observe the same external spike: that is because the potential change does not occur across the glial membrane, but in the external fluid. Thus, I do not believe that there is a net current across the glial membrane produced by the action potential. In mammals you can also see these phenomena.

F. Joo:

Can you give us some more informations on the uneven distribution of K+ channels in the glial cells?

R. Orkand:

These are the obervations by Erik Newman who found that in the retina the glial cells are distributed at the region of the astrocytes near the blood vessels. The K+ channels are mostly localized at the endfeet. Thus, his view is that buffer currents tend to dump the K+ in the regions outside the blood vessels and to help in this way the clearence.
Other studies have shown a good correlation between K+ accumulation and depolarization of glial cells, suggesting that K+ channels may be localized in the regions where the neurons extruse this ion.

FUNCTIONAL INTERACTIONS BETWEEN NEURONS AND GLIAL CELLS

Leif Hertz
Department of Pharmacology
University of Saskatchewan
Saskatoon, Saskatchewan
S7N OWO
Canada

INTRODUCTION

The fluid composition within the interstitial space in the central nervous system (CNS) is regulated partly by transport between neural cells and microvessels and partly by release and uptake in different types of neural cells. Interactions between neurons and glial cells (mainly astrocytes and oligodendrocytes) seem to play a major role in this respect. In most species glial cells outnumber neurons. Astrocytes contribute about 1/3 of the brain cortex volume in man, and oligodendrocytes another 1/5 (Pope, 1978). The involvement of oligodendrocytes in myelination is well established, but these cells may also have other functions, and the roles of astrocytes are still controversial. Astrocytes occupy strategically important sites in the CNS: cortical astrocytes form astrocytic endfeet around capillaries, surround synapses (Wolff, 1970), and form sheets between neurons (Palay and Chan-Palay, 1974), and white matter astrocytes participate in the formation of the node of Ranvier (Black and Waxman, 1988). The surface/volume ratio is extremely large (Wolff, 1970), making these cells particularly well suited for regulation of the extracellular milieu. The close anatomical contact with endothelial cells suggest on one hand that transport between astrocytes and capillaries, at least under some conditions might be important in the overall exchange between CNS and the vascular system; on the other hand it also provides the possibility that compounds released from astrocytes may exert a direct regulatory function on vessels.

Among the constituents of the interstitial fluid, neurotransmitters and neuromodulators as well as the two cations, potassium (K^+) and calcium (Ca^{2+}), play a major role for neuronal excitability. Interactions between neurons and glial cells contributing to the regulation of the extracellular concentrations of these compounds will be discussed in this review. It is not the intention to provide a comprehensive review of pertinent literature in this area. As far as possible, references will be made to other reviews, several by myself, quoting the original literature in much greater detail.

NATO ASI Series, Vol. H 33
Regulatory Mechanisms of Neuron to Vessel Communication in the Brain. Edited by F. Battaini et al.

NEUROTRANSMITTERS AND NEUROMODULATORS

Amino acid transmitters

Glutamate turnover. Evidence is found that glutamate is the major excitatory transmitter (Hertz et al., 1983a). It is released in large amounts by a potassium stimulated, calcium dependent mechanism from glutamatergic neurons such as cultured cerebellar granule cells (e.g., Hertz and Schousboe, 1986; Schousboe et al., 1988b). Although there is quite an intense reuptake of glutamate into these cells the uptake rate into astrocytes is even higher, reflecting a preferential uptake of glutamate into astrocytes in brain (Fig.1). This uptake has been demonstrated both in glial cells prepared by gradient centrifugation, in glial cells in vivo and in cultured astrocytes (Hertz, 1979; Hosli et al., 1986; Schousboe et al., 1988b). It may be essential for regulation of the interstitial glutamate concentration.

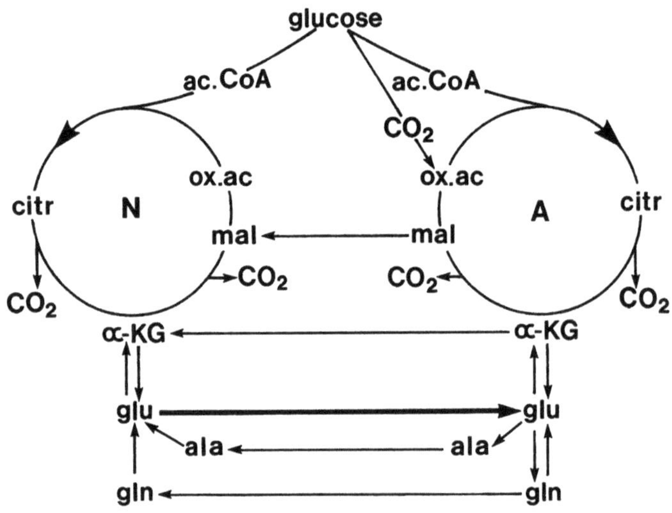

Fig.1. Glutamate (glu) is released from glutamatergic neurons (N) and intensely accumulated into astrocytes (A) (heavy line). Synthesis of glutamine (gln), an astrocyte-specific process, release of gln from astrocytes, its reaccumulation into neurons and hydrolysis to form glu contributes to replenish neuronal glu. However, astrocytic glu is also metabolized to CO_2 and its amino group is partly transaminated into other amino acids, e.g., alanine (ala). Unless there is a net import of gln into the CNS, also other glu precursors must therefore exist in glutamatergic neurons. Only astrocytes have pyruvate carboxylase activity, allowing a net synthesis of oxaloacetate (ox.ac) and other TCA cycle constituents, e.g., α-ketoglutarate (α-KG) and malate (mal) from glucose. These may be released from astrocytes, accumulated into neurons and used to form the carbon skeleton of glu. Incorporation of amino nitrogen from ala (and/or other amino acids) may supply the amino group.

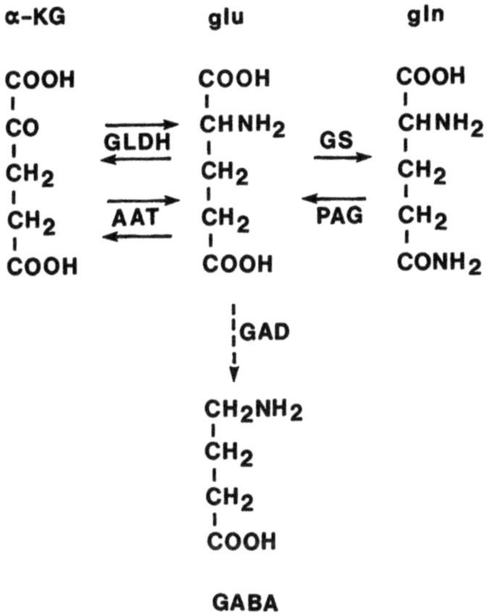

Fig.2. Structure of glutamate (glu) and its metabolites α-ketoglutarate (α -KG), glutamine (glu) and GABA. The α-KG-glu interconversion is either catalyzed by the aspartate aminotransferase (AAT), involving also another amino acid and its corresponding ketoacid (e.g. aspartate-oxaloacetate) or by the glutamate dehydrogenase (GLDH), leading to formation of α-KG by an oxidative deamination or formation of glutamate by a reductive amination. Synthesis of gln is catalyzed by the glutamate synthetase (GS), and formation of glu by the phosphate activated glutaminase (PAG). Formation of GABA (stippled arrow), catalyzed by the glutamate decarboxylase (GAD), does not occur in glutamatergic neurons but is restricted to GABA-ergic neurons.

Glutamate is not only accumulated into astrocytes, but also released from these cells. The release is enhanced by an elevation of the extracellular potassium concentration. Such a K^+ increase occurs in vivo during hypoxia (Hansen, 1985), which also impairs glutamate uptake into both neurons and astrocytes (Schousboe et al., 1988b). The resulting increase in interstitial glutamate may play a role in ischemic neuronal cell death.

It has been suggested (Benjamin and Quastel, 1975) that glutamate released from neurons is quantitatively converted to glutamine, catalyzed by the glutamine synthetase (GS), and that this glutamine is released to the interstitial fluid and reaccumulated into glutamatergic neurons, where it is hydrolyzed, catalyzed by phosphate activated glutaminase (PGA), to form glutamate (Figs 1-2). This concept is supported by the findings that GS is an astrocyte specific enzyme (Norenberg and Martinez-Hernandez, 1979), and that PAG is enriched in, although not restricted to, glutamatergic neurons (Hertz and Schousboe, 1986, 1988). However, experiments using astrocytes in primary cultures have shown that accumulated glutamate is only partly converted in astrocytes to glutamine. The major part is used as a metabolic substrate (Yu and Hertz, 1983; Hertz et al., 1988a) and oxidized, via α -ketoglutarate (α -KG) to CO_2 (Figs.1-2). Such a metabolic degradation of glutamate (or glutamine) to CO_2 has been observed also in non-cultured brain tissue (Yu and Hertz, 1983; Tildon, 1983). In addition, Yudkoff and coworkers (1986) have demonstrated that the amino group of glutamate is transaminated to several other amino acids in astrocytes, e.g. alanine (Fig.1). One implication of these findings is that the amount of glutamine which is synthesized in astrocytes is too small to replace the amount of glutamate accumulated into astrocytes (Hertz et al., 1983; Hertz and Schousboe, 1986, 1988). Moreover, glutamine is degraded to glutamate also in astrocytes and there is no preferential uptake of glutamine in neurons (Hertz et al. 1983b; Hertz and Schousboe, 1986,1988). The inevitable conclusion of these findings is that there must either be a net import of glutamine into the central nervous system (which was not observed by most authors but reported by Eriksson et al. (1983)), or other compounds must be utilized in addition to glutamine as precursors for transmitter glutamate in neurons. This applies to both the carbon skeleton and the amino group. Possible precursors for the carbon skeleton (Shank and Aprison, 1988) include tricarboxylic acid (TCA) cycle constituents like α-KG and malate (mal). These compounds can be synthesized from glucose in astrocytes (Fig. 1), due to the presence of pyruvate carboxylase activity (Yu et al., 1983; Shank et al., 1985), which is the major anaplerotic enzyme in brain (Shank and Aprison, 1988), and catalyzes the formation of the TCA cycle constituent oxaloacetate (ox.ac.). In contrast, this enzyme is absent in neurons (both glutamatergic and GABAergic), indicating that these cells can not perform a net synthesis of TCA cycle constituents from glucose. Shank and Campbell (1984) have found that both mal and α-KG are accumulated into synaptosomes, and we have demonstrated incorporation of radioactivity from mal into cellular glutamate in cultured cerebellar granule cells. However, the incorporation rate is low, which partly may be explained by isotope dilution in intracellular TCA cycle constituents, and it is abolished when transamination is inhibited (L. Hertz, A.C.H. Yu, M. Shargool, J. Drejer and A. Schousboe, unpublished).

Exogenous glutamine and exogenous α-KG may act as precursors for glutamate in very different manners (Palaiologos et al., 1988a,b). Thus, recent experiments using AAT inhibitors or inhibitors of the exchange of malate and α-KG across the mitochondrial membrane (the ketodicarboxylate carrier) have suggested that glutamine in cerebellar granule cells is converted to glutamate within the matrix of the mitochondrial membrane, crosses the inner mitochondrial membrane (in exchange with aspartate) and is converted to α-KG by an intramitochondrial transamination; this α-KG is subsequently transferred across the mitochondrial membrane, in exchange with malate, by the ketodicarboxylate carrier, to the cytoplasm where it is converted by another transamination to form transmitter glutamate (Fig.3).

Such a process requires aspartate aminotransferase (AAT) activity in both the cytoplasm and the mitochondria but does not lead to any net consumption of either oxaloacetate or aspartate. It resembles the malate-aspartate shuttle (see, e.g., Hertz et al., 1988b) with the exception that there is a constant supply of glutamate outside the inner mitochondrial membrane and a constant removal of glutamate (as a transmitter) from the cytoplasm. This sequence of events might explain the high activity of AAT in glutamatergic neurons observed both in vivo (Wenthold and Altschuler, 1983; Wenthold et al., 1986) and in culture (Hertz and Schousboe, 1986, 1988). A direct net synthesis of glutamate from aspartate is unlikely to explain the high AAT activity, because aspartate biosynthesis occurs from glutamate (by a transamination) and there is no net uptake of aspartate across the blood brain barrier (Eriksson et al., 1983). From Fig. 3 it can also be seen that there is a possibility that exogenous α-KG might be transaminated to trans-

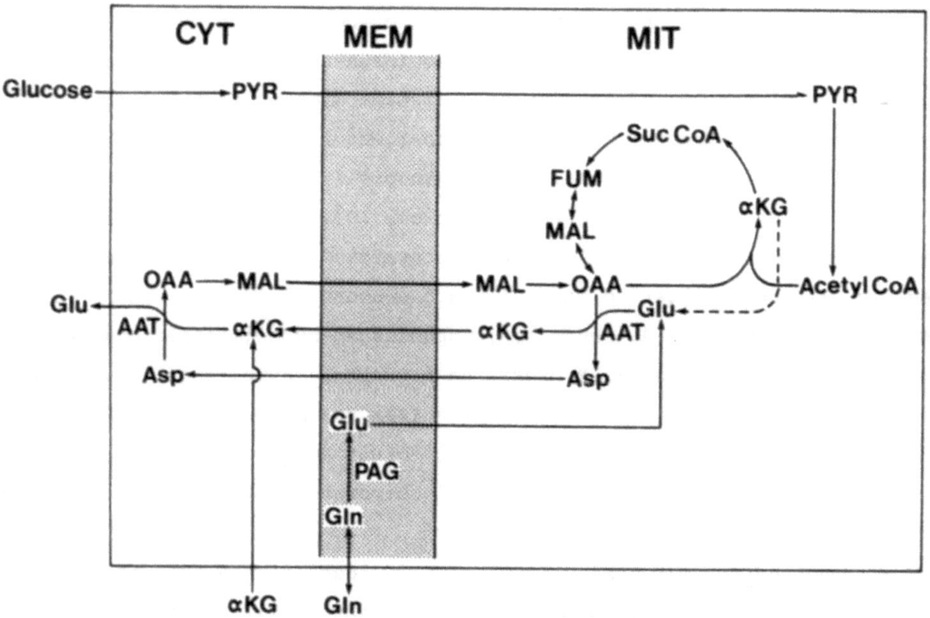

Fig.3. Schematic drawing of cytoplasmic (CYT) and mitochondrial (MIT) reactions which appear to be involved in biosynthesis of transmitter glutamate. Phosphate activated glutaminase (PAG) converts glutamine (Gln) to glutamate (Glu) within the mitochondrial membrane (MEM) matrix. Glu is transported across the inner mitochondrial membrane in exchange with aspartate (Asp), and transaminated to α-ketoglutarate (α-KG) in the mitochondria, accompanied by a transamination between oxaloacetate (OAA) and Asp. After transport of α-KG across the mitochondrial membrane, in exchange with malate (MAL), the opposite transamination occurs in the cytoplasm. Exogenous α-KG may be transaminated to Glu directly in the cytoplasm. Modified from Palaiologos et al., 1988b.

mitter glutamate directly in the cytoplasm. This would involve net utilization of an amino acid (e.g., conversion of alanine to pyruvate), taking part in the transamination. In contrast, mal cannot be used for net synthesis of transmitter glutamate in these cells, since this would require intramitochondrial glu formation from α-KG (stippled line in Fig.3) by a reductive amination, and this process is completely inhibited in cerebellar granule cells by AOAA (see above), indicating that it occurs by transamination only.

GABA turnover. GABA is formed in GABAergic neurons by decarboxylation of glutamate catalyzed by GAD (stippled arrow in Fig. 2). The release of GABA (per mg protein) from GABA-ergic neurons is quantitatively much less (about 1/10) than the glutamate release from glutamatergic neurons (Hertz and Schousboe, 1987; Schousboe et al., 1988b). Moreover, the GABA uptake rate in GABAergic neurons is at least as high as that in astrocytes. Thus, transfer of GABA from neurons to astrocytes during neuronal activity is quantitatively less important than the corresponding transfer of glutamate from glutamatergic neurons. After astrocytic accumulation, GABA is converted to the TCA cycle constituent succinate and from there via α-KG and glutamate to glutamine. Transport of glutamine (formed in this way and maybe also from glutamate) back to neurons (confer Fig. 1) might therefore conceivably suffice for neuronal GABA synthesis (the (GABA)-glutamate-glutamine cycle; (Van den Berg and Garfinkel, 1971)). It is consistent with this concept that no measurable glutamate synthesis from labelled malate (see Fig. 1) can be observed in GABA-ergic neurons (L. Hertz, A.C.H. Yu, M. Shargool, J. Drejer and A. Schousboe, unpublished experiments). Nevertheless, recent experiments (M. Yudkoff and L. Hertz, unpublished) have clearly shown that glutamine is not the only precursor for glutamate and GABA in GABAergic neurons, since amino nitrogen in other amino acids, e.g., alanine and leucine to a considerable extent is incorporated into glutamate and GABA in these cells.

Although astrocytic GABA accumulation is less intense than the corresponding accumulation of glutamate it may be functionally important because GABA accumulated into neurons can be reutilized as a transmitter (or metabolized), whereas GABA accumulated into astrocytes can be metabolized only. Thus, astrocytic GABA uptake diverts GABA from reuse. This may explain why astrocyte-specific GABA uptake inhibitors have been shown to possess antiepileptic activity in several animal models (Schousboe et al., 1988b).

GABA release from astrocytes is not stimulated by elevated potassium concentrations (Schousboe et al., 1988a).

<u>Glutamate and GABA receptors</u>. For the understanding of neuronal-glial interactions in the CNS and the consequences of altered transmitter concentrations in the interstitial fluid it is important to know to what extent different cell types are affected by alterations in transmitter concentrations, i.e., whether they possess receptors. GABA-ergic cerebral interneurons and glutamatergic cerebellar granule cells in primary cultures do, indeed, express receptors for both glutamate and GABA (for references, see Hertz and Schousboe, 1987). In immature cells, there is only a high-affinity binding site, but differentiated cells have both a high-affinity and a low-affinity binding site.

The functional role of GABA receptors on cerebral cortical neurons is unknown, whereas the function of these receptors on cerebellar granule cells has been studied in some detail. The low-affinity receptors are of the $GABA_A$ type and inhibit the release of glutamate evoked by 100 mM potassium. The glutamate receptors stimulate transmitter release, i.e., release of GABA from cerebral cortical neurons and release of glutamate from cerebellar granule cells. The cortical neurons possess all conventional glutamate, receptor subtypes, i.e., NMDA (N-methyl-D-aspartate), quisqualate-, and kainate-preferring receptors, whereas the glutamate receptor on the cerebellar granule cells is not markedly affected by any of these substances. It can, however, be blocked by γ-D-glutamylglycine, indicating that it is not part of the high-affinity uptake site (Hertz and Schousboe, 1987).

It has been debated whether astrocytes have genuine receptors for GABA or glutamate, since they do not show receptor binding (Ossola et al., 1980; Schousboe, 1980), whereas both glutamate and GABA (in relative high concentrations) do lead to an astrocytic depolarization (Bowman and Kimelberg, 1984; Kettenman et al., 1988). The demonstration that glutamate and GABA affect calcium fluxes in cultured astrocytes and that excitatory amino acid transmitters initiate the breakdown of membrane inositol phospholipids (Murphy and Pearce, 1987), are in support of a transmitter action.

Purines and Purine Derivatives

The role of adenosine and ATP as neurotransmitters or neuromodulators in the CNS has been extensively studied by Phillis and co-workers (e.g., Phillis and Wu, 1981). Adenosine, which besides its neuromodulatory activity also is a potent vasodilator, is accumulated more intensely into astrocytes than into neurons (Bender and Hertz, 1986). The subsequent metabolism in these two cell types differ (Matz and Hertz, 1989), with astrocytes converting adenosine mainly into nucleotides (AMP, ADP and ATP) and neurons mainly forming deaminated compounds (inosine and hypoxanthine). Both neurons and astrocytes are also capable of releasing adenosine (H. Matz and L. Hertz, unpublished). This provides the possibility that adenosine release from astrocytic endfeet may exert a direct vasodilatory effect on brain capillaries.

Adenosine receptors have repeatedly been demonstrated on cultured astrocytes; both A_1 and A_2 receptors may be present, and affect second messengers differently (Murphy and Pearce, 1987).

Monoamines

Uptake: It has long been recognized that the uptake of noradrenaline into neurons differs from that into peripheral tissues. Kimelberg and co-workers have reported a high affinity uptake of noradrenaline into astrocytes in primary cultures which resembles that in neurons, i.e., is concentrative, sodium dependent and inhibited by tricyclic antidepressants (Kimelberg, 1986). This is at variance with earlier findings that noradrenaline uptake in neurons and astrocytes have very different characteristics (Henn and Hamberger, 1971; Vernadakis, 1974; Hosli and Hosli, 1976). We have therefore reinvestigated noradrenaline uptake into primary cultures of astrocytes and found that the uptake can be resolved into two components, i.e., a saturable uptake of low affinity (compared to that reported for synaptosomes), which is not concentrative (at least not in the presence of a monoamine oxidase inhibitor) and another, non-saturable component. Neither of them is sodium dependent or inhibited by the antidepressant desmethylimipramine (Paterson and Hertz, 1989). These characteristics contrast not only neuronal adrenaline uptake but also uptake into brain endothelial cells, which is sodium dependent and inhibited by antidepressants (Abe et al., 1980; Joo, 1983). The functional relevance of this seems to be that it is only at relatively high interstitial noradrenaline concentrations that astrocytic uptake contributes significantly to noradrenaline clearance, and that subsequent metabolism (see below) might increase the uptake. Moreover, the differences between astrocytes and endothelial cells may be of methodological importance, since some preparations of astrocytes in primary cultures may be quite heavily contaminated with endothelial cells (see page 614 in Hertz et al., 1985). Such a contamination would explain the heterogeneous uptake of noradrenaline into astrocytes shown autoradiographically by Semenoff and Kimélberg (1985).

Kimelberg and co-workers have also reported a serotonin uptake into astrocytes which in principle is similar to the noradrenaline uptake described by this group (Kimelberg, 1986). However, we have in preliminary experiments failed to show any sodium dependence of serotonin uptake (page 333 in Hertz, 1982). In view of the methodological concern expressed above, this topic needs reinvestigation.

Metabolism: Monoamine oxidase (MAO), the major metabolizing enzyme for monoamines in brain tissue, has a relatively high activity in astrocytes in primary cultures, especially after prolonged culturing with dibutyryl cyclic AMP (Yu and Hertz, 1982). This procedure affects mainly the activity of MAO B. The physiological role of MAO A and B in astrocytes has not been clearly defined. Noradrenaline uptake might, at high interstitial concentrations, be quite intense, especially if accumulated noradrenaline is rapidly removed by metabolism, and/or uptake of other monoamines might also be followed by metabolic degradation. It is of pathological relevance that MPTP oxidation, leading to the neurotoxin MPP^+, occurs in astrocytes (Ransom, 1988).

Receptors. Astrocytes have a high receptor density, especially for β_1 adrenergic receptors (Murphy and Pearce, 1987; McCarthy et al., 1988) and may therefore, together with endothelial cells (Hartman, this volume), constitute the main target for a sympathetic nervous system within the brain. The β_1 receptor density on neurons may be considerably lower. Stimulation of astrocytic β receptors enhance glycolysis (Opler and Makman, 1972), and the release of the inhibitory amino acid transmitter taurine (Martin et al., 1988), at least in glioma cells. The astrocytic β receptors downregulate after chronic (weeks), but not acute (days) exposure to antidepressants (Hertz and Richardson, 1983). Astrocytes also express serotonin (Hertz et al., 1979; Whitaker-Azmitia et al., 1988; Murphy and Pearce, 1987) and dopamine (Hansson et al., 1984) receptors.

ENERGY METABOLISM.

It has already been mentioned that glutamate can be transported from neurons to astrocytes and subsequently used as an important metabolic substrate. Other substrates may also be transferred between neurons and astrocytes. Thus, a substantial CO_2 formation from lactate in both cultured neurons and non neuronal cells from sympathetic ganglia has been described by Larrabee (1983), so that "a lively exchange of lactate may go on behind the blood-brain barrier between brain cells both of the same and different

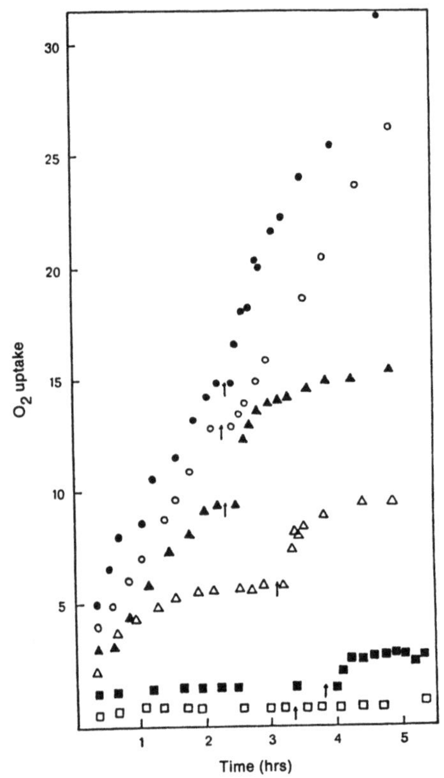

Fig.4. Cumulative oxygen uptakes (10^4 µl/sample) by individual samples of cat nerve cells from the cortex and mesencephalon (○ and ●, respectively), and neuroglia lumps from cat and rat (△ and ▲, respectively) compared to the apparent respiration in control divers without cells (□ and ■). O_2 uptake was measured for about 5 hrs from the start of the incubation. At the arrows, concentrated solutions of KCl were added to a final concentration of ≃ 50 mM. For graphical reasons the curves are consecutively displaced in a vertical direction. From Hertz, 1966.

kinds". Both lactate (Yu and Hertz, 1983) and pyruvate (Hertz et al, 1988a) are oxidized in cerebral cortical astrocytes and (at least pyruvate) also in cerebral cortical neurons and cerebellar granule cells.

The rate of oxygen uptake is high in astrocytes, i.e., comparable to average brain respiration in vivo. Cultured cerebellar granule cells consume oxygen at about the same rate, whereas cerebral cortical neurons are metabolically slightly less active (Hertz et al., 1988a). High concentrations of potassium stimulate respiratory activity in microdissected (Fig. 4) or cultured astrocytes, but not in microdissected neuronal perikarya (Fig. 4) or cultured neurons (Hertz, 1986b). This agrees with the observation by Sokoloff and coworkers (Sokoloff, this volume) that stimulation in vivo does not increase metabolic activity in neuronal perikarya. Lactate formation in cultured astrocytes is also enhanced in the presence of excess potassium, and is released to the medium (Walz and Mukerji, 1988).

CALCIUM

Calcium (Ca^{2+}) plays a very important role in the regulation of cellular functions in the central nervous system (CNS). Intracellular Ca^{2+} is involved in the control of a variety of physiological processes, including membrane permeability (e.g. calcium activated potassium channels), activation of second messengers and of proteinkinases (e.g., proteinkinase C), vascular tone (Van Neuten et al., 1988), and release of neurotransmitters (Janis et al., 1987; Miller, 1987; Turner and Goldin, 1988), as well as in pathological reactions, e.g., following ischemia, when the extracellular Ca^{2+} concentration in brain falls from about 1.0 to about 0.1 mM (Harris et al., 1981; Harris and Symon, 1984). Calcium enters cells via either voltage-sensitive (L, T, or N type) or receptor-operated Ca^{2+} channels. Of the available organic calcium-channel blockers, the 1,4-dihydropyridines interact specifically and potently with the voltage-sensitive Ca^{2+} channel of the L type (Janis et al., 1987; Miller, 1987), whereas the diphenylpiperazine flunarizine is less selective (Janis et al., 1987). Binding sites for dihydropyridines have been observed in isolated neurons (Miller, 1987), including primary cultures of cerebral cortical neurons (Table 1), which have both a high-affinity binding site and a binding site of lower affinity (Bender and Hertz, 1988). The 1-4-dihydropyridines are potent inhibitors of depolarization-induced uptake of calcium into neuronal cell lines (Miller, 1987) and primary cultures of cerebellar granule cells (Carboni et al., 1985), but are virtually ineffective in many other neuronal preparations (Miller and Freedman, 1984; Skattebol and Triggle, 1987) and in most primary cultures of neurons (Miller, 1987).

Table 1. Kinetic characteristics (K_D (nM) and B_{max} (femtomol per mg protein) for nitrendipine binding to intact astrocytes (cultured in the presence or absence of dibutyryl cyclic AMP (dBcAMP) and neurons in primary cultures. Values are from Bender and Hertz, 1989.

	Astrocytes		Neurons
	- dBcAMP	+ dBcAMP	- dBcAMP
K_D,nM		2.5 ± 0.5	1.1 ± 0.4
B_{max}		917 ± 132	485 ± 94
K_D,nM	218 ± 24	209 ± 31	449 ± 176
B_{max}	19,600± 1900	22,400 ± 1900	26,800± 8000

Microdissected Muller cells (Newman, 1985), which are retinal astroglial cells, and primary cultures of astrocytes that have been treated with dibutyryl cyclic AMP dBcAMP to obtain optimum differentiation, express voltage sensitive Ca^{2+} channels (MacVicar, 1984; Chun et al., 1986). Cultured astrocytes (Litzinger et al., 1986) treated with dBcAMP (Bender and Hertz, 1988) as well as astrocytes prepared by gradient centrifugation (L. Hertz, M. Farooq and W.T. Norton, preliminary experiments) also possess high-affinity nitrendipine binding sites (Table 1). Nevertheless, such channels could not be demonstrated in astrocytes in hippocampal slices (Walz and McVicar, 1988). Astrocytes also possess Ca activated K^{+} channels (Quandt and McVicar, 1986), and have a high activity of protein kinase C (Neary et al., 1986; Mobley et al., 1986).

Lazarewicz et al. (1977) and Albrecht et al. (1988) have observed a potassium-induced stimulation of uptake of ^{45}Ca in a human glioma cell line and/or astrocytes obtained by gradient centrifugation. Previous attempts to demonstrate a potassium-induced stimulation of calcium uptake in primary cultures of astrocytes were not successful (Barnes and Mandel, 1981; Walz and Wilson, 1986), but we have recently (Hertz et al., 1989) observed such an effect after prelabeling with ^{45}Ca (to saturate extracellular binding sites) slightly before the addition of excess potassium. As illustrated in Fig.5, an almost maximum effect was seen already at 10-15 mM K^{+}, i.e., the "ceiling level" of the interstitial K^{+} concentration during seizures (see "Potassium"). This observation is in contrast to potassium-stimulated ^{45}Ca uptake in neurons (Skattebol and Triggle, 1987), which increases linearly with the potassium concentration between 5 and 55 mM, and it possibly reflects the behavior of astrocytes as perfect potassium electrodes even at low extracellular potassium concentrations (see "Potassium"). Astrocytes which had not been treated with dBcAMP do not express high affinity nitrendipine binding sites (Table 1), and excess potassium did not enhance ^{45}Ca uptake.

The dihydropyridine calcium channel blocker nimodipine potently (IC_{50} 2.7 nM) abolished the potassium-induced Ca^{2+} uptake in astrocytes but had virtually no effect on non-stimulated Ca^{2+} uptake (Fig. 6). This strongly suggests that the potassium induced ^{45}Ca uptake into astrocytes occurs through voltage-dependent L-channels. The IC_{50} values is in very good agreement with the K_D value of 2.5 nM for nitrendepine binding (Table 1).

It is generally assumed that increased Ca^{2+} inward currents into neurons are responsible for the decreased extracellular Ca^{2+} levels observed

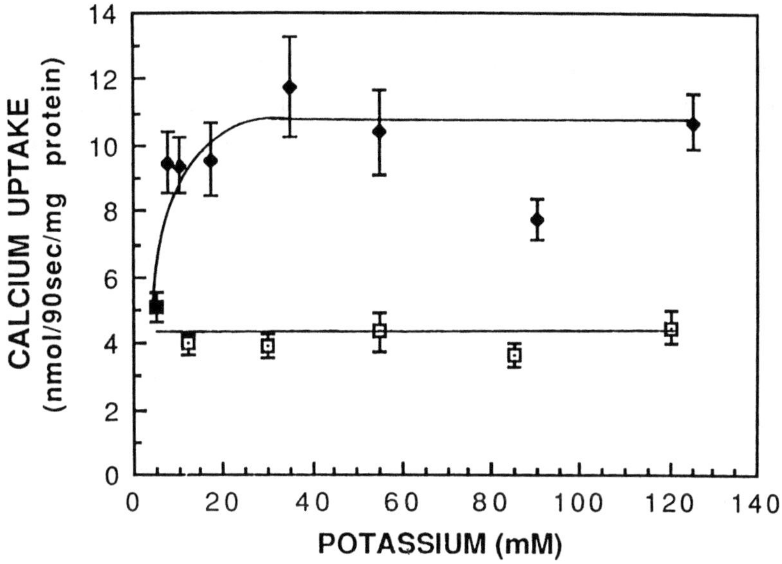

Fig.5. Uptake of ^{45}Ca into astrocytes in primary cultures as a function of the potassium concentration of the medium. Excess potassium was added either simultaneously with ^{45}Ca (□) or 60 sec. later (◆). In both cases ^{45}Ca uptake was terminated after 90 sec. From Hertz et al., 1989.

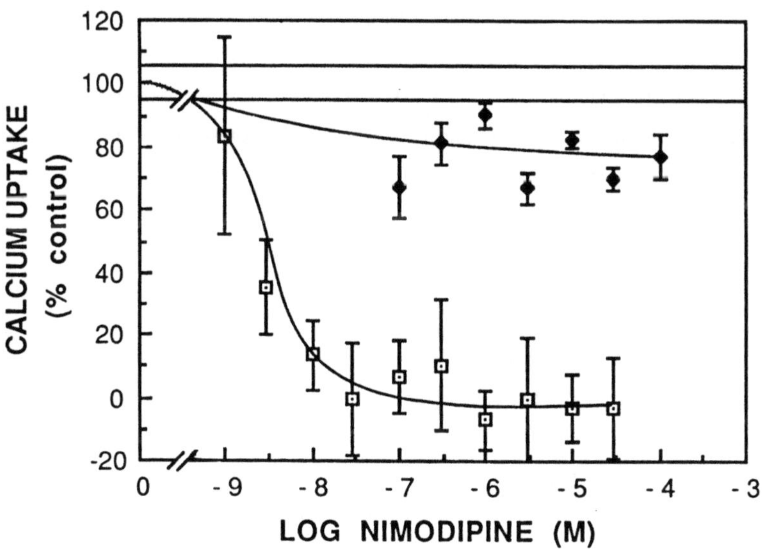

Fig.6. Inhibition of potassium-induced uptake of ^{45}Ca (i.e., uptake after addition of 50 mM K^+ (after 60 sec) to a medium with 5 mM K^+ minus uptake at 5 mM potassium) (□), or of non-stimulated uptake (5 mM K^+ only) (◆) as a function of the concentration of nimodipine. In both cases, ^{45}Ca uptake was terminated after 90 sec. From Hertz et al., 1989.

during increased CNS excitability (Hille, 1984). However, the findings described above suggest that during periods of increased neuronal activity, a large amount of Ca^{2+} may enter astrocytes, and that neuronally released potassium acts as the trigger for this effect. Moreover, stimulation of glutamate or GABA receptors has been reported to alter astrocytic Ca^{2+} uptake (see "Glutamate and GABA receptors"). The physiological consequences of this astrocytic response to neuronal activity have not been clearly defined. However, as briefly discussed above, Ca^{2+} plays a major role in regulation of several cell functions. In view of the fact that astrocytes constitute a large fraction of the brain volume (see "Introduction), they may also serve as a "calcium sink," thereby partly protecting neurons from the neurotoxic effects of Ca^{2+}, e.g., during ischemia. Under this condition the extracellular Ca^{2+} concentration in the brain in vivo decreases steeply, once the extracellular potassium concentration has reached a level of $\simeq$15 mM (in the rat) (Harris and Symon, 1984). Although this level is slightly higher than the K^+ concentration required to stimulate Ca^{2+} uptake in cultured astrocytes, it should be noted that a further increase in K^+ concentration has no additional effect, although neuronal Ca^{2+} uptake appears to be further enhanced when the K^+ concentration is increased beyond 20 mM. Further studies are clearly required to elucidate the contribution of each cell type to Ca^{2+} uptake under physiological and pathological conditions. Nevertheless, the observation that nimodipine is a potent blocker of potassium-induced Ca^{2+} influx into astrocytes demonstrates that these cells are accessible for pharmacological manipulation and, therefore, potential targets for drug therapy.

POTASSIUM.

Extracellular concentration

It has now been overwhelmingly well demonstrated that the potassium concentration in the extracellular space of the central nervous system increases measurably (from 3 to about 4 mM) during physiological stimulation, to a larger extent (from 3 up to a "ceiling level" of 12 mM) during seizures or direct stimulation of afferent pathways, and to exceedingly high values (up to 80-100 mM) during anoxia and spreading depression (Somjen 1979; Walz and Hertz, 1983). It is also well established that the potassium concentration in the extracellular space after stimulation, seizures, spreading depression or short-lasting anoxia rapidly returns to its resting level of 3 mM. Potential mechanisms for this potassium removal from the extracellular clefts would be that potassium either leaves the brain by accumulation into blood vessels or cerebrospinal fluid, or that potassium is redistributed within the brain tissue itself. Such a redistribution might either occur by diffusion within the extracellular space, or it might involve adjacent cells, be it neurons or astrocytes. Of these possiblities diffusion through the extracellular space or removal into blood vessels or cerebrospinal fluid can be ruled out as the major mechanism(s) for potassium removal from the mammalian brain in vivo, since the processes are not fast enough to explain the actually measured rates of potassium removal (for references, see Walz and Hertz, 1983). Thus, one or more kinds of cellular mechanisms must participate in the redistribution of potassium.

Role of glial cells.

Neuronal-glial interactions: Already in the mid-1960s, it was suggested that glial cells might be involved in potassium homeostasis at the cellular level of the central nervous system, and that this mechanism may be of major importance for brain function (Hertz, 1965; Orkand et al., 1966; Kuffler, 1967). Two different hypotheses were proposed, i.e., (1) that extracellular potassium is accumulated into glial cells by active uptake processes (Hertz, 1965), or (2) that an increased local concentration of extracellular potassium is redistributed through glial cells by a current-carried transport mechanism ("spatial buffering") (Orkand et al., 1966). The former of these hypotheses was suggested on the basis of findings in isolated rat or cat glial cells and the latter on the basis of biophysical membrane characteristics (selective potassium permeability and electrical coupling) of glial cells in intact non-mammalian nervous systems. During the past 20-25 years, evidence has accumulated supporting both of these hypotheses, and a substantial amount of information has been compiled of both potassium transport and membrane characteristics in isolated glial cells and, to a less extent, in isolated neurons.

These processes are outlined in Fig.7. Potassium transport through glial cells by a current carried mechanism (stippled line through all A's) removes a local excess of extracellular potassium, resulting from excitation of neurons, and redistributes this potassium to other regions (see Orkand, this volume). The process is passive, and the circuit is closed by return of extracellular sodium (the predominant extracellular ion) back to the origin. This redistribution of potassium is not directly dependent upon energy metabolism, although it does depend on membrane polarization which,

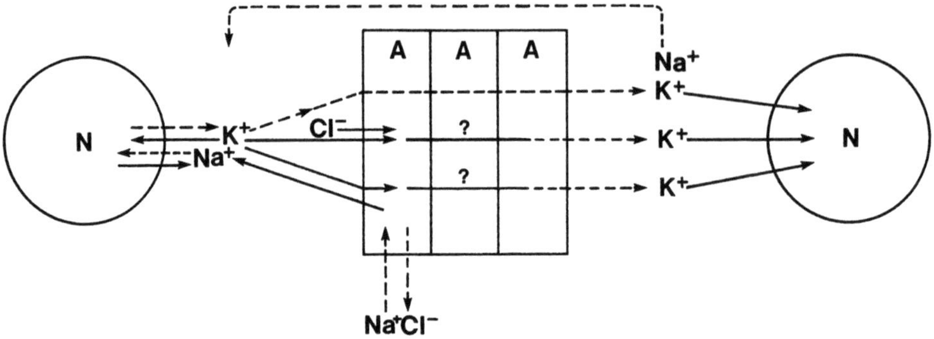

Fig.7. Interactions between neurons (N) and astrocytes (A) in potassium (K^+) homeostasis at the cellular level. Active transport processes are shown as fully drawn lines and passive processes as stippled lines. An excited neuron loses K^+ and gains sodium (Na^+). Inward active transport of K^+ and outward active transport of Na^+ in neurons partly restore prestimulated ion levels. However, in addition K^+ passes through electrically coupled astrocytes by a passive, current-carried transport (the "spatial buffer") from a region with elevated extracellular K^+ to a different extracellular location. The electrical circuit is closed by an extracellular transport of Na^+ back to the origin. Moreover, active transport of K^+, together with chloride (Cl^-) or in exchange with Na^+ occurs into astrocytes. It is unknown whether these K^+ ions can be further transported through astrocytes to different locations and whether they are subsequently released to the extracellular space or, maybe, transported directly from astrocytes to neurons. Na^+ enters astrocytes passively in spite of a low membrane permeability, due to a high electrochemical driving force, and Cl^- leaves passively.

in turn, is a result of energy metabolism. The resulting increase in intraglial potassium concentration is infinitesimally small. Conceptionally, this process shows similarities with the neuronal action potential in allowing a rapid transport of an ion, utilizing the energy residing in an uneven distribution of intracellular and extracellular ions. It does not remove potassium from the extracellular space to glial cells, since extracellular potassium after the process is completed, remains extracellular potassium. However, it does remove potassium ions from the neuron(s) which have suffered a potassium loss during excitation. By redistributing potassium ions and thus influencing neuronal membrane potential, it may be of importance in information processing. The pathway(s) for the redistribution will depend upon anatomical structure and membrane permeability for potassium at both the external membranes and at the gap junctions, coupling the individual astrocytes electrically. A possible regulation of the process by intra- and/or intercellular signals (see below) would provide the possi-

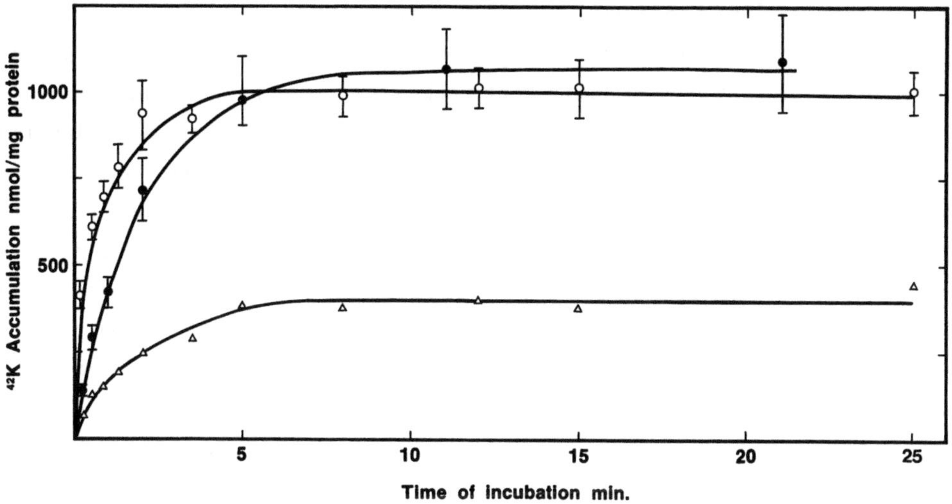

Fig.8. Time course of the uptake of ^{42}K into primary cultures of astrocytes measured either after potassium depletion (●) or at steady state (○) and into corresponding cultures of neurons, measured at steady state (△). From Hertz, 1986a.

bility of controlling the responsiveness of the system.

Active uptake into glial cells (fully draws lines into the first A) is conceptionally quite different. It is energy dependent and does transfer potassium ions from neurons to glial cells. Any potassium ions accumulated into glial cells will eventually have to be reaccumulated into neurons in order to prevent neuronal depletion of potassium. Thus, the entire process is, in principle, similar to that which appears to exist for glutamate turnover. There is, however, one major difference, i.e., that glutamate can be converted to other compounds (Fig.2), whereas potassium ions remain potassium ions.

Potassium uptake into isolated neurons and glial cells: Primary cultures of mouse astrocytes constitute a suitable preparation for studies of potassium turnover due to the high membrane potential in these cells (80 mV at 5.4 mM and 90 mV at 3 mM), their electrical coupling and their selective potassium permeability which makes them behave as "perfect potassium electrodes" down to a very low level (1 mM) of extracellular K^+ (Hertz, 1986a). These biophysical characteristics are similar to those observed for astrocytes in vivo. Some problems have been encountered with rat astrocytes which have been found to have a less selective potassium permeability (Walz and Kimelberg, 1985). However, it seems possible to amend this, at least partly, by altering the culturing conditions (Juurlink and Hertz, 1985).

Uptake of ^{42}K into cells at a steady state with respect to their potassium content (cells that have not been artificially depleted of potassium) may represent several different mechanisms, i.e., (1) an active uptake; (2) homoexchange, i.e., a direct coupling between uptake of labeled extracellular potassium ions and release of nonlabeled intracellular potassium ions (a process that would be of no physiological consequence); and (3) diffusion through the cell membrane, a process that can be expected to be fast due to the selective and large membrane permeability for potassium and a high negative membrane potential. This diffusion is a reflection of the characteristics allowing the current carried redistribution, but at a uniform extracellular potassium concentration the "spatial buffer" mechanism itself will not be operating. A relatively good distinction between these possibilities can be obtained by measuring potassium uptake into cells that by previous exposure (10-30 min) to potassium-free medium have been depleted of virtually all intracellular potassium, and thus also of most of the membrane polarization. The uptake ^{42}K into such cells at a potassium concentration of 5.4 mM is shown in Fig.8 (filled circles). The approximately rectilinear part of the uptake during the first 60 sec represents the initial uptake rate, i.e., an uptake of radioactivity that is only little obscured by the resulting increase in intracellular ^{42}K content. With time, the intracellular accumulation of ^{42}K leads to a steadily increasing release of radioactivity during the efflux of potassium, i.e., the subsequently less steep uptake observed during the next ≃5 min. Finally, after a constant intracellular potassium pool, having the same specific activity as potassium in the medium, has been established, there will be no further increase in the radioactivity of the cells. The level of ^{42}K at this time indicates the total pool of exchangeable potassium. Fig.8 shows that the level of exchangeable potassium is ≃1,000 nmol/mg protein and that the initial uptake rate, calculated during the first 30 seconds, is ≃600 nmol/min per mg protein (see also Table 2). With a water content of approximately 10 µl/mg protein, this corresponds to 100 µmol per ml intracellular water (100 mM) and ≃60 µmol per min per ml, respectively. This uptake rate almost exclusively represents an active uptake, and the uptake is almost completely abolished when the temperature is reduced to 1°C (Hertz, 1986). A decrease of the extracellular K^+ concentration from the normal 5.4 mM (usual concentration in the tissue culture medium) leads to a diminished uptake rate of potassium, whereas an increase causes an enhancement of the active uptake, following Michaelis-Menten kinetics with a K_m value of ≃15 mM. A large part of this uptake is inhibited by ouabain, i.e., occurs by a Na, K-ATPase activated exchange with sodium. The ouabain-sensitive uptake reaches its maximum at 10-15 mM K^+. This is also consistent with a high K_m value for Na^+,K^+-ATPase activity in glial cells obtained by gradient centrifugation (Grisar et al., 1979; Franck et al., 1983) and with an increase of Na^+, K^+-ATPase activity in isolated Muller cells from the rabbit when the extra-cellular K^+ concentration is elevated up to 10 mM (Reichenbach et al., 1985). In spite of a low membrane permeability towards sodium, Na^+ required for the exchange may enter the astrocytes passively on account of a high electrochemical driving force. Moreover, astrocytic K^+-Na^+ exchange seems to require only low levels of Na^+ (Reichenbach et al., 1987; Coles et al., 1986; L. Hertz, S.Y. Chow, H.S. White and D.M. Woodbury, unpublished experiments).

In astrocytes at steady state the uptake rate is considerably faster (2,000 nmol/min per mg protein) (Fig. 8; Table 2) but the equilibrated level is unaltered. The difference in uptake rate suggests that a very consider-able potassium uptake occurs either by homoexchange or by diffusion along an electrochemical gradient. Homoexchange seems to play little, if any role (Hertz, 1986a). Thus, it seems likely that the difference between potassium uptake at steady state and after potassium depletion is caused by an intense diffusion, occurring in either direction, which reflects the high potassium permeability of the astrocytic cell membranes. Over the last few years a substantial amount of information on K^+ permeability has been obtained using isolated retinal Muller cells (a modified astrocytic cell). An advantage of this preparation is that the Muller cell spans the entire stratified retina. After light stimulation, the extracellular K^+ concentration increases predominantly within the plexiform layers (Karwoski et al., 1985); the released K^+ ions therefore enter the Muller cells at this region as the initial part of a "spatial buffer mechanism". A very high potassium conductance at the endfeet of the Muller cell (Newman, 1985; Nilius and Reichenbach, 1988) strongly suggests that the K^+ ions subsequently travel through the cells to be released across the endfeet, i.e., into the vitreous humor (Newman, 1985; Nilius and Reichenbach, 1988). Similar "hot spots" for K^+ conductance may conceivably also exist on astrocytic endfeet lining the capillaries of the cerebral hemispheres. Potassium release here may cause a local vasodilation and thus contribute to the regulation of blood flow (Paulson and Newman, 1987). However, as stated above, K^+ dissipation into the vascular system is too slow to account for the major part of K^+ homeo-stasis at the cellular level of the brain (Walz and Hertz, 1983).

Low concentrations of barium (≃50 μM), which are known to inhibit potassium conductance, cause a drastic decrease of potassium influx into astrocytes in primary cultures at steady state (Walz et al., 1984) and a less pronounced augmentation of input resistance (Walz, 1986). This concentration of barium has little effect upon the ouabain-sensitive uptake of potassium and on Na^+,K^+-ATPase activity, whereas higher concentrations (e.g. 1-5 mM) of barium also inhibit these parameters (Walz et al., 1984). A similar inhibition is exerted by nanomolar concentrations of phorbol 12-myristate 13 acetate (PMA) (Fig.9). Its effect on active uptake is orders of magnitude less potent. It is possible that the PMA effect on K^+ permeability could be due to an inhibition of voltage-sensitive calcium channels as a result of the stimulation of protein-kinase C. Such an effect has been described in dorsal root ganglia neurons (Kaczmarek, 1987). McVicar et al. (1987) have observed that active phorbolesters cause oscillations of membrane potential in astrocytes, accompanied by an increase in input resistance. They have suggested that this may be due to oscillating levels of intracellular Ca^{2+} which, in turn, affect K^+ permeability and thus extracellular K^+ concentration and neuronal excitability. The proposed alterations of intracellular Ca^{2+} concentration could, however, not be attributed to opening or closure of voltage-sensitive Ca^{2+} channels, since the membrane potential oscillations were not abolished by dihydropyridines (McVicar et al., 1987). It is in keeping with this conclusion that we recently have observed that the calcium channel blocker nimodipine at a concentration of 100 μM has very little effect on K^+ permeability and that the calcium channel agonist Bay K 8646 has a similar, not an opposite effect (Fig.10). The effects of these huge concentrations are probably non-specific (Janis et al., 1987), and active uptake of K^+ is at least equally much affected (Fig.10). The less specific Ca^{2+} blocker flunarizine is more potent (Fig.10), causing 25% inhibition at 10 μM. At higher concentrations it also inhibits active uptake. Diazepam, at a concentration of 10 μM, inhibits active uptake of potassium (Fig.10). At 1 μM it has no effect on K^+ permeability although other evidence suggests that benzodiazepines may interact with Ca^{2+} channels in astrocytes (Bender and Hertz, 1985, 1988; Hertz and Bender, 1988).

In addition to a high potassium permeability and a ouabain-inhibited active uptake, astrocytes in primary cultures also possess a furosemide-sensitive uptake system, which is activated only at elevated K^+ concentrations (Walz and Hertz, 1984), and is abolished after preincubation in potas-

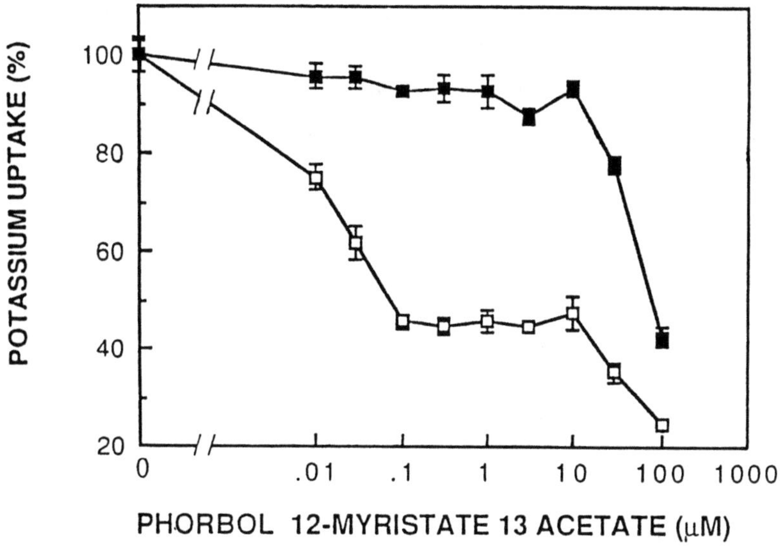

Fig. 9. Inhibition of uptake of ^{42}K into astrocytes in primary cultures by phorbol 12-myristate 13 acetate (PMA) under steady state conditions (□), i.e., reflecting mainly K^+ permeability and in previously K^+-depleted cells (■), i.e., reflecting mainly active uptake. From Bender et al., 1989.

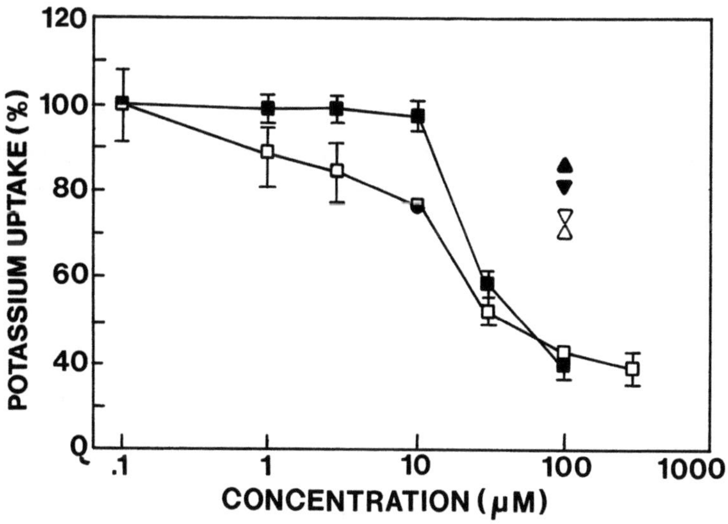

Fig. 10. Inhibition of ^{42}K uptake into astrocytes in primary cultures at steady state (□ , △ , ▽), i.e., mainly K^+ permeability, and after previous K^+ depletion (■, ▲ , ▼ , ●), i.e., mainly active uptake, by flunarizine (□ and ■), nimodipine (△ and ▲), Bay K 8646 (▽ and ▼), and diazepam (●). L. Hertz, A. Bender, D.M. Woodbury and H.S. White, unpublished experiments, except for o which is from Finegan et al., 1987.

Table 2. Uptake rates for ^{42}K in primary cultures of mouse astrocytes, sheep oligodendrocytes and mouse neurons at 5.4-6.0 mM K^+.

	Steady state (including permeability)	K^+ depleted (active)
Astrocytes[1)]	1500 - 2500	200 - 600
Oligodendrocytes[2)]	100	50
Neurons (GABA-ergic)[1)]	100 - 250	

1) From Hertz, 1986
2) L. Hertz, S. Szuchet and E. Hertz, unpublished experiments

sium free media (L. Hertz, unpublished experiments). In all likelihood it represents a cotransport of K^+ and Na^+ with Cl^- (Fig.7). The energy for this transport system may reside in the Na^+ gradient (Walz, 1986). Accumulated Cl^- is probably passively released.

Mature ovine oligodendrocytes cultured from white matter show lower potassium uptake rates (Table 2) and equilibrated levels than astrocytes (L. Hertz, S. Szuchet and E. Hertz, unpublished experiments). However, again the uptake is faster in cells at steady state than in potassium depleted cells (Table 2), suggesting that part of the uptake reflects diffusion into the cells along one or more types of channels. Electrophysiological studies have clearly shown that similar oligodendrocytes do express different potassium channels (Soliven et al., 1988), and both a high potassium conductance and active potassium uptake have been observed in cultured oligodendrocytes from the embryonic mouse spinal cord (Kettenmann, 1986). Moreover, oligodendrocytes cells form gap junctions with astrocytes (Mugnaini et al., 1987).

The equilibrated level of potassium is lower in primary cultures of neurons than in astrocytes (Fig.8). This probably indicates a lower intracellular potassium concentration, since there are no major differences between cultured neurons and astrocytes in water content per mg protein (H.S. White, S.Y., Chow, D.M. Woodbury and L. Hertz, unpublished). A somewhat lower membrane potential in cultured neurons than in astrocytes (at 5.4 mM K^+) is consistent with this conclusion but could, of course, also reflect a less selective potassium permeability.

In contrast to astrocytes, there is little, if any, difference between potassium uptake rates in neurons at steady state and in potassium-depleted cells (Hertz, 1986a). The uptake rates are distinctly lower than in astrocytes (Table 2). About one-half of the potassium uptake into neurons in primary cultures is inhibited by ouabain, and the potency of ouabain is the same in neurons and in astrocytes (Hertz, 1986a). Furosemide has no effect on the uptake, even at elevated potassium concentrations, and an increase of the potassium concentration above 5 mM does not enhance ouabain-sensitive potassium uptake (L. Hertz, S.Y. Chow, H.S. White and D.M. Woodbury, unpublished results). This is in agreement with a lack of stimulation of neuronal Na^+,K^+-ATPase activity by elevated potassium concentration (Grisar et al., 1979; Franck et al., 1983).

Kinetic evidence is found that there may be a direct transfer of potassium from astrocytes to neurons. Latzkovits et al. (1974) concluded that in mixed neuronal-astrocytic chick cultures there is, besides uptake into either cell type from the medium, also a transfer of potassium between astrocytes and neurons. This conclusion was based upon the finding of an exponential uptake curve for ^{42}K in astrocytic cultures versus a complex uptake curve in mixed neuronal-astrocytic cultures and does not allow any distinction between transport from neurons to astrocytes and from astrocytes to neurons. More recently, the complex K^+ uptake in mixed cultures has been confirmed in mouse cultures (L. Latzkovits, W. Walz and L. Hertz, unpublished experiments), but in these cells there also seemed to be a transport between individual cells in pure astrocytic cultures.

Cellularly intact preparations

Coles and Tsacopoulos (1981) have elegantly shown that physiological stimulation of the intact honey bee retina by light leads to a decrease of K^+ activity in neurons and an increase in K^+ activity not only in the extracellular space but also in glial cells (Fig. 11). Both passive and active mechanisms seem to be involved (Coles et al., 1986). Similarly, Schlue and Wuttke (1983) have demonstrated an enhanced K^+ activity in leech neuropile glial cells exposed to elevated concentrations of potassium, and Kettenmann (1986) has observed a similar increase in oligodendrocytes in explant cultures. Grafe and Ballanyi (1987) concluded that both active uptake and Donnan forces contribute to stimulus-induced K^+ uptake into glial cells in rat olfactory brain cortex slices. They attempted to estimate the relative role of the active uptake of potassium in slices exposed to barium. However, the barium concentration used (0.5 mM) may have affected both active uptake and membrane permeability (see Walz et al., 1984). Moreover, K^+ uptake into brain slices is relatively slow (Hertz and Franck, 1978).

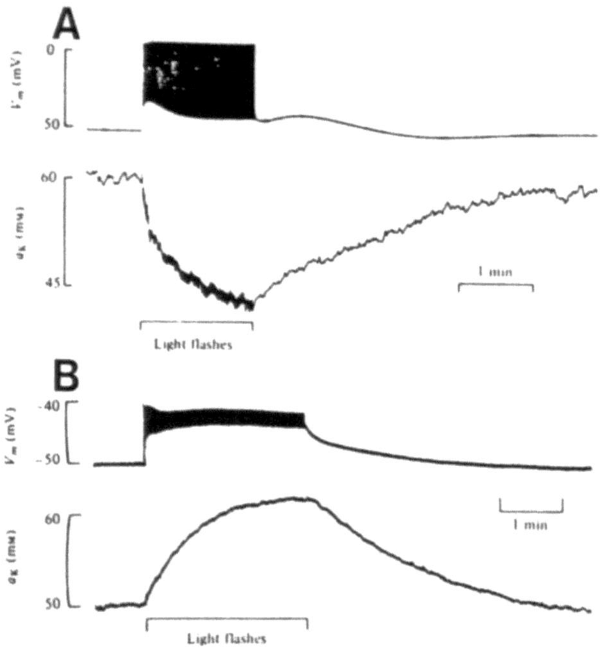

Fig. 11. Changes in intracellular potassium activity in a photoreceptor and a glial cell in the drone retina during and after a train of light flashes. Recordings by means of double-barrelled potassium selective microelectrodes. (A) Receptor potential (upper trace) and potassium signal (lower trace) from a photoreceptor. (B) Membrane potential (upper trace; the electrical responses to the individual flashes appear fused) and potassium signal (lower trace) from a glial cell (from Coles and Tsacopoulos, 1981).

CONCLUDING REMARKS.

This review has focused on neuronal-glial interactions involved in the regulation of, and the response(s) to, interstitial contents of a multitude of compounds, including noradrenaline, potassium ions, calcium ions, glutamate, glutamine, and maybe also pyruvate or lactate. Mutual interaction between neurons and glial cells, two autonomous cell types, may play a potentially very important role in CNS function under normal conditions, and aberrations in these interactions could contribute in an essential manner to metabolic and degenerative disease in the CNS. Several steps in such interactions may be regulated by enzyme, carrier or channel activity. This greatly enhances the responsiveness of the system by creating a multitude of "check-points," which each can be regulated separately. In a previous paper on the role of neuronal-glial interactions in potassium homeostasis (Hertz, 1965), it was suggested that K^+ ions, released from an active neuron, may be transported through active, enzyme-catalyzed transport mechanisms through glial cells to a previously resting neuron; it was indicated that "the flexibility of the proposed system within certain limits seems to represent its major advantage for transfer of impulses in the central nervous system. It provides a physiological basis of giving external stimuli an integrated response, which is dependent on heredity (anatomical arrangement), previous experience (enzyme induction), simultaneous or immediately preceding activity in neighbouring areas (local concentrations of potassium) and the total state of the individual (hormonal balance, drug effects)". In spite of an increasing amount of evidence for a role of astrocytes in cellular homeostasis of potassium during the intervening years the mechanisms for returning K^+ ions to neurons remain to be resolved. In this context it may also be important that both active uptake (see above) of potassium into astrocytes and potassium permeability (Fig.9) in astrocytes can be regulated by physiologically occurring processes. This may create conditions (e.g., low extracellular potassium concentration, protein kinase stimulation) under which neuronal uptake predominates. Elevated extracellular potassium concentrations were found to affect not only potassium uptake but also calcium uptake into astrocytes. In view of the widespread regulatory functions exerted by intracellular calcium, this may represent a very important aspect of neuronal-astrocytic interactions. Both calcium and potassium may play critical roles in information processing, and a possible role of a non-synaptic system in "higher mental activity" has been suggested by Galambos (1961) and by Schmitt (1984). It is in agreement with this concept that astrocytic potassium homeostasis mechanisms tend to increase in complexity in phylogenetically more highly developed species (Hertz and Nissen, 1976; Nilius and Reichenbach, 1988).

It has also become quite clear that interactions in ion homeostasis represent just one facet of regulable neuronal-glial interactions. Some aspects, like glutamate production, release and metabolism are directly related to neuronal excitation, whereas metabolic effects may be more indirectly involved in brain function. Hartman (this volume) has provided solid evidence for the exciting concept that an autonomic sympathetic nervous system, extending from locus coeruleus across the entire brain, controls various aspects of cerebral microvascular function, and that similar serotonergic and dopaminergic systems may also exist. He has pointed out that regulation of microvascular tone and permeability is only one part of autonomic functions in brain, and that other functions include establishment of "baseline tone" of various neuronal systems regulating and modulating such "autonomic" functions as temperature control, sleep-wake cycle, food intake and concentration. The findings reviewed in the present paper suggest that astrocytes may be critically involved in such actions and constitute a second target for autonomic regulation. Some astrocytic responses to β-adrenergic stimulation, like an increase in glycolysis, have already been demonstrated, suggesting yet another link between neuronal excitation and energy metabolism in astrocytes.

Aberrations in neuronal-astrocytic interactions with respect to K^+ and Ca^{2+} may play a major role in epilepsy and ischemic brain disease. It is of interest that Ca^{2+} blockers seem to be of therapeutic value in both conditions (for references, see Hertz et al., 1989), and evidence was presented that part of the effect of these drugs may be exerted on astrocytes. Aberrations in glutamate formation, release and metabolism are more likely to be related to neurodegenerative disease, involving excitotoxins. It was also shown that antidepressant drugs alter noradrenergic receptor function; moreover, lithium in a pharmacological concentration inhibits astrocytic K^+ uptake (Walz and Hertz, 1982). Thus, manic-depressive illness may well be a result of a cerebral "autonomic dysregulation", paralleling "psychosomatic" conditions caused by "autonomic dysregulation" of, e.g., the cardiovascular and gastrointestinal systems. Since both dopaminergic and serotonergic autonomic may co-exist with the noradrenaline system in brain, and astrocytes also have receptors for these transmitters, other psychiatric diseases might similarly involve astrocytic abnormalities. Therefore, not only the greatness and diversity of the human mind but also the devastating consequences of severe disease involving the CNS may to a large extent be based upon neuronal-astrocytic interactions.

REFERENCES

Abe T, Abe K, Rausch WD, Klatzo I, Spatz M (1980): Characteristics of some monoamine uptake systems in isolated cerebral capillaries. In: The cerebral microvasculature, Eisenberg HM, Suddith RL (eds.). Plenum Publishing Corp, 45-55.

Albrecht J, Hilgier W, Lazarewics JW, Rafalowska U, Wysmyk-Cybula U (1988): Astrocytes in acute hepatic encephalopathy: metabolic properties and transport functions. In: The Biochemical Pathology of Astrocytes, Norenberg MD, Hertz L, Schousboe A (eds.). Alan R. Liss Inc., New York, 465-476.

Barnes EM Jr, Mandel P (1981) Calcium transport by primary cultured neuronal and glial cells from chick embryo brain. J Neurochem, 36: 82-85.

Bender A, Hertz L (1985) Pharmacological evidence that the non-neuronal diazepam binding site in primary cultures of glial cells is associated with a calcium channel. Europ J Pharmacol, 110: 287-288.

Bender A, Hertz L (1986) Similarities of adenosine uptake systems in astrocytes and neurons in primary cultures. Neurochem Res, 11: 1507-1524.

Bender AS, Hertz L (1988) Astrocytic benzodiazepine receptors: Correlation with a calcium channel and involvement in anxiety and convulsions. In: Biochemical Pathology of Astrocytes, Norenberg MD, Hertz L, Schousboe A (eds.). Alan R. Liss Inc., New York, 599-610.

Bender A, Hertz L (1989) [^{3}H]Nitrendipine binding in neurons and astrocytes in primary cultures, submitted for publication, Brain Res.

Bender AS, White HS, Woodbury DM, Hertz L (1988) Potent inhibition of potassium permeability in astrocytes in primary cultures by phorbol esters, submitted for publication.

Benjamin AM, Quastel JH (1975) Metabolism of amino acids and ammonia in rat brain cortex slices in vitro: A possible role of ammonia in brain function. J Neurochem, 25: 197-206.

Black JA, Waxman SG (1988) The perinodal astrocyte. Glia, 1:169-183.

Bowman CL, Kimelberg HK (1984) Excitatory amino acids directly depolarize rat brain astrocytes in primary culture. Nature, 311: 656-659.

Carboni E, Wojcik WJ, Costa E (1985) Dihydropyridines change the uptake of calcium induced by depolarization into primary cultures of cerebellar granule cells. Neuropharmacol, 24: 1123-1126.

Chun LLY, Barres BA, Corey DP (1986) Inducation of calcium channel in astrocytes by cAMP. Abst Soc Neurosci, 12: 1346.

Coles JA, Tsacopoulos M (1981) Ionic and possible metabolic interactions between sensory neurons and glial cells in the retina of the honey bee drone. J Exp Biol, 95: 75-92.

Coles JA, Orkand RK, Yamate CL, Tsacopoulos M (1986) Free concentration of Na, K, and Cl in the retina of the honey bee drone: Stimulus-induced redistribution and homeostasis. Ann New York Acad Sci, 481: 303-317.

Eriksson LS, Law DH, Hagenfeldt L, Wahren J (1983) Nitrogen metabolism of the human brain. J Neurochem, 41: 1324-1328.

Finegan BA, Weiler R, Hertz L (1987) Effects of diazepam, ketamine, morphine and naloxone on the uptake of potassium into astrocytes. Br J Anaesth, 59: 655P-656P.

Franck G, Grisar T, Moonen G (1986): Mechanisms of K^+ regulation in brain passive or active process? In: Dynamic Properties of Glial Cells, Cellular and Molecular Aspects, Grisar T, Franck G, Hertz L, Norton WT, Sensenbrenner

M, Woodbury DM (eds.). Pergamon Press, 137-144.

Grisar T, Frere JM, Franck G (1979). Effects of K^+ ions on kinetic properties of the (Na^+,K^+)ATPase (EC 3.6.1.3) of bulk isolated glia cells, perikarya and synaptosomes from rabbit brain cortex. Brain Res, 165: 87-103.

Hansen AJ (1985) Effect of anoxia on ion distribution in the brain. Physiol. Rev., 65: 101-148.

Hansson E, Ronnback L, Sellstrom A (1984) Is there a "dopaminergic glial cell?" Neurochem Res, 9: 679-689.

Harris RJ, Symon L (1984) Extracellular pH, potassium and calcium activities in progressive ischaemia of rat cortex. J Cerebral Blood Flow Met, 4: 178-186.

Harris RJ, Symon L, Branston NM, Bayhan M (1981) Changes in extracellular calcium activity in cerebral ischaemia. J Cerebral Blood Flow Met, 1: 203-209.

Henn FA, Hamberger A (1971) Glial cell function: Uptake of transmitter substances. Proc Nat Acad Sci USA, 68: 2686-2690.

Hertz L (1965) Possible role of neuroglia: A potassium-mediated neuronal-neuroglial-neuronal impulse transmission system. Nature, 206: 1091-1094.

Hertz L (1966) Neuroglial localization of potassium and sodium effects on respiration in brain. J Neurochem, 13: 1373-1387.

Hertz L (1979) Functional interactions between neurons and astrocytes. I. Turnover and metabolism of putative amino acid transmitters. Progr Neurobiol, 13: 277-323.

Hertz L (1982) Astrocytes. In: Handbook of Neurochemistry, 2nd Ed, Vol 1, Lajtha A (ed.). Plenum Press, New York, 319-355.

Hertz L (1986a) Potassium transport in astrocytes and neurons in primary cultures. Ann NY Acad Sci, 481: 318-333.

Hertz L (1986b) Potassium as a signal in metabolic interactions between neurons and astrocytes. In: Dynamic Properties of Glial Cells, Cellular and Molecular Aspects, Grisar T, Franck G, Hertz L, Norton WT, Sensenbrenner M, Woodbury DM (eds.). Pergamon Press, 215-224.

Hertz L, Bender AS (1988): Astrocytic benzodiazepine receptors: Their possible role in regulation of brain excitability. In: Glial Cell Receptors, Kimelberg HK (ed.). Raven Press, 159-181.

Hertz L, Franck G (1978) Effect of increased potassium concentrations on potassium fluxes in brain slices and in glial cells. In: Dynamic Properties of Glial Cells, Schoffeniels E, Franck G, Hertz L, Tower DB (eds.). Pergamon Press, Oxford, 383-388.

Hertz L, Nissen C (1976) Differences between leech and mammalian nervous sytems in metabolic reactions as an indication of differences in potassium homeostasis mechanism, Brain Res. 110: 182-188.

Hertz L, Richardson JS (1983): Acute and chronic effects of antidepressant drugs on β-adrenergic function in astrocytes in primary cultures - an indication of glial involvement in affective disorders? J Neurosci Res, 9: 173-183.

Hertz L, Schousboe A (1986): Role of astrocytes in compartmentation of amino acid and energy metabolism. In: Astrocytes, Fedoroff S, Vernadakis A (eds.). Academic Press, New York, 2: 179-208.

Hertz L, Schousboe A (1987) Primary cultures of GABAergic and glutamatergic neurons as model systems to study neurotransmitter functions. I. Differentiated cells. In: Model Systems of Development and Aging of the Nervous System, Vernadakis A, Privat A, Lauder JM, Timiras PS, Giacobini E (eds.). Martinus Nijhoff Publishing, Boston, 19-31.

Hertz L, Schousboe A (1988) Metabolism of glutamate and glutamine in neurons and astrocytes in primary cultures. In: Glutamine and Glutamate in Mammals, Kvamme E (ed.). CRC Press, Boca Raton, FL., 2: 39-55.

Hertz L, Baldwin F, Schousboe A (1979) Serotonin receptors on astrocytes in primary cultures: Effects of methysergide and fluoxetine. Can J Physiol Pharmacol, 57: 223-226.

Hertz L, Kvamme E, McGeer EG, Schousboe A (eds.)(1983a) Glutamate, Glutamine and GABA in the Central Nervous System. Alan R. Liss, New York.

Hertz L, Yu ACH, Potter RL, Fisher TE, Schousboe A (1983b) Metabolic fluxes from glutamate and towards glutamate in neurons and astrocytes in primary cultures. In: Glutamine, Glutamate and GABA in the Central Nervous System, Hertz L, Kvamme E, McGeer EG, Schousboe A (eds.). Alan R. Liss Inc., New York, 327-342.

Hertz L, Juurlink BHJ, Szuchet S (1985) Cell cultures. In: Handbook of Neurochemistry, Lajtha A (ed.). Plenum Press, New York, Vol 8, 603-661.

Hertz L, Drejer J, Schousboe A (1988a) Energy metabolism in glutamatergic neurons, GABAergic neurons and astrocytes in primary cultures. Neurochem Res, 13: 605-610.

Hertz L, Murthy ChRK, Schousboe A (1988b) Metabolism of glutamate and related amino acids. In: Biochemical Pathology of Astrocytes, Norenberg MD, Hertz L, Schousboe A (eds.). Alan R. Liss, New York, 395-406.

Hertz L, Bender AS, Woodbury D, White HS (1989) Potassium induced calcium uptake in astrocytes and its potent inhibition by a calcium channel blocker. J Neurosci Res, in press.

Hille B (1984) Ionic channels of excitable membranes, Sunderland: Sinauer Assoc. J.

Hosli E, Hosli L (1976) Autoradiographic studies on the uptake of ^{3}H-noradrenaline and ^{3}H-GABA in cultured rat cerebellum, Exp Brain 26: 319-324.

Hosli E, Schousboe A, Hosli L (1986) Amino acid uptake. In: Astrocytes, Fedoroff S, Vernadakis A (eds.). Academic Press, New York, 2: 133-153.

Janis RA, Silver PJ, Triggle DJ (1987) Drug action and cellular calcium function. Adv Drug Res, 16: 309-591.

Joo F (1983) The blood-brain barrier in vitro: Ten years of research on microvessels isolated from the brain. Neurochem, 7: 1-25.

Juurlink BHJ, Hertz L (1985) Plasticity of astrocytes in primary cultures: An experimental tool and a reason for methodological caution. Dev Neurosci, 7: 263-277.

Kaczmarek LK (1987) The role of protein kinase C in the regulation of ion channels and neurotransmitter release. Trends Neuro Sci, 10: 30-34.

Karwoski CJ, Newman EA, Shimazaki H, Proenza LM (1985) Light-evoked increases in extracellular K^+ in the plexiform layers of amphibian retina. J Gen Physiol, 86: 189-213.

Kettenmann H (1986) Oligodenrocytes control extracellular potassium by active uptake and spatial buffering. In Dynamic Properties of Glia Cells II - Cellular and Molecular Aspects, Grisar T, Franck G, Hertz L, Norton WT, Sensenbrenner M, Woodbury DM (eds), Pergamon Press, Oxford 155-163.

Kettenmann H, Backus KH, Schachner M (1988) GABA receptors on cultured astrocytes. In: Glial Cell Receptors, Kimelberg HK (ed.). Raven Press Ltd., New York, 95-106.

Kuffler SW (1967) Neuroglial cells: Physiological properties and a potassium mediated effect of neuronal activity on the glial membrane potential. Proc R Soc Series B, 168: 1-21.

Kimelberg HK (1986) Catecholamine and serotonin uptake in astrocytes. In: Astrocytes: Biochemistry, Physiology and Pharmacology of Astrocytes, Fedoroff S, Vernadakis A (eds.). Academic Press, Orlando, Fl, 2: 107-131.

Larrabee MG (1983) Lactate uptake and release in the presence of glucose by sympathetic ganglia of chicken embryos and by neuronal and nonneuronal cultures prepared from these ganglia. J Neurochem, 40: 1237-1250.

Latzkovits L, Sensenbrenner M, Mandel P (1974) Tracer kinetic model analysis of potassium uptake by dissociated cell culture: glial-neuronal interrelationship. J Neurochem, 23: 193-200.

Lazarewicz JW, Kanje M, Sellstrom A, Hamberger A (1977) Calcium fluxes in cultured and bulk isolated neuronal and glial cells. J Neurochem, 29: 495-502.

Litzinger MJ, Foster G, Brenneman DE (1986) [^{3}H]-Nitrendipine binding in non-neuronal cell cultures. Biochem Biophys Res Comm, 136: 783-788.

MacVicar BA (1984) Voltage-dependent calcium channels in glial cells. Science, 226: 1345-1347.

MacVicar BA, Chrichtow SA, Burnard DM, Tse F (1987) Membrane conduction oscillations in astrocytes induced by phorbol esters. Nature, 329: 242-243.

Martin DL, Shain W, Madelain V (1988) Receptor-mediated release of taurine from glial cells and signaling between neurons and glia. In: Glial Cell Receptors, Kimelberg HK (ed.). Raven Press, New York, 183-195.

Matz H, Hertz L (1988) Adenosine metabolism in neurons and astrocytes in primary cultures, J. Neurosci, in press.

McCarthy KD, Salm A, Lerea LS (1988) Astroglial receptors and their regulation of intermediate filament protein phosphorylation. In: Glial Cell Receptors, Kimelberg HK (ed.). Raven Press, New York, 1-22.

Miller RJ (1987) Multiple calcium channels and neuronal function. Science, 235: 46-52.

Miller RJ, Freedman SB (1984) Are dihydropyridine binding sites voltage sensitive calcium channels? Life Sci, 34: 1205-1221.

Mobley PL, Scott SL, Cruz EG (1986) Protein kinase C in astrocytes: a determinant of cell morphology. Brain Res, 398: 366-369.

Mugnaini E, Massa PT, Szuchet S (1987) Specialized junctions of glial cells in situ and in vitro. In: Model Systems of Development and Aging of the Nervous System, Vernadakis A, Privat, R, Lauder JM, Timiras PS, Giacobini E (eds.). Martinus Nijhoff, Boston, 139-170.

Murphy S, Pearce BR (1987) Functional receptors for neurotransmitters on astroglial cells - a commentary. Neuroscience, 22: 381-394.

Neary JT, Norenberg LOB, Norenberg MD (1986) Calcium-activated, phospholipid- dependent protein kinase and protein substrates in primary cultures of astrocytes. Brain Res, 385: 420-424.

Newman EA (1985) Voltage-dependent calcium and potassium channels in retinal glial cells. Nature, 317: 809-811.

Nilius B, Reichenbach A (1988) Efficient K^+ buffering by mammalian retinal glial cells is due to cooperation of specialized ion channels. Eur J Physiol, 411: 654-660.

Norenberg MD, Martinez-Hernandez A (1979) Fine structural localization of glutamine synthetase in astrocytes of rat brain. Brain Res, 161: 303-310.

Opler LA, Makman MH (1972) Mediation by cyclic AMP of hormone-stimulated glycogenolysis in cultured rat astrocytoma cells. Biochem Biophys Res Comm, 46: 1140-1145.

Orkand RK, Nicholls JG, Kuffler SW (1966) Effect of nerve impulses on the membrane potential of glial cells in the central nervous system of amphibia. J Neurophysiol, 29: 788-806.

Ossola L, DeFeudis FV, Mandel P (1980) Lack of Na^+-independent binding of [^{3}H-GABA or [^{3}H-muscimol to particulate fractions of cultured astroblasts. J Neurochem, 34: 1026-1029.

Palaiologos G, Hertz L, Schousboe A (1988a) Evidence that aspartate aminotransferase activity and ketodicarboxylate carrier function are essential for biosynthesis of transmitter glutamate. J Neurochem, 51:

317-320.

Palaiologos G, Hertz L, Schousboe A (1986b) Role of aspartate aminotransferase and mitochondrial dicarboxylate transport for release of endogenously and exogenously supplied neurotransmitter in glutamatergic neurons, Neurochem Res, submitted for publication.

Palay SL, Chan-Palay V (1977) Cerebellar Cortex: Cytology and Organization, Springer, Berlin.

Paterson IA, Hertz L (1989) Effects of monoamines on the uptake of [^{3}H]-noradrenaline by rodent astrocytes in primary cultures. J Neurocsci Res, in press.

Paulson, OB, Newman EA (1987): Does the release of potassium from astrocytic endfeet regulate cerebral blood flow. Science 237, 896-898.

Phillis JW, Wu PH (1981): The role of adenosine and its nucleotides in central synaptic transmission. Prog Neurobiol, 16: 187-239.

Pope A (1978): Neuroglia: Quantitative aspects. In: Dynamic Properties of Glia Cells, Schoffeniels E, Franck G, Hertz L, Tower DB (eds.). Pergamon, Oxford, 13-20.

Quandt FN, MacVicar BA (1986): Calcium activated potassium channels in cultured astrocytes. Neurosci, 19: 29-41.

Ransom BR (1988) The role of astrocytes in MPTP-induced Parkinsonism. In: The Biochemical Pathology of Astrocytes, Norenberg MD, Hertz L, Schousboe A (eds.). Alan R. Liss, NY, 203-213.

Reichenbach A, Dettmer D, Bruckner G, Neumann M, Birkenmeier G (1985) Morphological variability, lectin binding and Na^{+}, K^{+} activated adenosine triphosphatase activity of isolated Muller (glial) cells from the rabbit retina shows a K^{+} dependence similar to that of brain astrocytes. Neurosci Lett, 59: 281-284.

Reichenbach A, Dettmer D, Reichelt W, Eberhan W (1987) High Na^{+} affinity of the Na^{+} K^{+} pump in isolated rabbit retinal matter (glial) cells. Neurosci Lett, 75: 157-162.

Schlue WR, Wuttke, W (1983) Potassium activity in leech neuropil cell changes with external potassium concentration, Brain Res 270, 368-372.

Schousboe A (1980) Primary cultures of astrocytes from mammalian brain as a tool in neurochemical research. Cell Mol Biol, 26: 505-513.

Schousboe A, Larsson OM, Krogsgaard-Larsen P, Drejer J, Hertz L (1988a) Uptake and release processes for neurotransmitter amino acids in astrocytes. In: Biochemical Pathology of Astrocytes, Norenberg MD, Hertz L, Schousboe A (eds.). Alan R. Liss, New York, 381-394.

Schousboe A, Drejer J, Hertz L (1988b) Uptake and release of glutamate and glutamine in neurons and astrocytes in primary cultures. In: Glutamine and Glutamate in Mammals, Kvamme E (ed.). CRC Press, Boca Raton, Fl, 2: 21-38.

Semenoff D, Kimelberg HK (1985) Autoradiography of high affinity uptake of catecholamines by primary astrocyte cultures. Brain Res, 348: 125-136.

Shank RP, Aprison MH (1988) Glutamate as a neurotransmitter. In: Glutamine and Glutamate in Mammals, Kvamme E (ed.). CRC Press, Boca Raton, Fl, 2: 3-19.

Shank RP, Campbell GLeM (1984) α-Ketoglutarate and malate uptake and metabolism by synaptosomes: Further evidence for an astrocyte to neuron metabolic shuttle. J Neurochem, 42: 1153-1161.

Shank RP, Bennett GS, Freytag SD, Campbell GL (1985) Pyruvate carboxylase: an astrocyte-specific enzyme implicated in the replenishment of amino acid neurotransmitter pools. Brain Res, 329: 364-367.

Schmitt FO (1984) Molecular regulators of brain function: a new view, Neurosci, 13: 991-999.

Skattebol A, Triggle DJ (1987) $^{45}Ca^{2+}$ uptake in rat brain neurons: absence of sensitivity to the Ca^{2+} channel ligands nitrendipine and Bay K 8644. Can J Physiol Pharmacol 65: 344-347.

Soliven B, Szuchet S, Arnason BGW, Nelson DJ (1988) Voltage-gated potassium currents in cultured ovine oligodendrocytes. Neurosci, 8: 2131-2141.

Somjen GG (1979) Extracellular potassium in the mammalian central nervous system. Ann Rev Physiol, 41: 159-177.

Tildon JT (1983) Glutamine: A possible energy source for the brain. In: Glutamine, Glutamate and GABA in the Central Nervous System, Hertz L, Kvamme E, McGeer EG, Schousboe A (eds.). Alan R. Liss, New York, 415-429.

Turner TJ, Goldin SM (1988) Do dihydropiridine-sensitive calcium channels play a role in neurosecretion in the central nervous system? Ann NY Acad Sci, 522: 278-283.

Van den Berg CJ, Garfinkel D (1971) A simulation study of brain compartments. Metabolism of glutamate and related substances in mouse brain. Biochem J, 123, 211-218.

Van Neuten JM, Janssens WJ, Vanhoutte P, (1988) Calcium antagonism and vascular smooth muscle. Ann NY Acad Sci, 522: 234-247.

Vernadakis A (1974) Neurotransmission: a proposed mechanism of steroid hormones in the regulation of brain function. In: Psychoneuroendocrinology, Hatotani N (ed.). Workshop conference of the International Society for Psychoneuroendocrinology, S. Karger AG, Basel, 251-258.

Walz W (1986) Potassium channels and carriers in glial cell membranes. In: Dynamic Properties of Glial Cells, Cellular and Molecular Aspects, Grisar T, Franck G, Hertz L, Norton WT, Sensenbrenner M, Woodbury DM (eds.). Pergamon Press, 145-154.

Walz W, Hertz L (1982) Acute and chronic effects of lithium in therapeutically relevant concentrations on potassium uptake into astrocytes. Psychopharmacology 78: 309-313.

Walz W, Hertz L (1983) Functional interactions between neurons and astrocytes II. Potassium homeostasis at the cellular level. Progr. Neurobiol. 20: 133-183.

Walz W, Hertz L (1984) Intense furosemide-sensitive potassium accumulation into astrocytes in the presence of pathologically high extracellular potassium levels. J Cerebr Blood Flow and Metab 4: 301-304.

Walz W, Kimelberg HK (1985) Differences in cation transport properties of primary astrocyte cultures from mouse and rat brain. Brain Res, 340: 333-340.

Walz W, MacVicar BA (1988) Electrophysiological properties of glial cells: comparison of brain slices with primary cultures. Brain Res, 443: 321-324.

Walz W, Mukerji S (1988) Lactate production and release in cultured astrocytes. Neurosci Lett, 86: 296-300.

Walz W, Wilson DC (1986) Calcium entry into cultured mouse astrocytes. Neurosci Lett, 67: 301-306.

Walz W, Shargool M, Hertz L (1984) Barium-induced inhibition of K^+ transport mechanisms in cortical astrocytes - its possible contribution to the large Ba^{2+}-evoked extracellular K^+ signal in brain. Neurosci, 13: 945-969.

Wenthold RJ, Altschuler RA (1983) Immunocytochemistry of aspartate aminotransferase and glutaminase. In: Glutamine, Glutamate and GABA in the Central Nervous System, Hertz L, Kvamme E, McGeer EG, Schousboe A (eds.). Neurology and Neurobiology, Alan R Liss, NY, 7: 33-50.

Wenthold RJ, Skaggs KK, Altschuler RA (1986): Immunocytochemical localization of aspartate aminotransferase and glutaminase immunoreactivities in the cerebellum. Brain Res, 363: 371-375.

Whitaker-Azmitia PM (1988) Astroglial serotonin receptors. In: Glial Cell Receptors, Kimelberg, HK (ed) Raven Press, NY, 107-120.

White HS, Bender A, Chow SY, Woodbury D and Hertz L (1988) Effect of anticonvulsant drugs and pentylenetetrazol (PTZ) on potassium (K^+) regulation of cultured asrocytes. In: The Biochemical Pathology of Astrocytes, Norenberg MD, Hertz L, Schousboe A (eds) Alan R. Liss, NY, 537-539.

Wolff JR (1970): Quantitative aspects of astroglia. VIth Intern Congr Neuropathology (Paris), 352: 327-333.

Yu ACH, Hertz L (1983): Metabolic sources of energy in astrocytes. In: Glutamine, Glutamate and GABA in the Central Nervous System, Hertz L, Kvamme E, McGeer EG, Schousboe A (eds.). Alan R. Liss, NY, 431-439.

Yu ACH, Drejer J, Hertz L, Schousboe A (1983): Pyruvate carboxylase activity in primary cultures of astrocytes and neurons. J Neurochem, 41: 1484-1487.

Yu PH, Hertz L (1982): Differential expression of type A and type B monoamine oxidase of mouse astrocytes in primary cultures. J Neurochem, 39: 1493-1495.

Yudkoff M, Nissim I, Hummeler K, Medow M, Pleasure D (1986): Utilization of (^{15}N)-glutamate by cultured astrocytes. Biochem J, 234: 185-192.

DISCUSSION

Chaired by Joo F. and Carruba M.O.

R. Pumain:

What was the K^+ concentration that you have used to induce Ca2+ uptake.

L. Hertz:

50 mM K^+. It has been observed that the decrease in extracellular Ca2+ in vivo occurs at a K+ concentration around 20 mM; the question is: does this coincide with Ca uptake into the neurons or also into the astrocytes?

S. Govoni:

You have shown the existence of ^{3}H-Nitrendipine binding sites at the level of astrocytes. The high affinity component appears only when the cells are incubated with dibutyrryl cyclic AMP. Is there any physiological conditon in which you expect the astrocytes to show the high affinity component?

L. Hertz:

It has been greatly disputed whether the high affinty component is an artifact. Several authors have demonstrated K+ sensitive Ca channels in cultures of astrocytes, always in the presence of dibutyrryl cAMP.
In brain slices of hippocampus not treated with this substances, it has not been possible to detect K+ sensitive Ca channels.
The reason of this is not clear. We have approached this problem by using binding studies and the results suggest the existence of two components.

R. Pumain:

Did you measure Ca^{2+} uptake in glial cells not treated with dibutyrryl cAMP?

L. Hertz:

Yes. We have observed that with low K^+ concentration the presence of dibutyrryl cAMP does not influence Ca2+ uptake, whereas with higher K+ levels the presence of the compound is essential to measure this parameter.

ALTERATION OF THE BLOOD-BRAIN-BARRIER BY COMPONENTS OF THE IMMUNE SYSTEM

Ute Traugott, M.D.

Department of Pathology (Neuropathology) and Neurology
and the Rose F. Kennedy Center for Research in Mental Retardation and Human Development

Albert Einstein College of Medicine

The Bronx, New York, U.S.A.

ABSTRACT

Under normal conditions, blood is separated from CNS parenchyma by a barrier made up of a layer of endothelial cells which are joined by continuous tight junctions and possess only a few vesicles. Solutes can cross the barrier by ultrafiltration, facilitated diffusion or active and sometimes highly-selective energy-consuming transport mechanisms. Normal CNS endothelial cells are Class I MHC-positive but do not express Class

---

Key Words: Blood-brain barrier (BBB); endothelial cells; astrocytes; Class I and Class II major histocompatibility complex (MHC) antigen; interferon (IFN)-gamma; activated T cells.

NATO ASI Series, Vol. H 33
Regulatory Mechanisms of Neuron to Vessel Communication in the Brain. Edited by F. Battaini et al.

II MHC. They express Fc receptors, transferrin receptors and complement receptors and various adhesion molecules. In inflammatory diseases of the CNS, such as in autoimmune demyelination, interferon gamma induces Class II MHC expression on endothelium and enhances the density of adhesion molecules. CNS endothelial cells thus acquire the ability to present antigen locally to T cells. Cytokines released by activated T cells facilitate lymphocyte-endothelial cell interactions and increase the permeability of the BBB for plasma proteins and hematogenous cells, leading to vasogenic edema and infiltration of the CNS parenchyma by inflammatory cells. In the CNS, the disease process is propagated locally by Ia+ astrocytes. Methods interfering with the homing mechanism of lymphocytes to the CNS, such as blocking of adhesion molecules by monoclonal antibodies might prove useful for treatment of immune-mediated CNS diseases such as MS.

HISTORIC OVERVIEW

The concept of a barrier between blood and central nervous system (CNS) dates back to 1898, when Briedl and Kraus observed that brain and spinal cord tissue remained unstained by bile acids in icteric patients (for detailed Review see Bradbury, 1979). In 1902, Ehrlich and co-worker noticed that CNS and fat tissue had similar staining properties and attributed this to the presence of cholesterol and lecithine in both organs. That CNS tissue was stained by a few basic but not by acid dyes and by those containing sulphate groups was interpreted as reflecting their special affinity to CNS tissue. Based upon the lack of specific morphologic features at the light microscopic level, Ehrlich rejected the idea that CNS endothelium might have selective properties. However, the presence of a barrier which was either related to CNS endothelial cells or directly to neurons and glial cells was postulated by other investigators from similar experiments. From extensive studies on the tissue distribution of the vital stain trypan blue, Goldmann, one of Ehrlich's associates, concluded that the choroid plexus and the placenta protect CNS tissue and the fetus, respectively from blood. That trypan blue injected directly into the subarachnoid space resulted in seizures and staining of superficial parts of the CNS whereas there was no staining when given systemically lead Lewandowsky to postulate a barrier between the CNS parenchyma and blood. In 1933, presence of a blood-brain-barrier (BBB) and a blood-cerebrospinal fluid (CSF) barrier was clearly established by Walter and Spatz and these were located to the inner lining of cerebral blood vessels and to vessels of the choroid plexus and leptomeninges, respectively. The importance of capillaries for the supply of nutrients and ion exchange and a selective function of the barrier were clearly recognized. In 1945 Krogh defined the basic characteristics of the BBB as they are accepted today. He demonstrated that ions and non-electrocytes of low molecular weight penetrate the BBB slowly; that lipid soluble substances easily pass through the BBB and that concentrations of certain organic compounds such as glucose and urea are kept at constant levels in the CSF even in the presence of significantly increased blood values. Moreover, Krogh documented an extracellular space of 20%-30% and a similar composition of CSF and interstitial fluid of the brain. In 1965, Davson and Oldendorf suggested that astrocytic foot processes induce the features special for CNS endothelium, a conclusion which was confirmed recently by tissue culture studies (Tao-Cheng et al., 1987).

Demonstration of a morphologic structure which could account for the observed physiologic barrier function of the cerebral vasculature was impossible at the light microscopic level and depended upon the introduction of the electron microscope. In early electron microscopic studies, CNS endothelial cells were found to lack intracellular or intercellular fenestrations, to be completely surrounded by a basement membrane and to be covered on the abluminal side by glial processes. Cerebral capillaries were thus classified as type A, I, beta (Bradbury, 1979). While physiologic data had provided convincing evidence for the presence of an extracellular space in the CNS, such a space could not be documented by light or early electron microscopic studies. Based upon this, the concept of an "illusion of a barrier" was advocated in the early sixties which suggested that neurons and glial cells themselves exerted a barrier function. After improving fixation procedures, Van Harreveld et al. succeeded in 1965 to document appreciable extracellular space in CNS tissue which disappeared after asphyxia (Bradbury, 1979). A detailed electron-microscopic study describes CNS endothelium as 100-150 nm thick cells with elongated nuclei the cytoplasm of which is rich in filaments, ribosomes and mitochondria but contains only a few pinocytotic vesicles. Where endothelial cells overlapp, they form a 10-15 nm wide cleft which is interrupted by tight junctions (zona occludens) forming a continuous seal around endothelial cells. The introduction of horse-radish peroxidase (HRP; MW:40,000) as electron-dense marker by Graham and Karnovsky in 1966 facilitated studies on the BBB at the electron microscope level. After intraveneous administration, HRP was found in the lumen of CNS capillaries. From there it penetrated into the intercellular cleft where it was stopped by tight junctions. In contrast, in heart and muscle where endothelial cells are not connected by tight junctions, HRP was demonstrable in vesicles of endothelial cells and passed readily through intercellular clefts. After cisternal application of HRP, it readily penetrated between ependyma, pial cells, glial cells, and the basement membrane and was prevented from entering the capillary from the abluminal side by interendothelial tight junctions. In addition, tight junctions were also observed between epithelial cells of the choroid plexus and between ependymal cells overlying the median eminence and the area postrema. Astrocytes and ependymal cells were joined by 2-3 nm wide gap junctions which permit free penetration of HRP throughout the extracellular space of the brain. These results clearly demonstrated that in most parts of the brain, CNS parenchyma and CSF are separated from blood by a barrier of continuous tight junctions on capillary endothelium and on epithelium of the choroid plexus and the outer layer of the arachnoid, respectively. By both physiologic and morphologic criteria, the resistance of the BBB is higher than that of the blood-CSF barrier. This is also reflected by the large number and complexity of strands forming tight junctions between endothelial cells im comparison to their relatively simple arrangement between ependyma cells as shown by freeze fracture studies (Connell and Mercer, 1974, see Bradbury, 1979).

PHYSIOLOGY OF THE BLOOD-BRAIN-BARRIER

Continuous tight junctions between CNS endothelial cells, together with the lack of pinocytotic vesicles, provide an effective barrier which separates blood from CNS parenchyma. This allows a constant internal milieu to be maintained within the CNS which is essential for the proper functioning of neurons. Short-term local fluctuations in concentrations of electrolytes and neurotransmitters within the CNS are regulated by glial cells. This epithelial-like barrier function of the CNS endothelium necessitates the presence of transport mechanisms for nutrients and metabolic by-products (Crone, 1986). The BBB can be crossed by ultrafiltration of macromolecules via pinocytotic vesicles and channels; by facilitated diffusion or by active transport mechanisms some of which are highly specific (Bradbury, 1979; Essner 1987; Cunningham, 1986). Although radioactively-labelled test substances have been observed in higher concentrations in grey than in white matter, this does not reflect a qualitative difference of the BBB but is due to the greater capillary area in grey matter. The BBB is completely absent in the neurohypophysis, the median eminence of the hypothalamus (tuber cinereum), the area postrema, the paraphysis (intercolumnar tubercle, subfornical body), the pineal gland and the preoptic areas. Capillaries in these areas are fenestrated and the perivascular space is relatively large. These modified vessels might be important for the release of neuroendocrine products or might function as chemoreceptors. The vasculature of the retina and the optic nerve has BBB-like properties. There is no barrier in vessels of the extramedullary part of dorsal roots and their ganglia.

In general, entry of highly lipid-soluble substances into the CNS occurs rapidly and is mainly limited by blood flow. Penetration of lipid insoluble compounds is slower and depends upon their charge, size and shape, whereby permeability becomes more restricted with increase in molecular size. A variety of substances require active transport mechanisms similar to those described previously for erythrocyte membranes for crossing of the BBB (Bradbury, 1979).

Electrolytes and Ions: Concentrations of Ca^{++}, K^+ and Mg^{++} are kept constant in the CSF even in the presence of significantly altered blood levels. Maintaining homeostasis of these compounds is faciliated by the steep increase in efflux with increasing CSF concentrations and the insensitivity of the transport system to changes in plasma levels. Na^+ and K^+ are distributed asymetrically in CNS endothelium, in that Na^+ is predominantly located on its luminal and K^+ on its abluminal side

(Bradley, 1979). K+ transport is ATP-ase dependent and can be associated with secretion of interstitial fluid. Exchange rates between blood and CNS range from 2h for Na+ to 30-40h for K+ (Bradbury, 1979).

Monosaccharides cross via sodium- and insulin-independent facilitated diffusion. D-glucose is transported at a rate of 1-3 umol/min/g brain. Rapid adaption of the blood-flow rate can be observed during hypoglycemia. Fructose and L-glucose cross only to a limited extent (Bradbury, 1979; Cunningham, 1986).

Insulin is transported across the BBB via ATPase-dependent receptor-mediated endo-, trans- and exocytosis which shows self inhibition. Insulin receptors on CNS endothelium contain sialic acid residues and are thus distinct from those on neurons.

Amino Acids. Transport of amino acids can occur against a gradient from CSF to blood and is characterized by Na+ independency, stereo-specificity, saturation and competitive inhibition (Smith and Takasato, 1986). Under normal conditions, this transport mechanism is 96% saturated. It is specific for neutral, basic and dicarboxylic amino acids. Essential amino acids (phenylalanine, methionine, tryptophan, histidine) cross rapidly and inhibit strongly. Smaller neutral amino acids (threonine, serine, alanine and glycin) enter more slowly and inhibit only weakly. Self-inhibition and competition has not been documented for proline, alanine and glycine.

Oranigic Acids: Specific mechanisms are also involved in the transport of acetate, butyrate, lactate, pyruvate, keton bodies (beta-hydroxphenlyic acid and aceto-acetate), nucleic acid precursors and choline (Bradbury, 1979). Transport of fatty acids can be probenecid-sensitive (octanic acid, myristic acid) or probenecid insensitive (linoleic acid; Spector, 1988).

Specific carriers exist on the BBB for ribonucleosides (thiamine, uridine) and on the choroid plexus for deoxyribonucleosides (thymidine, deoxycytidin; Bradbury 1979; Spector, 1986).

Transport of neuropeptides occurs via ATP-ase dependent receptor-mediated endo- and exocytosis which is specific, sterospecific, saturable and can be mediated by amino acids (Bradbury, 1979; 1986; Banks and Kastin, 1987). Neuropeptides can exert their effect directly or by triggering a second messenger signal. Some neuropeptides can modulate BBB permeability (Partdridge, 1983; Young et al.; 1986). Substrate-limited transport has been reported for serotinin, dopamin and histamin.

Enzymes not detectable in significant amounts on endothelium outside the CNS are gamma glutamyl transpeptidase, DOPA decarboxylase, pseudocholinesterase and GABA-transaminase/succine semialdehyde dehydrogenase (Bradbury, 1979). Furthermore, aminopeptidase, peptidyl-dipeptidase, alkaline phosphatase and Na+K+ ATPase are present in CNS endothelium (Bradbury, 1979; Lather et al., 1987). Enzymatic degradation of proteins or peptides in CNS endothelial cells can act as "functional BBB" (Goldstein and Betz, 1986).

Receptors have been documented on CNS endothelium for insulin; insulin-like growth factor (ILGF)-I and ILGF II, for peptides, neurotransmitters, catecholamines, atriopeptin, prostaglandin, transferrin, the third complement component (CR3), the Fc fragment of immunoglobulin (Ig) and for lectins (Bradbury, 1979; Vorbrodt et al.; 1986, Joo, 1986; Goldberg and Betz, 1986; Steado and Nathanson, 1987).

DAMAGE OF THE BLOOD-BRAIN-BARRIER

The BBB can be damaged by increased tension of the vessel wall and by physical, chemical, infectious, allergic, neoplastic and radiation insults (Bradbury, 1979). Osmotic opening of the BBB can be achieved by hypertonic solutions (Bradbury, 1979; Rapoport and Robinson, 1986). This can be due to increased pinocytotic activity, to the opening of tight junctions or to both and leads to vasogenic edema. Vasogenic edema is less common in grey than in white matter where it spreads rapidly. Edema can be reduced by decreasing intracappillary hydrostatic or by increasing colloid osmotic pressure.

In summary, the BBB plays an important role in protecting the brain from exogeneous and endogenous "toxins" and prevents release of neurotransmitters into the blood. It provides specific transport mechanisms for various nutrients and metabolic by-products and can adopt to the metabolic needs of the brain. The BBB allows maintainance of a separate internal milieu within the CNS with constant levels of Ca++, Mg++ and K+ and prevents fluctuation in their serum concentrations from interfering with normal membrane potentials and function of neurons.

THE BLOOD-BRAIN-BARRIER IN INFLAMMATION

In recent years, particular attention has focussed on the role of endothelial cells in inflammatory disease processes (Cotran, 1986). Even under normal conditions, lymphocytes constantly recirculate through most organs to exert immunologic surveillance and re-enter the blood in lymphoid tissue. For recirculation, close interaction between lymphocytes and endothelial cells is required. This involves various adhesion molecules such as leukocyte function associated antigen (LFA), interleukin (IL)-1, intercellular adhesion molecules (ICAM-1), Class I (HLA-ABC) and Class II (HLA-DR; Ia) major histocompatibility (MHC) antigens and the T cell epitopes CD2, CD4 and CD8 (Butcher, 1986). Expression of some of these molecules can be induced or enhanced by cytokines such as e.g. interferon (IFN)-gamma which is produced by activated T cells (Butcher, 1986; Doyle and Strominger, 1987). Moreover, receptors have been demonstrated on lymphocytes which recognize high endothelial venules (HEV) in certain organs. Specific receptors have been documented for high endothelial venules of lymph nodes, for gut associated lymphoid tissue and for vessels in synovia in rheumatoid arthritis (Butcher, 1986; Jalkanen et al., 1987) but not yet for the CNS.

For inflammation to develop, hematogeneous cells have to migrate in considerable numbers through vessels into the parenchyma. That endothelial cells are not only innocent bystanders in inflammation but can actively participate in a T cell-mediated immune response was documented by Burger et al. in 1981. He reported that activated T cells can induce expression of Class II MHC on endothelial cells which then become capable of presenting antigen to T cells. More detailed studies revealed IFN-gamma to be the mediator for this phenomenon (Pober et al., 1983). Even in the absence of specific antigen, Class II MHC was found to facilitate adhesion of T cells, CD4+ (helper/inducer) T cells in particular to and migration through endothelial cells (Masuyama et al., 1986). IFN-gamma-induced Ia expression on endothelium becomes detectable after about 24h, is dose dependent, persists as long as IFN-gamma is present and disappears after its removal with a half-life time of 30h (Male et al., 1987). Like most nucleated cells of the body, endothelial cells normally express Class I MHC (HLA-ABC). Its expression can be rapidly enhanced by IFN-alpha/beta as well as by IFN-gamma. Class I MHC peaks at 48h and decreases thereafter with a half-life time of 40h. mRNA for MHC molecules become detectable in endothelium about 12h prior to their surface expression (Male and Pryce, 1988). For T cell activation, specific antigen has to be recognized in combination with MHC molecules on the surface of antigen presenting cells (APC). In blood and non-CNS tissues, antigen is primarily presented by monocytes/macrophages. This involves uptake of the specific antigen by APC, intracellular processing and re-expression of antigen fragments in combination with MHC molecules on their surface. This antigen-MHC complex is then recognized by specific T cell receptors and interaction with the CD3 epitope leads to transduction of the activating signal. This is followed by expression of activation markers such as Ia antigen, transferrin receptors, and interleukin-2 (IL-2) receptors; by production of IL-2, endoglycosidase and interferons; by histamine and serotonin release from mast cells and by T cell proliferation. Macrophage-derived interleukin (IL)-1 plays also an important role in T cell activation. Overall, endothelial cells are more effective in inducing interleukin-2 (IL-2) production by T cells while macrophages are better antigen presentors for B cells (Male et al., 1986).

It has remained an enigma why inflammatory cells preferrentially accumulate in certain organs during various disease processes. Endothelial cells have been postulated to play a key role in this highly selective homing of leukocytes. This could be related to organ specific epitopes on endothelium which are recognized by certain leukocytes as described above for HEV in lymphoid organs. It could also involve local antigen presentation by endothelial cells or the tendency of sensitized lymphocytes to migrate predominantly through organs they had been activated in (Butcher, 1986). In most tissues, except the nervous system, endothelial cells are normally involved in pinocytotic transport of macromolecules and low numbers of leukocytes can leave the blood stream through intercellular clefts. In inflammation, interaction between leukocytes and endothelial cells is facilitated by an IFN-gamma-induced increase in adhesion molecules which enhances migration of leukocytes. Moreover, release of vasoactive amines from mast cells, triggered by activated T cells leads to an increase in the permeability of vessels for serum proteins and cells. In the parenchyma, the inflammatory response is enhanced and perpetuated by resident tissue macrophages which can function as antigen presenting cells. Immune-mediated tissue damage can be due to a direct attack by cytotoxic (CD8+) T cells; to antibody-mediated cellular cytotoxicity (ADCC) the effector cells of which are macrophages; or to antibody mediated complement-dependent cytolysis.

In contrast to other organs, the CNS is isolated from the circulation by a tight barrier which prevents recirculation of and immunologic surveillance by lymphocytes. Together with the lack of resident MHC-positive facultatively antigen presenting cells and the absence of MHC molecules from neurons and glial elements, this provides the basis for the CNS having being considered an immunologic priviledged site (Fontana and Grob, 1984; Wekerle et al., 1986). CNS endothelium shares many properties with endothelial cells in other organs. Normal CNS endothelial cells are Class I MHC-positive but do not express Class II MHC. Presence of Class I MHC on CNS endothelium and its known preferrential interaction with CD8+ (suppressor) T cells might be important for maintaining immunologic tolerance to auto-antigens such e.g. myelin basic protein (MBP) which has been demonstrated, although at low levels in the circulation (Wekerle, 1986). Similar to other organs, Ia expression can also be induced on CNS endothelium by IFN-gamma in a dose-related fashion. Ia+ CNS endothelial cells become capable of presenting antigen such as MBP to specifically-sensitized T cells (McCarron et al., 1987). By modulating their membrane properties or opening tight junctions, vasoactive amines increase the permeability of CNS endothelial cells for serum components. Endoglycosidase produced by activated T cells facilitates their migration even through normal endothelium (Wekerle, 1986; Joo, 1987).

Once activated T cells have crossed the BBB, they can interact with glial elements. Tissue culture studies have demonstrated that similar to endothelial cells, astrocytes can also be induced by IFN-gamma to express Class II MHC and can thus acquire the ability to function as APC (Fontana and Grob, 1984). Cultured Ia+ astrocytes were able to present MBP effectively to specifically-sensitized T cells or T cell lines (Fontana et al., 1983; Fontana and Grob, 1984). In one of the studies, killing of antigen presenting astrocytes by T cells of the helper phenotype was observed (Sun and Wekerle, 1984). The relevance of this phenomenon, however, for the pathogenesis of immune-mediated demyelinating diseases such as multiple sclerosis (MS) and its animal model experimental autoimmune encephalomyelitis (EAE) remains obscure since astrocyte hypertrophy and not destruction is a hallmark of these diseases (Raine, 1982). The relevance of these tissue culture findings on endothelial cells and astrocytes to events occurring in vivo will be discussed below for some inflammatory CNS diseases such as MS, EAE, subacute sclerosing panencephalitis (SSPE), human immuno-defficiency virus (HIV)-related myelopathy and lymphocytic choriomeningitis in mice.

MULTIPLE SCLEROSIS

Normal CNS endothelium expresses Class I but not Class II MHC. In CNS tissue from MS subjects, however, some Ia+ endothelial cells could be detected and these were randomly distributed throughout grey and white matter. Their overall frequency was somewhat higher in active than in inactive MS but showed no significant differences between grey and white matter or between lesions and uneffected white matter (Traugott et al., 1983 and 1985). Interestingly, CNS endothelium in MS was less consistently reactive for Class I MHC than in controls, particularly in the astrocytic scar tissue (Traugott et al., 1987). This suggested Class II MHC-dependent down regulation of Class I MHC expression. The random distribution of Ia+ endothelial cells throughout grey and white matter is in accord with the concept of induction of Class II MHC expression on endothelium by circulating activated T cells. Restriction of Ia antigen to endothelium indicated that the inducing signal was not transferred to the closely-related astrocytic foot processes. In addition to a decrease in CD8+ cells, activated T cells have, indeed, been demonstrated, although in low numbers in the blood of MS patients while they are absent from normal controls (Hafler et al., 1985). Presence of myelin antigens such as MBP and galactocerebroside (GC) on some endothelial cells further supported the possibility that specific antigen can be presented locally within the CNS in vivo. Facilitation of lymphocyte migration through Ia+ endothelial cells was indicated by the diffuse presence of low numbers of CD2+ (pan T cells), CD4+ T cells, CD8+ T cells and Ia+ cells (B cells, macrophages, activated T cells) in perivascular areas in grey matter and in uneffected white matter parenchyma (Traugott et al., 1983). Enhanced Ia expression on CNS endothelium and more pronounced migration of lymphocytes into white matter could have been the cause for the increase in the exacerbation rate following therapeutic application of IFN-gamma (Panitch et al., 1987) and intercurrent viral infections (Sibley et al., 1985). In accord with the known preferrential interaction between Class II MHC and helper/inducer T cells, CD4+ cells were found to predominate in highly active chronic MS lesions (Traugott et al., 1983). CD4+ cells were mainly seen in the normal-appearing white matter close to the lesion edge where IL-2R+ cells could also be detected, indicating activation of some of the CD4+ cells (Traugott et al., 1985). Ia+ astrocytes were detectable at the edge of active chronic lesions and in the adjacent normal-appearing white matter. The similar distribution pattern of activated CD4+ cells and Ia+ astrocytes in active chronic MS lesions indicated that Ia expression on astrocytes in vivo can also be induced by IFN-gamma (Traugott et al., 1985). This was recently confirmed by the demonstration of IFN-gamma on astroglia in active MS lesions (Traugott and Lebon, 1988). While Ia+ endothelial cells seem to be important for initiation of lesion development in MS, antigen presentation by Ia+ astrocytes contributes to continuous growth of lesions from the periphery.

EXPERIMENTAL AUTOIMMUNE ENCEPHALOMYELITIS (EAE)

Immunocytochemical studies on CNS tissue from animals with EAE provided results similar to those obtained from MS material. When the time course of immunopathologic events was studied in acute EAE in SJL/J mice, a few Ia+ endothelial cells were detected as early as 12h post-inoculation (PI) and their frequency remained low throughout the latent period. At 12h PI, low-grade infiltration of the CNS parenchyma by T cells and Ia+ cells was seen and this was associated by a mild degree of albumin and Ig deposits indicating some increase in the permeability of the BBB for serum proteins. At 10 days PI, 2 days before onset of clinical signs, Ia+ endothelial cells increased in numbers. This coincided with the appearance of dense cellular infiltrates in the CNS parenchyma and extensive perivascular albumin and Ig deposits (Oldstone and Dixon, 1968; Traugott et al., 1985; Juhler et al., 1986), indicating vasogenic edema. While in acute EAE in mice, cellular infiltration was not preceeded by extravasation of serum proteins this was the case in previous studies on EAE in Strain 13 guinea pigs (Traugott et al., 1982) and rabbits (Simmons et al., 1987). However, Ia+ endothelial cells were not only seen in the CNS but also in non-CNS tissue such as heart, liver and kidney. This random distribution of Ia+ endothelial cells throughout the body suggested non-specific induction of Ia expression by circulating factors. Since at 12h PI, activated T cells and IFN-gamma were most likely not present in sufficient amounts, Ia expression at this early timepoint might have been due to the effect of adjuvant as has been described previously for a streptococcal-induced arthritis model (Wilder et al., 1987). This was supported by similar findings in mice which had received complete Freund's adjuvant (CFA) only. As the immune response developed, activated T cells and IFN-gamma become available later during the latent period and might then have been the primary inducer for Class II MHC molecules on CNS endothelium. In

EAE, the concept of local antigen presentation was also supported by the presence of MBP and GC on some CNS endothelial cells. At 10 days PI, extensive inflammation was only seen in CNS tissue from mice inoculated for acute EAE but not in those given CFA only or in non-CNS organs. This suggested that early generalized non-specific enhanced recirculation of lymphocytes was followed by immunopathologic changes which were highly selective for the target organ of the disease process, the CNS. While a role of endothelial cells is implied, the mechanisms involved in selective accumulation of inflammatory cells in certain organs have yet to be defined in detail. In chronic EAE, relapses were associated with recurrent vasogenic edema (Traugott et al., 1986). In addition to albumin and Ig deposits, the third complement component could also be demonstrated in lesions and white matter (Traugott et al., 1982 and 1986). This opening of the BBB was accompanied by influx of inflammatory cells, L3T4+ cells in particular. During remission, when vasogenic edema was less pronounced, Lyt-2+ (CD8+) T cells predominanted (Hickey et al., 1984; Traugott et al., 1986). While endothelial cells are not irreversibly damaged after a single episode of EAE, after repeated attacks, the physiologic astrocyte-endothelial cell interaction might be permanently disturbed leading to fenestrated blood vessels and chronic vasogenic edema.

<u>VIRAL-INDUCED INFLAMMATION OF THE CNS</u>

In <u>SSPE,</u> immunopathologic changes similar to those seen in acute MS lesions were found. In lesion areas and normal CNS tissue, endothelial cells were positive for both Class I and Class II MHC. Ia+ astrocytes were present throughout lesions and all types of inflammatory cells (CD2+, CD4+, CD8+, Ia+, Ig+, IL-2R+ cells) were randomly distributed whereby CD8+ cells were more numerous than CD4+ cells.

In <u>HIV-related myelopathy,</u> inflammatory demyelinating lesions were seen in the spinal cord by routine morphology. Immunocytochemical analysis revealed pronounced expression of Class I MHC on endothelial cells and on some astrocytes, whereas Class II MHC was rarely detectable on endothelium, astrocytes or macrophages. Inflammatory infiltrates mainly contained CD8+ and only a few CD4+ cells. Similar immunopathologic changes with enhanced Class I MHC expression and predominance of CD8+ cells in infiltrates have been described previously in <u>lymphocytic choreomeningitis</u> in mice (Doherty et al., 1986).

These findings demonstrate that while CNS is normally an immunologic priviledged site, in inflammatory diseases, endogeneous CNS elements such as endothelial cells and astrocytes can be induced to function as APC which enhance inflammation and localize it to the CNS.

ACKNOWLEDGEMENTS

The author thanks Dr. L.C. Scheinberg, M.B. Bornstein and J. Prineas for providing some of the MS material and Dr. C.S. Raine for helpful discussion. The excellent technical assistance of I. Martinez and N. Sambamurty and the secretarial help of M. Briggs are appreciated.

Supported in part by research grants from the National Multiple Sclerosis Society RG-1664-B3; and by grants from the Public Health Service NS 11920 and NS 08952.

REFERENCES

Banks WA, Kastin AJ (1987): Saturable transport of peptides across the blood-brain-barrier. Life Sci, 41:1319-1338
Bradbury M (1979): The concept of a blood-brain-barrier. John Wiley & Sons New York
Bradbury MWB (1986): Introductory remarks: The developing experimental approach to the idea of a blood-brain-barrier. In: The Neuronal Microenvironment, Cserri HF (ed). Ann NY Acad Sci, 481:137-141
Breitmeyer JB (1987): How T cells communicate. Nature, 329:760-761
Burger DR & al. (1981): Endothelial cell presentation of antigen to human T cells. Human Immunol, 3:209-230
Butcher EC (1986): The regulation of lymphocyte traffic. Curr Topics in Microbiol and Immunol, 128:85-122
Cotran RS (1987): New roles for the endothelium in inflammation and Immunity. Am J Pathol, 129:407-413
Crone C (1986): The blood-brain-barrier as a tight epithelium: where is information lacking? In: The Neuronal Microenvironment, Cress HF (ed). Ann NY Acad Sci, 481:174-185
Cunningham VJ (1986): The influence of transport and metabolism on brain glucose content. In: The Neuronal Microenvironment, Cress HF (ed). Ann NY Acad Sci, 481:161-173
Doherty PC & al. (1987): Characteristics of the CSF inflammatory exudate in murine lymphocytic choriomeningitis (1987). In: Cellular and Humoral Immunological Components of Cerebrospinal Fluid in Multiple Sclerosis, Lowenthal A, Raus J (eds). Nato ASI Series Life Sci, 129:351-360
Doyle C, Strominger JL (1987): Interaction between CD4 and Class II MHC molecules mediates cell adhesion. Nature, 330:256-259
Essner E (1987): Role of vesicular transport in breakdown of the blood-retinal barrier. Lab Invest, 56:457-460
Fontana A & al. (1984): Astrocytes present myelin basic protein to encephalitogenic T cell lines. Nature, 307:273-276
Fontana A, Grob PJ (1984): Lymphokines and the brain. Springer Sem in Immunopathol, 7:375-386
Goldstein E, Betz L (1986): The blood-brain-barrier. Scientific America, 255:74-83
Halfer DA & al. (1985): Investigation of in vivo activated T cells in multiple sclerosis and inflammatory central nervous system disease. Clin Imm and Immunopath, 37:163-171
Hickey WF, Gonatas NK (1984): Suppressor T lymphocytes in the spinal cord

of Lewis rats recovered from acute experimental allergic encephalomyelitis. Cell Immunol, 85:284-288

Jalkanen S & al. (1986): A distinct endothelial cell recongition system that controls lymphocyte traffic into inflammed synovia. Science, 233:556-558

Joo F (1986): New aspects to the function of the cerebral endothelium. Nature, 321:197

Juhler M & al. (1986): The distribution of immunoglobulins and albumin in the central nervous system in acute experimental allergic encephalomyelitis. Acta Neurol Scand, 73:119-124

Latker CH & al. (1987): Differential localization of alkaline phosphatase in brain tissues of the frog and rat nervous systems: a cytochemical and biochemical study. J Comparative Neurol, 264:291-302

Male D, Pryce G (1988): Kinetics of MHC gene expression and mRNA synthesis in brain endothelium. Immunol, 63:37-42

Male D & al. (1987): Antigen presentation in brain: MHC induction on brain endothelium and astrocytes compared. Immunol, 60:453-459

Martz E (1987): LFA-1 and other accessory molecules functioning in adhesion of T and B lymphocytes. Human Immunol, 18:3-37

Masuyama J & al. (1986): Mechanisms of lymphocyte adhesion to human vascular endothelial cells in culture. J Clin Invest, 77:1596-1605

McCarron RM & al. (1986): Interaction between myelin basic protein-sensitized T lymphocytes and murine cerebral vascular endothelium cells. J Immunol, 137:3428-3435

Meuer SC & al. (1982): Clonal analysis of human cytotoxic T lymphocytes: T4+ and T8+ effector T cells recognize products of different major histocompatibility complex regions. Proc Nat Acad Sci USA, 79:4395-4399

Oldstone MBA, Dixon FJ (1968): Immunohistochemical study of allergic encephalomyelitis. Am J Pathol, 52:251-257

Panitch HS & al. (1987): Exacerbations of multiple sclerosis in patients with gamma-interferon. The Lancet, 8538:893-895

Pardridge WM (1986): Mechanisms of neuropeptide interaction with blood-brain-barrier. In: The Neuronal Microenvironment, Cress HF (ed), Ann NY Acad Sci, 481:231-249

Pober JS & al. (1983): Ia expression by vascular endothelium is induced by activated T cells and by human gamma-interferon. J Exp Med, 158:1339-1353

Raine CS (1982): Multiple sclerosis and chronic relapsing EAE: Comparative ultrastructural neuropathology. In: Multiple Sclerosis: The Patient, the Treatment and the Disease, Hallpike JF, Adams CWM, Tourtellotte WW (eds). London: Chapman & Hall, 411-458

Rapoport SI, Robinson PJ (1986): Tight-junctional modification as the basis of osmotic opening of the blood-brain-barrier. In: The Neuronal Microenvironment, Cserr HF (ed). Ann NY Acad Sci, 481:250-267

Sibley WA & al. (1985): Clinical viral infections and multiple sclerosis. Lancet, 1313:13-15

Simmons & al. (1987): Simultaneous visualization of vascular permeability change and leukocyte egress in the central nervous system during autoimmune encephalomyelitis. Acta Neuropath, 74:191-194

Smith QR, Takasato Y (1986) Kinetics of amino acid transport at the blood-brain-barrier studied using an in situ brain perfusion technique. In: The Neuronal Microenvironment, Cress HF (ed). Ann NY Acad Sci 481:186-201

Sobel RA, Colvin RB (1986): Responder strain-specific enhancement of endothelial and mononuclear cell Ia in delayed hypersensitivity

reactions in (Strain 2 X Strain 13) F_1 guinea pigs. J Immunol, 137:2132-2138
Spector R (1988): Fatty acid transport through the blood-brain-barrier. J Neurochem, 50:639-643
Spector R (1986): Nucleoside and vitamin homeostasis in the mammalian central nervous system. In: The Neuronal Microenvironment, Cress HF (ed). Ann NY Acad Sci, 481:221-230
Steardo L., Nathanson JA (1987): Brain-barrier tissues: end organs for atriopeptins. Science, 235:470-473
Sternberger NH, Sterberger LA (1987): Blood-brain-barrier protein recognized by monoclonal antibody. Proc Natl Acad Sci USA, 84:8169-8173
Sun D, Wekerle H (1986): Ia-restricted encephalitogenic T lymphocytes mediating EAE lyse autoantigen-presenting astrocytes. Nature, 320:70-72
Tao-Cheng JH & al. (1987): Tight junctions of brain endothelium in vitro are enhanced by astroglia. J Neurosci, 7:3293-3299
Traugott U (1987): Multiple sclerosis: relevance of Class I and Class II MHC-expressing cells to lesion development. J Neuroimmunol, 16:283-302
Traugott U & al. (1986): Immunopathology of the lesion in chronic relapsing experimental autoimmune encephalomyelitis in the mouse. Cell Immunol, 99:395-410
Traugott U & al. (1985): Acute experimental allergic encephalomyelitis in the mouse. Immunopathology of the developing lesion. Cell Immunol, 91:240-254
Traugott U & al. (1983): Multiple sclerosis: distribution of T cells, T cell subsets and Ia-positive macrophages in lesions of different ages. J Neuroimmunol, 4:201-221
Traugott U & al. (1985): On the presence of Ia-positive endothelial cells and astrocytes in multiple sclerosis lesions and its relevance to antigen presentation. J Neuroimmunol, 8:1-14
Traugott U & al. (1982): Chronic relapsing experimental allergic encephalomyelitis: identification and dynamics of T and B cells within the central nervous system. Cell Immunol, 68:261-275
Traugott U & al. (1982) Acute experimental autoimmune encephalomyelitis: T and B cell distribution within the target organ. Cell Immunol, 70:345-356
Vorbrodt AW (1986): Ultrastructural localization of lectin receptors on the luminal and abluminal aspects of brain micro-blood vessels. J Histochem Cytochem, 34:251-261
Wekerle H (1986): Cellular immune reactivity within the CNS. Trends in Neurosci, 9:271-277
Wilder RL & al (1987): Thymus-dependent and independent regulation of Ia antigen expression in situ by cells in the synovium of rats with streptococcal cell wall-induced arthritis. J Clin Invest, 79:1160-1171
Young AR & al. (1986): Recent studies on the serotonergic innervation of the cerebral circulation: a review. In: Neural regulation of brain circulation. Owman C, Hardebo JE (eds). Amsterdam: Elsevier Science Publishers, 195-217

DISCUSSION

Chaired by Joo F. and Carruba M.

M. De Sousa:

The acute MS is characterized by positive Ia astrocytes and positive Ia endothelial cells. You have shown that the lymphocytes have a role in determining such properties. In this respect I have two specific questions:
1) The astrocytes in culture are known to influence endothelial cell junctions. Can you further comment this phenomenon? I think that this would help us to understand the fact that Ia positivity for endothelial cells is necessary to allow the lymphocytes to cross the endothelium.
2) What's the importance of the mast and the polymorphonucleocyte cells in the events that you have illustrated?

U. Traugott:

In acute multiple sclerosis, Ia+ astrocytes and endothelial cells are present simultaneously even in the absence of any inflammatory cells in the parenchyma. In my opinion, in this particular instance the expression of Ia was most likely induced by interferon gamma produced by circulating activated T cells. In the acute stage this process reaches such a level that there might be transduction of the activating signals from the endothelium to the astrocyte processes which are closely attached to it. There is also damage of the BBB, which likely enables gamma interferon to get there just by diffusion. The level of activated T cells in the circulation increases up to 30% in MS patients with intercurrent infections. When stimulated "in vitro", MS lymphocytes produce more gamma interferon than T cells of control subjects. Probably this is a prerequirement in acute MS to have high Ia expression. In contrast, in chronic MS I did not observe any Ia-positive astrocytes adjacent to Ia+ endothelial cells. I could find Ia+ astrocytes only in the presence of activated T cells in active chronic lesions. In regard to polymorphonuclear leukocytes they are not present in lesions of MS patients and in guinea pigs with experimental allergic encephalomyelitis, an animal model for MS. However, in lesions in chronic or acute EAE in the mouse, there are plenty of polymorphonuclear leukocytes. Thus, by electonmicroscopy the pathology of the guinea pig model is quite different from that of the mouse model: the first is characterized by selective destruction of the myelin sheath and of the oligodendrocytes, whereas the axons are preserved. In the mouse model, on the contrary, an hyperacute disease develops in which demyelination is accompanied by axonal destruction and presence of polymorphonuclear leukocytes.
Also the mast cells are important in the chain of pathological events leading to the expression of the lesion: lymphokines from activated T cells trigger mast cells to

release vasoactive amines. In MS, this event occurs very early during the imune process and is important for the initial opening of the BBB or other endothelia outside the CNS.

A. Betz:

Do viral particles, like AIDS virus, enter the CNS inside the lymphocytes or are they able of penetrating by themselves?

U. Traugott:

The concept for AIDS viral particles is, at present, that they enter via monocytes which have CD4 receptors on the surface. Once penetrated, they are released and infect locally the glial cells. A problem which has not yet been solved is the fact that the message for CD4-epitopes has been demonstrated in various CNS cells whereas its expression has never been detected. Since it appears that the presence of these epitopes on the surface of cells is a requisite for the virus to enter, it is not clear, at the time, how AIDS encephalopathy develops.

RELATIONSHIPS BETWEEN NEURONAL ACTIVITY, ENERGY METABOLISM AND CEREBRAL CIRCULATION

Vincent J Cunningham, Jill E Cremer and Richard J Hargreaves [1]

MRC Cyclotron Unit, Hammersmith Hospital, Du Cane Road, London W12 0HS

Simple motor and cognition tasks are associated with regionally specific and rapid changes in cerebral blood flow. Examination of these phenomena in man has been made possible by Positron Emission Tomography. Improvements in the spatial and temporal resolution of the tomograph, together with the development and application of dynamic tracer modelling techniques, have in particular allowed repeat studies to be carried out in the same individual, so that comparisons can be made between resting and activated states over short periods. For example it can be shown that wilful movements of the hand and arm are accompanied by increased blood flow in the ipsilateral cerebellum, supplementary motor area and in the contralateral motor cortex. Specific regional changes in flow are also characteristic of visual stimulation and of the performance of mental tasks. The simplest interpretation of these phenomena is that they are in some way related to increased energy demands arising from increased neuronal activity, and they illustrate the rapidity with which metabolic and physiological adaptation can occur

Regional cerebral blood flow is intimately related to the delivery of glucose and oxygen which are required to satisfy energy demands arising from the maintenance of the ionic environment and transport processes within the tissue, and the removal of waste metabolic products. In this paper we consider these relationships principally from the point of view of the regulation of the supply of glucose for glycolysis and oxidation occuring at the level of the blood-brain barrier .

Glucose Transport Proteins

The facilitated diffusion of glucose from the capillary plasma to sites of phosphorylation within the brain requires its transport across the luminal and abluminal membranes of the capillary endothelial cells and across the plasma membranes of the neuronal and glial cells. Glucose transport proteins are present in most cell types and much can be inferred

---

1 Address - Merck, Sharp and Dohme, Neuroscience Research Centre, Terlings Park, Harlow, Essex CH20 2QR. U.K.

NATO ASI Series, Vol. H 33
Regulatory Mechanisms of Neuron to Vessel Communication in the Brain. Edited by F. Battaini et al.

as to the structure and function of those in brain by taking advantage of homologies with transport proteins in other tissues.

Birnbaum et al. (1986) reported the cloning and characterization of a cDNA encoding rat brain glucose transporter, from which the amino acid sequence of the primary translation product was deduced. Identification of the clone was based on the use of antibodies raised against purified human erythrocyte glucose transporter, and these authors drew attention to a striking sequence homology of greater than 97% between the rat and human proteins despite the differences in tissue and species.

Mueckler et al.(1985) have reported the sequence and structure of a human glucose transporter derived from a hepatoma cell line, again with a high homology with that from erythrocytes. Structural analysis of the 492 residue sequence indicated 12 membrane spanning domains which they suggested can be arranged so as to form a transmembrane channel through which glucose is transported. Supportive evidence for this basic structural model has come from a variety of physicochemical, ligand binding and inhibitor studies (see Walmsley, 1988). This constitutive glucose transporter is an extensively and heterogeneously glycosylated protein of molecular weight 55000, and has been shown in studies using polyclonal antibodies to be widely distributed in many cell types. Many of its properties are consistent with an alternating conformation model of transport, with mutually exclusive inward and outward facing conformations, allowing access to a membrane embedded glucose binding site (Cairns et al. 1987). This model will form the basis of our later consideration of the kinetics of glucose transport across the blood-brain barrier.

Tissue specific differences in glucose transporters are however apparent. Birnbaum et al. (1986) reported in their studies that mRNA, coding for a primary translation product of 38000, from insulin sensitive tissues was indistinguishable from that expressed in brain, but did not exclude the possibility that differences may occur between tissues as a result of post-translational modifications. Wang (1987) has shown that tissue specific differences exist in the extent of transporter glycosylation. Differing antibody reactivities further suggested that there are differences between the transporter in insulin sensitive tissues (such as skeletal muscle and adipocytes) and that in brain and red cells independent of differences in the extent of glycosylation. The situation has been clarified by a recent report by James et al. (1988) in which a monoclonal antibody was used to identify an insulin regulatable protein in muscle and adipose tissue, which is not present in rat brain or human erythrocytes. Thus it is likely that insulin sensitive tissues contain a characteristic transporter in addition to the ubiquitous constitutive transporter. The level at which the distinction between the types arises, genetic or post-translational, is not yet known.

Whilst being aware therefore of differences between insulin sensitive tissues and the brain, modulation of glucose transport capacity across cell membranes is an important feature of both, and it is of value to consider mechanisms. Cushman and Wardzala (1980) and Suzuki and Kono (1980) first reported translocation of glucose transporter proteins from an intracellular pool to cell membranes of adipocytes. Since then this process has been extensively investigated in insulin responsive systems. Blok et al. (1988) have recently demonstrated this in an electron microscopic study employing an antibody labelling technique. This work shows translocation of glucose transporter from an intracellular compartment located at the Golgi complex and its insertion in the plasma membrane in response to insulin treatment. Removal of insulin caused a rapid reversal of the effect, consistent with control being exerted on the rates of dynamic cycling of the transport protein between functional and reserve pools. This phenomenon probably involves the tissue specific transport protein more than the constitutive transporter. This is indicated by James et al. (1988) who observed a more marked degree of translocation for the specific insulin regulatable protein, than that detected using less specific polyclonal antibodies against glucose transport proteins.

What then is the situation in brain? The constitutive glucose transporter is present, and, as will be discussed later, regional cerebral glucose transport is modulated in response to local metabolic rate and plasma glucose concentration in a manner which is consistent with changes in total glucose transport capacity across the blood-brain barrier. Does this modulation involve translocation of the constitutive glucose transporter or is there an additional brain specific glucose transport protein?

Homologies and analogies between the glucose transporters from different tissues may be exploited in studies of various aspects of cerebral function: In the use of cross reacting antisera in the anatomical mapping of the brain transporter; in examining the kinetics of glucose transport across the blood-brain barrier; and in considering the possible mechanisms. involved in the modulation of cerebral glucose transport capacity.

Distribution of Glucose Transporter Protein in Rat Brain

The distribution of glucose transporter in rat brain has been investigated using two techniques; autoradiographic localization of radiolabelled ligand binding, and the use of specific antibodies.

Glucose transport proteins are characterised by a specific D-glucose inhibitable binding site for the high affinity ligand cytochalasin B. In brain, binding sites have been identified in high concentration in preparations of cerebral microvessels and the choroid plexus, and in particulate subcellular fractions obtained from cortical and cerebellar homogenates (Baldwin et al. 1985; Dick and Harik, 1986; Matthaei et al. 1986). Autoradiographical localization of tritiated cytochalasin B binding in fresh frozen sections of rat brain has further allowed the regional distribution of the glucose transporter to be examined (Tucker and Cunningham, 1988). Marked intra- and interregional differences in the extent of binding of the ligand were observed. Within the hippocampus, for example dense binding was associated with the synaptic zones and in particular with the medial molecular layer of the fascia dentata, whereas layers corresponding to the pyramidal cell bodies showed a lower degree of binding. The implied distribution of glucose transport proteins within this region correlates with glucose phosphorylation rates as can be measured using the autoradiographic deoxyglucose technique (see also below).

In contrast, the interregional comparisons of Cytochalasin B binding did not show the expected correlation with glucose phosphorylation rates. This was most striking in the case of the inferior colliculus which displayed low binding relative to adjacent sections of the cerebellum, whereas deoxyglucose studies show characteristically high phosphorylation rates in the inferior colliculus. Since there is an interregional correlation between glucose transport and phosphorylation rates (Cremer et al. 1983; Hawkins et al. 1983, Planas and Cunningham, 1987), it is therefore possible that the density of Cytochalasin B binding is related to the total available pool of glucose transport proteins, only a fraction of which is functionally active. If this is so then the capacity for modulation of glucose transport by translocation from storage to functional pools may differ between regions and show no simple relationship with transport and metabolic rates in the resting state.

A more detailed anatomical mapping of glucose transporter protein in rat brain has been made possible by the use of antisera raised against the protein from human erythrocytes. This work,using an immunogold technique, is to be reported by Bagley et al. (1988). Silver enhancement of antibody binding to perfusion fixed paraffin sections of rat brain was used to allow a detailed examination of the tissue at the light microscopic level. In this work the distribution of the enzyme pyruvate dehydrogenase was also studied using a similar technique. The relationships between the distribution of the transport protein and the enzyme will be considered later.

The most dense transporter immunoreactivity was associated with capillaries and with the epithelial cells of the choroid plexus, extending the observations using Cytochalasin B binding as cited above. This study also clearly demonstrates extravascular staining in the neuropil throughout the brain. Neuronal cell bodies and apical dendrites displayed little glucose transporter immunoreactivity. A heterogeneous distribution of transporter was observed within the neuropil, again raising the question as to the relationships between the localization of glucose transport proteins and the subsequent metabolism of glucose. Bagley et al. (1988) have compared the relative densities of glucose transporter immunostaining in different regions of the hippocampus and dentate gyrus, with the ranking of phophorylation rates in these structures as determined by Wree et al. (1988) using 2-deoxyglucose autoradiography. Whilst a close correlation was observed within this region, as with Cytochalasin B binding, no simple interregional correlations with phosphorylation rates were found. This also supports an hypothesis of reserve pools of transport proteins with this reserve capacity differing between brain structures.

Kinetics of Glucose Transport across the Blood-Brain Barrier

The facilitated diffusion of glucose from capillary plasma across the blood-brain barrier has conventionally been interpreted in terms of a simple saturable Michaelis-Menten relationship;

$$J_{in} = \frac{T_{max} \cdot S_p}{K_t + S_p} \qquad \text{---- 1}$$

where J_{in} represents unidirectional flux from plasma to brain of glucose at concentration S_p in plasma, T_{max} the maximal transport rate, and K_t a half saturation constant. This form implies a linear relationship between, for example, J_{in} and J_{in} / C_p (an Eadie-Hofstee plot). Experimentally observable deviations in such plots have been interpreted as due to a non-saturable free diffusion component, or to the presence of a second transport system. As written, equation 1 represents an irreversible process, and it is therefore required that the efflux of glucose from brain to plasma be described in terms of an equivalent relationship between efflux (J_{out}) and the tissue glucose concentration (C_t). Although the double membrane nature of the barrier (the luminal and abluminal endothelial cell membranes) can be accomodated in this kinetic scheme (Pappenheimer and Setchell, 1973; Gjedde and Christensen, 1984) it is clear that this simple formulation requires examination and justification if it is to be considered as an adequate representation of the more complex kinetics describing carrier systems.

We have previously suggested an alternative double membrane analysis of glucose transport across the barrier, based on the reversible Michaelis-Menten equation. (Cunningham et al. 1986) This emphasized a need to take into account glucose concentrations in the endothelial cell and the tissue when considering influx from plasma to brain, and could account for an apparent 'non-saturable' component of glucose transport. This scheme however is kinetically equivalent to simple stereospecific pores, which have been rejected as a description of glucose transport in the human erythrocyte (Stein,1986) and may therefore be considered doubtful for brain.

Both the above schemes can be shown to be special cases of a more general model for transport across a double membrane system (Cunningham and Hargreaves, 1988). It is our purpose here to take advantage of the high degree of homology between the much studied erythrocyte glucose transporter and that in brain, so as to explore blood to brain glucose transport kinetics in terms of this general model.

The kinetics of a simple single transmembrane carrier have been described in detail by Stein (1986) and it is illustrated diagrammatically in Fig 1 with associated rate constants. In this scheme the carrier alternates between outward and inward facing conformations (1 and 2) and is either unloaded (E) or loaded with substrate (ES).

Fig 1 Diagramatic Representation of Simple Carrier

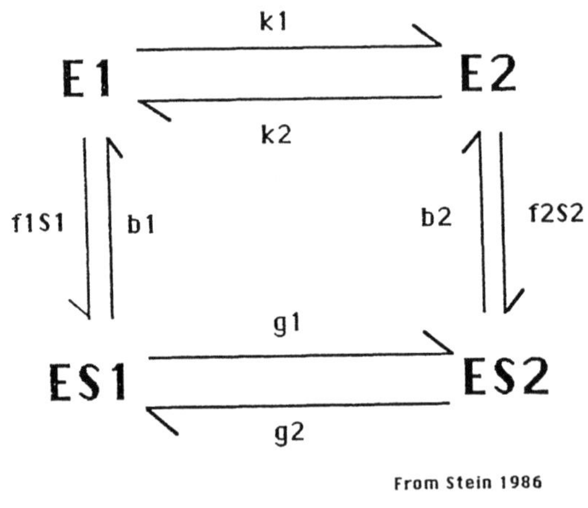

From Stein 1986

When the association and dissociation of substrate and carrier is rapid compared with the reorientations of the transporter, as will be assumed to be the case with glucose, then resistance parameters are given by (see Stein, 1986);

$$n.R_{12} = g_1^{-1} + k_2^{-1}$$
$$n.R_{21} = g_2^{-1} + k_1^{-1}$$
$$n.R_{ee} = g_1^{-1} + g_2^{-1}$$
$$n.R_{00} = k_1^{-1} + k_2^{-1}$$

where n is the total transporter concentration, R_{ee} and R_{00} are resistance parameters relating to the movement of the loaded and unloaded carriers respectively, and R_{12} and R_{21} are resistance parameters which relate to the movement and breakdown of loaded carrier on one side of the membrane and its return so that free carrier again becomes available on the other side.

The steady-state unidirectional transfer of substrate from side 1 to side 2 of a single membrane (J_{12}) is then given by;

$$J_{12} = \frac{KS_1 + S_1S_2}{K^2R_{00} + KR_{12}S_1 + KR_{21}S_2 + R_{ee}S_1S_2} \qquad \text{---- 2}$$

where K is an intrinsic dissociation constant and S_1 and S_2 are the concentrations of the substrate at side 1 and at side 2 respectively.

Equation 2 may be rearranged to the form of equation 1, to give expressions for the apparent Tmax and apparent Kt for influx, which are then functions not only of the intrinsic parameters but also of the trans substrate concentration (S_2);

$$T_{max}(app) = \frac{K + S_2}{KR_{12} + R_{ee}S_2}$$

$$K_t\ (app) = \frac{K^2R_{00} + KR_{12}S_2}{KR_{12} + R_{ee}S_2}$$

When R_{00} is greater than R_{ee} then the phenomenon of trans acceleration occurs in which increasing trans concentrations of substrate lead to increased influx. When R_{ee} is greater than R_{00} the kinetics approach those of a simple pore in which increasing trans concentrations competitively inhibit influx.

The blood-brain barrier may be considered as the conjunction of two membrane transport systems sharing a relatively small and well mixed endothelial cell compartment. The apparent unidirectional flux of substrate across the barrier as a whole therefore requires that four expressions of the form of equation 2 be used to define transport between the three compartments - plasma (p) endothelial cell (e) and tissue (t) - to give fluxes J_{pe}, J_{ep}, J_{et}, and J_{te}. The apparent unidirectional flux from plasma to brain across the barrier as a whole (J_{in}) is then given by ;

$$J_{in} = \frac{J_{pe} \cdot J_{et}}{J_{et} + J_{ep}}$$

with a corresponding expression for Jout. This model therefore allows unidirectional fluxes to be calculated from given plasma and tissue concentrations and estimates of the rate constants in fig 1 (The concentration of substrate in the endothelial cell is also required, but can be obtained readily in calculations by adjusting it until the steady state condition ; $J_{pe} - J_{ep} = J_{et} - J_{te}$. is obtained)

Simulations based on such a double membrane model show that only when the resistance parameters for loaded and unloaded carriers are equal, does equation 1 apply to describe influx from plasma to brain tissue. This is the case even when tissue substrate concentrations are held constant. When these resistance parameters are unequal deviations from the form of equation 1 arise because variations in the substrate concentration within the endothelial cell compartment accompany changes in plasma substrate concentration. This results in deviations from linearity in kinetic plots such as the Eadie Hofstee (Cunningham and Hargreaves, 1988)

In order to apply this general model model to glucose transport across the blood-brain barrier we must lean heavily on studies using the single membrane of the erythrocyte for estimates of the appropriate kinetic constants .Although the precise kinetics of erythrocyte glucose transport are still a matter of some debate, we base our treatment on the data of Lowe and Walmsley (1986) (see also Walmsley 1988) who have presented estimates of the rate constants governing the reorientation of the loaded and unloaded forms of the carrier (see fig 1) for human erythrocytes at 0'C, together with their respective activation energies. The Arrhenius equation can be used to convert these estimates to values at 37'C from which resistance parameters may be calculated as described above. These are summarised in Table 1. In addition an estimate of the intrinsic dissociation constant (K) may be obtained from the value for the half saturation constant under zero-trans conditions (K^{zt} in 1 - 2 direction,) at 37'C from the relationship;

$$K = K^{zt}.R_{12}/R_{00} \quad \text{(Stein, 1986).}$$

As can be seen from the results presented in Table 1, the marked assymetry of the rate constants observed at 0'C is much reduced with increasing temperature. It can be calculated that at 37'C a fairly even distribution of the transporter between outward and inward facing conformations occurs and at this temperature the affinity for glucose is about 2.5 times greater outside than inside (Lowe and Walmsley, 1986).

Table 1

Human Red Cells - Glucose Transport Rate Constants

Data based on Lowe and Walmsley (1986)
Nomenclature Stein (1986)

rate constant	0'C (sec^{-1})	Activation Energy (Joules. mol^{-1})	37'C (sec^{-1})
g_1	1113	31700	5895
g_2	90.3	88000	9234
k_1	12.1	127000	9620
k_2	0.726	173000	6484

Zero - trans Half Saturation Constant (K^{zt} 1 -2) (mM at 37'C) = 7.24

Intrinsic Dissociation Constant (K, mM) = 9.08

Transporter concentration (n) x Resistance parameters at 37'C (sec)

$nR_{12} = 3.24 \times 10^{-4}$
$nR_{21} = 2.12 \times 10^{-4}$
$nR_{00} = 2.58 \times 10^{-4}$
$nR_{ee} = 2.78 \times 10^{-4}$

In extrapolating these results to a double membrane system representative of blood-brain barrier transport, we proceed with the simulation in placing two such transporters back to back. Thus for the first transporter, the outward facing side 1 is plasma and side 2 corresponds to the endothelial cell environment. For the second transporter side 2 is the endothelial cell and side 1 the tissue interstitium.

In these simulations a total transporter protein concentration of 40 pmoles per gm of brain was required to give an apparent T_{max} of 2.4 umoles glucose transported per gram (cf Cortical values in Cunningham et al. 1986) This figure is an order of magnitude higher than estimates of the amount of transporter in microvessels as based on Cytochalasin B binding (Morin et al. 1988). In view of some uncertainties regarding enrichment of transporter in microvessel preparations (see discussion Morin et al. 1988) and of approximations in the present approach, we regard this as a reasonable agreement.

Apparent K_t's for glucose influx from plasma to brain are given in Table 2. The values are in excellent agreement with estimates obtained in vivo (Hawkins et al. 1979; Gjedde and Rasmussen, 1980; Pardridge et al. 1982; Cunningham et al, 1986). Furtheremore it is of interest to note that assymetries in the luminal/abluminal distribution may account for the small interregional differences in K_t that have been observed in these studies. (see also Gjedde and Christensen, 1984). The minimal effect of tissue glucose concentration on the apparent K_m for influx (Table 2) results from the similar values for the resistance parameters relating to loaded and unloaded carrier (Table 1). This is in agreement with the rejection by Christensen et al. (1982) of a putative accelerative exchange diffusion of glucose. This model thus provides a firm mechanistic basis for the interpretation of apparent transport constants obtained in vivo using the simple form of equation 1.

Table 2

Apparent K_t's for glucose transport from plasma to brain based on double membrane model (mM)

Tissue Glucose Concentration	0.5	<----------->			5.0
K_t	8.6	<----------->			8.7
Luminal / Abluminal Asymmetry in Transporter Density	1:10	<-->	1:1	<-->	10:1
Km	7.4	<-->	8.7	<-->	10.4

The similarities between results obtained in the present simulation and those obtained in blood-brain barrier transport studies in vivo support the extrapolation of work on erythrocyte glucose transport kinetics to those of the barrier. It is therefore interesting to speculate that the erythrocyte might provide a convenient model for studies of control and disturbances in cerebral glucose transport!

Modulation of Glucose Transport Capacity across the Blood-Brain Barrier

By glucose transport capacity is meant the apparent Tmax for transport expressed per weight or volume of the brain region of interest. Since under normal physiological conditions, plasma glucose concentrations range over values close to the apparent K_t for transport (9mM), then the system is not close to saturation (unlike amino acid transport for example). Differences in transport capacity are thus indicated by transport rates being maintained despite changes in plasma glucose concentration as well as by variation in transport rates independent of changes in plasma glucose concentration (Hargreaves et al.1986).

Within control states in the rat an interregional correlation exists between the rate of glucose utilization and the rate of glucose influx (Cremer et al. 1983; Hawkins et al. 1983, Planas and Cunningham, 1987) The unidirectional influx of glucose is approximately double the rate of phosphorylation. Studies over a range of physiological states indicate that two main factors affect the local modulation of glucose transport capacity,. These are the plasma glucose concentration and the local metabolic rate for glucose in the tissue. These phenomena may also be illustrated in a simple comparison of fed and fasted, and of awake and anaesthetised rats (Hargreaves et al. 1986).

We have employed a short term dual label 2-deoxyglucose technique to obtain simultaneous regional estimates of glucose influx and rates of glucose phosphorylation in the rat. This technique involves direct assays of glucose, and of labelled 2-deoxyglucose and 2-deoxyglucose phosphate concentrations in discrete brain regions following timed sequential intravenous injections of tritiated and ^{14}C labelled deoxyglucose. Estimates of glucose influx obtained using this technique are in close agreement with those using labelled D-glucose as a tracer. A comparison of fed and fasted conscious rats using this technique shows that in general not only are regional rates of glucose phosphorylation similar in the two groups but that also regional influxes are similar. This is the case despite a marked difference in plasma glucose concentration, with little difference in regional plasma flow rates. A comparison of conscious and anaesthetised fasted rats showed differences in both plasma flow rate and rates of phosphorylation of glucose. Corresponding differences in glucose influx were apparent although plasma glucose concentrations were similar in these two groups. This and other studies therefore indicate that regional rates of influx commensurate with the rate of phosphorylation can be set despite wide variations in plasma glucose concentrations. What factors might be involved in these effects?

a) Plasma flow rate. The arteriovenous extraction of glucose under normal conditions is low and thus changes in flow rate alone has little effect on the mean capillary glucose concentration relative to that in arterial plasma. There is however increasing evidence for the possibility of differences in the number of perfused capillaries under different physiological conditions which needs to be taken into account particularly when differences in plasma flow occur. Such a phenomenon would involve changes in the capillary surface area and hence the number of available glucose transporters.

We have previously shown that in conscious control rats and in rats displaying local increases in glucose metabolic rate that a highly significant interregional correlation exists between blood volume and Tmax, but a weaker correlation of Tmax with blood flow (Cremer et al. 1983) In a recent study in which both blood flow and capillary transit times were measured, Shockley and LaManna (1988) emphasize the need to consider changes in blood volume as well as flow, and further present evidence that in the normoxic and normocapnic anaesthetised rat 'considerably less than 100% of the cerebral capillaries were actively perfusing the tissue' and estimate this to be about 30%. This strongly suggests that capillary recruitment can occur in brain and is a major factor governing blood-brain transport phenomena. Such a mechanism would account for the increased influx of glucose in conscious vs anaesthetised fasted rats at the same plasma glucose concentration.

b) Transporter density. The permeability and surface area components governing influx across the blood-brain barrier cannot be distiguished in kinetic analyses of blood-brain glucose transfer in vivo. Furthermore, there has been, as yet, no direct demonstration of changes in permeability arising from modulation of the endothelial membrane concentration of glucose transport proteins. Nonetheless, indirect arguments point to this possibility. The regional distribution of glucose transport proteins as visualized using Cytochalasin B, or by antibody labelling, summarised above shows no simple correlation with the regional 'flow/transport/metabolism' couples which have been established in the resting state, which is strongly suggestive of spare transport capacity. A quantitatve analysis of the distribution of transport protein between cell types would however be required to establish this. Analogies might be drawn with the insulin responsive systems in which modulation by translocation has been directly demonstrated. Although insulin sensitive transport protein iis absent from brain, the presence of a brain specific regulatable protein cannot be excluded. Observations such as the repression of glucose transfer in chronic hyperglycaemia (Gjedde and Crone, 1981) and the increased regional transport capacity of fasted rats relative to fed (above) in which plasma glucose concentration appears as the major factor influencing transport capacity may involve such a translocation mechanism.

Pyruvate Dehydrogenase Distribution in Brain

The cerebral metabolic rate for glucose, as measured using glucose analogues, relates to its phosphorylation. The detailed distribution of extravascular glucose transport protein in the neuropil shows similarities to that of hexokinase (Bagley et al. 1988; Wilkin and Wilson, 1977), consistent with a tight coupling between the supply of glucose and its phophorylation. The subsequent metabolism of glucose, under normal resting conditions, is predominantly oxidative (Siesjo, 1979) and involves the entry of pyruvate into the TCA cycle via pyruvate dehydrogenase.

Pyruvate dehydrogenase, as visualized using the immunogold technique (Bagley et al. 1988) is predominantly located in the neuronal cell bodies and proximal cell processes. The neuropil generally shows much less immunoreactivity, except in certain specific nuclei. This distribution contrasts markedly with that of the glucose transporter, leading to clear distinctions between the two, particulary in hippocampal cell layers. These results indicate that the neuropil and glia are richly endowed with the capacity to transport and phosphorylate glucose, but suggest that the local diffusion of lactate and pyruvate between cell types and their oxidation in neuronal cell bodies is an important feature of brain glucose metabolism under normal conditions.

There is increasing evidence that local temporal mismatches or uncoupling of glycolytic versus oxidative metabolism are associated with physiological activations such as those refered to in the introduction. Fox and Raichle (1986) described an uncoupling of cerebral flow and oxidative metabolism during somatosensory stimulation in man. More recently, this group (Fox et al. 1988) also report, from parallel measures of flow, oxygen consumption and the metabolic rate for glucose, during visual stimulation that 'transient increases in neural activity cause a tissue uptake of glucose in excess of that measured by oxidative metabolism'. Furthermore Hossman and Linn (1987) reported a decrease in glucose and significant increases in lactate in the somatosensory cortex after forepaw stimulation in the rat. Collins et al. (1987) also attribute an increased loss of label from [6-^{14}C]-glucose relative to phosphorylation of 2-deoxyglucose, over a 10 minute period of visual stimulation in the rat, as being primarily due to loss of lactate.

These observations raise the question as to how the excess lactate is handled under conditions of physiological activation - whether this is a temporal phenomenon, involving diffusion and eventual oxidation in the tissue, or whether significant local egress of lactate can occur. Lactate transport across the blood-brain barrier occurs via facilitated diffusion, but in adult rats its maximal transport capacity is very low (Oldendorf, 1972; Cremer et al. 1979). This does not however exclude the possibility of assymetries in its transport, such that a concentration gradient from brain to blood could be dissipated more rapidly than inferred from net blood to brain transfer. Egress phenomena under normal physiological conditions is an area which needs further investigation.

Acknowledgement. We are indebted to Peter R Bagley for stimulating discussions.

References

Baldwin SA, Cairns MT, Gardiner RM, Ruggier RA (1985) D-glucose-sensitive cytochalasin B binding component of cerebral microvessels. J. Neurochem. 45: 650-652

Bagley PR, Tucker SP, Nolan C, Davies A, Baldwin SA, Cremer JE, Cunningham VJ (1988) Anatomical mapping of glucose transporter protein and pyruvate dehydrogenase in rat brain: An immunogold study: manuscript in preparation

Birnbaum MJ, Haspel HC, Rosen OM (1986) Cloning and characterisation of a C-DNA encoding the rat brain glucose transporter protein. PNAS 83: 5784- 5788

Blok J, Gibbs EM, Leinhard GE, Slot JW, Geuze HJ (1988) J. Cell Biol. 106: 69-76

Cairns MT, Alvarez J, Panico M, Gibbs AF, Morris HR, Chapman D, Baldwin SA (1987) Investigation of the structure and function of the human erythrocyte glucose transporter by proteolytic dissection. Biochim. Biophys. Acta. 905: 295-310

Christensen O, Lund-Andersen H, Betz AL, Gilboe D (1982) Transport of glucose across the blood-brain barrier: re-evaluation of the accelerative exchange diffusion. Acta Physiol. Scand. 115: 233 - 238

Collins RC, McCandless DW, Wagman IL (1987) Cerebral glucose utilization: Comparison of [^{14}C]deoxyglucose and [6-^{14}C]glucose quantitative autoradiography. J. Neurochem. 49: 1564 - 1570

Cremer JE, Cunningham VJ, Pardridge WM, Braun LD, Oldendorf (1979) Kinetics of blood-brain barrier transport of pyruvate, lactate and glucose in suckling, weanling and adult rats. J. Neurochem. 33: 439 - 445

Cremer JE, Cunningham VJ, Seville MP (1983) Relationships between extraction and metabolism of glucose, blood flow and tissue blood volume in regions of rat brain. J. Cereb. Blood Flow Metabol. 3:291-302

Cunningham VJ, Hargreaves RJ (1988) Kinetic analyses of carrier mediated blood-brain barrier transport with reference to amino acids. In: *Peptide and Amino Acid Transport Mechanisms in the Central Nervous System.* pp 219 - 228 (Eds) Rakic L, Begley DJ, Davson H, Zlokovik BV. MacMillan Press.London / Serbian Acadamy of Science.

Cunningham VJ, Hargreaves RJ, Pelling D, Moorhouse SR (1986) Regional blood-brain glucose transfer in the rat: a novel double-membrane kinetic analysis. J. Cereb. Blood Flow Metabol. 6: 305 -314

Cushman SW, Wardzala LJ (1980) Potential mechanisms of insulin action on glucose transport in the isolated rat adipose cell. J. Biol. Chem. 255: 4758-4762

Dick APK, Harik SI (1986) Distribution of the glucose transporter in the mammalian brain. J. Neurochem 46: 1406-1411

Fox PT, Raichle ME (1986) Focal physiological uncoupling of cerebral blood flow and oxidative metabolism during somatosensory stimulation in human subjects. PNAS 83: 1140 - 1144

Fox PT. Raichle ME, Mintun MA, Dence C (1988) Nonoxidative glucose

consumption during focal physiologic neural activity. Science 241:462 - 464
Gjedde A, Crone C (1981) Blood-brain glucose transfer: repression in chronic hyperglycaemia. Science 214: 456 - 457
Gjedde A, Christensen O (1984) Estimates of Michaelis-Menten constants for the two membranes of the brain endothelium. J. Cereb. Blood flow metabol. 4:241-249
Gjedde A, Rasmussen M (1980) Pentobarbital anaestesia reduces blood-brain glucose transfer in the rat. j. Neurochem. 35: 1382 - 1387
Hargreaves RJ, Planas AM, Cremer JE, Cunningham VJ (1986) Studies on the relationship between cerebral glucose transport and phosphorylation using 2-deoxyglucose. J. Cereb Blood Flow Metabol. 6: 708-716
Hawkins RA, Hass WK, Ransohoff J (1979) Measurement of regional brain glucose utilization in vivo using [2-14C]-glucose. Stroke 10: 690 - 703
Hawkins RA, Mans AM, Davis DW, Hibbard LS, Lu DM (1983) Glucose availability to individual cerebral structures is correlated to glucose metabolism. J. Neurochem. 40: 1013-1018
Hossman KA, Linn (1987) Regional energy metabolism during functional activation of the brain. J. Cereb. Blood Flow Metabol. 7, Suppl 1: s297
James DE, Brown R, Navarro J, Pilch PF (1988) Insulin-regulatable tissues express a unique insulin-sensitive glucose transport protein. Nature 333: 183-185
Lowe AG, Walmsley AR (1986) The kinetics of glucose transport in human red blood cells. Biochim. Biophys. Acta. 857: 146-154
Matthaei S, Horuk R, Olefsky JM (1986) Blood-brain glucose transfer in diabetes mellitus: decreased number of glucose transporters at blood-brain barrier. Diabetes 35: 1181-1184
Morin AM, Dwyer BE, Fujikawa DG, Wasterlain CG (1988) Low [^{3}H]Cytochalasin B binding in the cerebral cortex of newborn rat. J. Neurochem. 51: 206 - 211
Mueckler M, Caruso C, Baldwin SA, Panico M, Blench I, Morris HR, Allard WJ, Leinhard GE, Lodish HF (1985) Sequence and structure of a human glucose transporter. Science 229: 941-945
Oldendorf WH (1972) Blood-brain Barrier permeability to lactate. Eur. Neurol. 6:49 - 55
Pappenheimer JR, Setchell BP (1973) Cerebral glucose transport and oxygen consumption in sheep and rabbits. J. Physiol. 233: 529-551
Pardridge WM, Crane PD, Mietus LJ, Oldendorf WH (1982) Kinetics of regional blood-brain glucose transport and brain phosphorylation of glucose and 2-deoxyglucose in the barbiturate-anaesthetized rat. J. Neurochem. 38: 560 - 568
Planas AM, Cunningham VJ (1987) Uncoupling of cerebral glucose supply and utilization after hexane-2,5-dione intoxication in the rat. J, Neurochem 48: 816-823
Shockley RP , LaManna JC (1988) Determination of rat cerebral corical blood volume changes by capillary mean transit time analysis during hypoxia, hypercapnia and hyperventilation. Brain Res. 454: 170-178
Siesjo BK (1979) *Brain Energy Metabolism* . John Wiley and Sons

Stein WD (1986) *Transport and diffusion across cell membranes.* Academic Press, New York.

Suzuki R, Kono T (1980) Evidence that insulin causes translocation of glucose transport activity to the plasma membrane from an intracellular storage site. PNAS 77: 2542-2545

Tucker SP, Cunningham VJ (1988) Autoradiography of [^{3}H]cytochalasin B binding in rat brain. Brain Res. 450: 131-136

Walmsley AR (1988) The dynamics of the glucose transporter. TIBS 13: 226 -231

Wang C (1987) The D-glucose transporter is tissue specific. J. Biol. Chem. 262: 15689-15695

Wilkin GP, Wilson JE (1977) Localization of hexokinase in neural tissue: light microscopic studies with immunofluorescence and histochemical procedures. J. Neurochem. 29: 1039 - 1051

Wree A, Schleicher A, Zilles K, Beck T (1988) Local cerebral glucose utilization in the Ammon's horn and dentate gyrus of the rat brain. Histochemistry 88:415-426

DISCUSSION

Chaired by Siesjo B.K. and Lenzi G.L.

G.L. Lenzi:

It is important to underline that the intimate association between basic science and more clinically oriented research is absultely essential in order to understand what is going on during the "cogito ergo sum".
I have a question that comes from my clinical experience: glucose transport to the brain may probably vary in conditions of hyperglycemia or hypoglycemia: what is the situation in diabetic conditions?

V. Cunningham:

As far as I know, there is no consistent evidence that glucose transport across the BBB is modulated directly by insulin, although there are changes associated with long term hyper or hypoglycemia. This makes sense: in certain conditions the utilization and storage of glucose is encouraged, for example at the level of adipocytes, muscles etc., whereas when the plasma glucose concentration decreases, the rest of the body tends to spare it. With respect to the brain, its behaviour is the opposite: under conditions of low blood glucose levels the capacity for glucose cram part to the neurones increases. It appears that the signals for the regulation of glucose transport capacity are its changes in blood level and modifications in CBF are involved. We have also failed to find an effect of glucagon on glucose transport across the barrier.

B.K. Siesjo:

Do you agree with Dr. L. Anderson who stated that glucose transport across the BBB is slow, while across the other cell membranes it is rather rapid?

V. Cunningham:

I agree on that. There is a very high capacity for glucose transport across some membranes, particularly in the neuropil.
Although the density of glucose transport proteins in the capillaries is even higher, the surface area of the barrier is much less than that of the neurones and glia.

B.K. Siesjo:

You have spoken of recruitment of capillaries and increase in glucose transport capacity. There are two foundamentally different possibilities:
(1) additional number of capillaries which open in parallel with the increase in cerebral activity or

(2) increased number of capillaries opening per unit of time. Can you further comment on this phenomenon?

V. Cunningham:

Recent data suggest that 30% of the capillaries are open or "used" in the anaesthetized state. This suggests a possibility for a dramatic increase in the potential surface area in the awake state; however, I do not know how to experimentally distinguish between the twinkling on and off of the capillaries and the increase in capillary number. It is not yet known which of the two possibilities is more likely.

B.K. Siesjo:

The presynaptic and postsynaptic elements are functional units endowed of many transporters but with little piruvate dehydrogenase (PDH) activity, in contrast with the cell bodies where the ratio is inverted; thus, we are possibly dealing with lactate production at the level of the neuropil and oxydative metabolism at the level of cell bodies. I wonder why nature has developed the machinary for ATP production at a site where it is apparently less convenient for the use. For example, in the cell bodies the ionic traffic is rather low and therefore there is less need for oxydative metabolism as compared with the neuropil. Can the measure of PDH be an artifact?

V.Cunningham:

This phenomenon was very surprising also to us; however I do not think that it represent an artifact. What I suggest is that the synapses are quite capable of oxydizing glucose, as shown by several authors, and that there is also a net movement of metabolites from the neuropil and the glia into the cell bodies.
It is possible that PDH activity in the terminals is sufficient for the process localized at this level.

L. Hertz:

Concerning the observation by Siesjo on the localization of PDH activity and the machinary for ATP production, I think that nature has developed in the cell different "check points". Instead of one single cellular location determining metabolic rate, there are also various transport processes and enzyme activities. For example, in the case of glutamate metabolism, the neurotransmitter released from the terminals in large amount is uptaken by astrocytes, converted to different metabolites and sent back to the neurons. This means that there are interactions between the different cells which may represent a tuning or "check point" in cell activity and metabolism.

B.K. Siesjo:

I agree. However, what amazes me is not the fact that PDH is high in the cell body, since this reflects not only transport on activity but also several other processes, such as macromelecule synthesis, but the low level of PDH in the terminals, where the ionic traffic is very intense.

L. Sokoloff:

How stereospecific or substrate-specific is the glucose transporter? Is it possible that the difference between the fed and the fasted animals in the influx may be not only due to the modulation of the amount of the transporter but also to the presence of inhibitory substances in the diet, competitive with the binding of glucose to its carrier.

V.J. Cunningham:

It is certainly possible that some physiological signals may modulate glucose transport via competitive mechanisms.

R. Orkand:

Glucose carrier does have affinity for other types of sugars: the affinity for glucose is the highest, for galactose about half, while lactose has very little affinty for the transporter. Furthermore, the concentration of sugars other than glucose never reaches in the blood levels which have a practical influence on this system.
There is evidence in fibroblast cultures of the possibility of up and down regulating of the number of carriers in an insulin independent manner by changing the glucose concentration in the medium. This indicates that the sugar level may be a signal that allows the cells to adjust the number of carriers in the membrane. There are suggestions that similar mechanisms are also operative at the level of the BBB. According to some authors the kinetic of glucose transport changes in chronically hyperglycemic animals. However, these findings have been contradicted by other studies. Do you have any comment on that?

V.J. Cunningham:

If the carriers are transported from the intracellular compartments, such as Golgi, to the endothelium cell membrane in response to the environmental conditions, this movment may be detectable by electromycroscopic studies. The point is that it is the number of exposed transporters that counts. This may explain the discrepancies in the reported studies.

MEASUREMENT OF REGIONAL HEMODYNAMIC AND METABOLIC CHANGES IN THE CENTRAL NERVOUS SYSTEM WITH IMAGING TECHNIQUES.

Louis Sokoloff, M.D.
Laboratory of Cerebral Metabolism, National Institute of Mental Health
9000 Rockvbille Pike: Building 36, Room 1A05
Bethesda, Maryland 20892. U.S.A.

INTRODUCTION

The modern era of quantitative studies of cerebral blood flow and energy metabolism in vivo began approximately 40 years ago with the development of suitable methods that were applicable to conscious unanesthetized animals and man. These methods were global in nature and provided only average values for blood flow and/or metabolism for the brain as a whole, but much of our current knowledge of the physiology and pharmacology of the cerebral circulation and metabolism was derived from applications of these methods. They have, furthermore, provided reliable quantitative data that are the basis by which all subsequent methods, global and/or regional, are evaluated.

MEASUREMENT OF OVERALL CEREBRAL BLOOD FLOW AND METABOLISM

Methods

In a steady state the tissue concentration of any substance utilized or produced by the brain is by definition constant. When this substance is exchanged between brain and blood, the difference in its rates of delivery to the brain in the arterial blood and removal in the venous blood is in the steady state equal to the net rate of its utilization or production by the brain. This relation can be expressed as follows:

$$CMR = CBF\ (C_a - C_v) \qquad [1]$$

where $(C_a - C_v)$ is the steady state difference in concentrations of the substance in arterial and representative cerebral venous blood, CBF is the rate of cerebral blood flow, and CMR (cerebral metabolic rate) is the rate of utilization or production of the substance by the brain. If both CBF and the arteriovenous difference are known, the net rate of utilization or production of the substance by the brain can then be calculated by Eq. (1). This relationship has been the basis of most quantitative studies of cerebral metabolism in vivo.

The most reliable method for determining CBF is the inert gas method of Kety and Schmidt (1948). It was originally designed to be used in conscious man, but it has been adapted for use in animals. The method is based on the Fick Principle and utilizes low concentrations of a freely diffusible,

NATO ASI Series, Vol. H 33
Regulatory Mechanisms of Neuron to Vessel Communication in the Brain. Edited by F. Battaini et al.

chemically inert gas as the tracer substance. The original gas used was N_2O, but subsequent modifications have substituted other gases, e.g., ^{85}Kr, ^{79}Kr, ^{133}Xe, or H_2, which can be measured more conveniently in blood. During constant inhalation of 15% N_2O in air timed arterial and cerebral venous blood samples are withdrawn at intervals and analyzed for their N_2O contents. CBF (in ml/100 g of brain tissue/min) is then calculated from the equation:

$$CBF = 100\ \lambda\ C_v(T)\ /\int_0^T (C_a - C_v)dt \qquad [2]$$

where C_a and C_v = arterial and cerebral venous blood concentrations of tracer, respectively

$C_v(T)$ = concentration Of N_2O in venous blood at time T

λ = partition coefficient for N_2O between brain tissue and blood (i.e., ratio of tissue to blood concentrations at equilibrium)

T = total duration of inhalation (min)

t = variable time (min)

$\int_0^T (C_a - C_v)dt$ = integrated arteriovenous difference in N_2O concentration over T min of inhalation

CBF = cerebral blood flow in ml/100 g brain tissue/min.

The brain-blood partition coefficient (λ) for N_2O is approximately 1.0. At least 10 min of inhalation is required to approach equilibrium, but it has become customary to extend T to 15-20 min or longer to achieve more complete equilibrium and accuracy. At the end of this interval, the N_2O concentration in brain tissue is then approximately equal to the cerebral venous blood concentration at the same time.

Because the N_2O method requires sampling of both arterial and cerebral venous blood, it lends itself readily to the simultaneous measurement of arteriovenous differences of substances involved in cerebral metabolism, and their rates of utilization or production can then be calculated by Eq. [1]. This method and its modifications have provided most of our current knowledge of the rates of substrate utilization or product formation by the brain *in vivo*.

Normal Rates of Blood Flow and Energy Metabolism in Brain as a Whole

The brain utilizes oxygen and glucose at very rapid rates and is absolutely dependent on uninterrupted oxidative carbohydrate metabolism for maintenance of its functional and structural integrity. Under anaerobic conditions glycolysis is limited to the conversion of glucose to lactate and provides two molecules of ATP for each molecule of glucose consumed. Under fully aerobic conditions glucose is metabolized to pyruvate which is then metabolized via the tricarboxylic cycle to carbon dioxide and water with a net production of 36 molecules of ATP per molecule of glucose consumed. The gain in ATP production is the result of oxidative phosphorylation that is associated with oxygen consumption. The availability of oxygen inhibits the rate of glycolysis and reduces net lactate production to zero or near zero levels, an effect generally known as the Pasteur effect. There is a very large Pasteur effect in brain, and glycolysis can be increased as much as 10-fold under anaerobic conditions. Even at its maximal rate, however, anaerobic glycolysis is unable to provide sufficient ATP to meet the brain's energy demands. Since the stores of oxygen and glucose in brain are small compared with their rates of utilization, the brain requires their continuous replenishment by the circulation. If cerebral blood flow is completely interrupted, consciousness is lost within less then 10 sec, or the amount of time required to consume the oxygen contained within the brain tissue and its blood content.

The cerebral blood flow must be able to satisfy the brain's avaricious appetite for oxygen and glucose. The average rate of blood flow in the brain as a whole in normal conscious adult man is about 57 ml/100 g tissue per minute (TABLE 1). Blood flow to the whole brain is almost 800 ml/min, or approximately 15% of the total basal cardiac output. This level must be maintained within relatively narrow limits, for the brain cannot tolerate any major decline in its perfusion rate. A fall in cerebral blood flow to approximately half of its normal rate is sufficient to cause loss of consciousness in normal, healthy young men.

Cerebral oxygen consumption in normal, conscious, young men is approximately 156 μmoles/100 g brain per minute (TABLE 1); the rate is similar in young women. The rate of oxygen consumption by an entire brain of average weight (1,400 g) is then approximately 2.2 mmoles/min. The magnitude of this rate can be more fully appreciated when it is compared with the metabolic rate of the body as a whole. The average man weighs 70 kg and consumes approximately 11.2 mmoles oxygen per minute in the basal state. Therefore, the brain alone, which represents only 2% of total body weight, accounts for 20% of the resting total body oxygen consumption.

TABLE 1. Cerebral blood flow and metabolic rate in normal young adult human males

Function	Rate	
	Per 100 g brain tissue	Per whole brain[a]
Cerebral blood flow (ml/min)	57	798
Cerebral O_2 consumption (μmol/min)	156	2,188
Cerebral glucose utilization (μmol/min)	31	428

[a] Brain weight assumed to be 1,400 g. From Sokoloff (1981a).

Oxygen is normally utilized in brain almost entirely for the oxidation of carbohydrate (Kety, 1957; Sokoloff, 1960, 1981a). On the basis of the caloric value of oxygen in the oxidation of carbohydrate, the energy equivalent of the cerebral metabolic rate can be estimated to be approximately 20 watts, or 0.25 kcal/min. If it is assumed that this energy is utilized mainly for the synthesis of high-energy phosphate bonds, that the efficiency of the energy conservation is approximately 20%, and that the free energy of hydrolysis of the terminal phosphate of ATP is 7 kcal/mole, then it can be estimated that this energy expenditure supports the steady turnover of close to 7 mmoles (approximately 4×10^{21} molecules) of ATP per minute in the entire brain.

There has been considerable speculation about the nature of the functions responsible for the brain's energy demands. The brain does not do mechanical work like cardiac and skeletal muscle or osmotic work, as the kidney does in concentrating urine. It does not have the complex, energy-consuming metabolic functions of liver, and, despite the synthesis of hormones, neuropeptides, neurotransmitters, and a variety of lipids, it is not noted for its biosynthetic activities. Clearly the functions of nervous tissues are mainly excitation, inhibition, and conduction, which are reflected in the unceasing electrical activity of the brain. The electrical energy is ultimately derived from chemical processes, and it is likely that most of the brain's energy consumption is used for active transport of ions to sustain and restore membrane potentials discharged during the processes of excitation, inhibition, and conduction.

In contrast to most other tissues, which exhibit considerable flexibility with regard to the nature of the foodstuffs extracted and consumed from the blood, the brain is normally restricted almost exclusively to glucose as the substrate for its energy metabolism. Under normal circumstances no other potential energy-yielding substance has been found to be extracted by the brain from the blood in more than relatively trivial amounts compared to glucose. The stoichiometry of glucose utilization and oxygen consumption is summarized in TABLE 2.

TABLE 2. Relationship between cerebral oxygen consumption and glucose utilization in normal young adult man

Function	Value[b]
Oxygen consumption (μmol/100 g brain tissue/min)	156
Glucose utilization (μmol/100 g brain tissue/min)	31
Oxygen/glucose ratio (μmol/μmol)	5.5
Glucose equivalent of oxygen consumption (μmol glucose/100 g brain tissue/min)	26[a]
CO_2 production (μmol/100 g brain tissue/min)	156
Cerebral respiratory quotient (R.Q.)	0.97

[a] Calculated on the basis of 6 mol oxygen required for complete oxidation of 1 mol glucose.
[b] Values are the median of the values reported in the literature. From Sokoloff (1960, 1981a).

The normal, conscious human brain consumes oxygen at a rate of 156 μmoles/100 g tissue per minute. Carbon dioxide production is the same, leading to an R.Q. (ratio of CO_2 production to oxygen consumption) of approximately 1.0, further evidence that carbohydrate is the ultimate substrate for oxidative metabolism. Oxygen consumption and CO_2 production are equivalent to a rate of glucose utilization of 26 μmoles/100 g/min, assuming 6 μmoles of oxygen consumed and carbon dioxide produced for each μmole of glucose completely oxidized to CO_2 and H_2O. The glucose utilization actually measured is 31 μmoles/100 g/min, however, which indicates that glucose consumption is not only sufficient to account for the total oxygen consumption but is in excess by 5 μmoles/100 g/min. For the complete oxidation of glucose, the theoretical oxygen/glucose utilization ratio is 6.0 μmoles/μmole; the excess glucose utilization is responsible for the measured ratio of only 5.5 μmoles of oxygen per micromole of glucose. The fate of this excess glucose that is consumed is unknown, but it is probably distributed in part in lactate, pyruvate, and other intermediates and products of carbohydrate metabolism that are released from the brain into the blood, each in an insufficient amount to be detectable in arteriovenous differences. Some of the glucose is also utilized not for the production of energy but for the synthesis of other chemical constituents of the brain such as glycolipids, glycoproteins, phospholipids, acetylcholine, etc.

The combination of a cerebral R.Q. of unity, an almost stoichiometric relationship between oxygen uptake and glucose consumption, and the absence of any significant arteriovenous difference for any other energy-rich substrate is strong evidence that the brain normally derives its energy from the oxidation of glucose. In this respect, cerebral metabolism is unique because no other tissue, with the possible exception of the testis, has been found to rely only on carbohydrate for energy. It should be noted, however, that this situation applies only to the normal state. In ketotic states ketone bodies may replace glucose, at least in part, as oxidizable substrates of energy metabolism (Owen et al., 1967; Krebs et al., 1971).

MEASUREMENT OF LOCAL CEREBRAL BLOOD FLOW

The rates of blood flow and metabolism presented in TABLES 1 and 2 and discussed above represent average values in the brain as a whole. The brain is a heterogeneous organ, however, and is composed of almost innumerable structural elements that often function independently or even inversely with respect to one another. Correspondingly, their local rates of blood flow and energy metabolism vary widely but under physiological conditions are closely correlated with one another (Sokoloff, 1978).

Measurement of regional rates of cerebral metabolism will be discussed below. Most of the quantitative methods for the measurement of local rates of cerebral blood flow have been based on the principles of inert gas exchange between blood and tissues developed by Kety (1951) and utilize freely diffusible, chemically inert radioactive tracers that exchange freely between blood and cerebral tissues. They are often referred to as "tissue indicator equilibration" techniques and are in principle essentially the same, differing mainly with respect to the species of radioactive tracer utilized, whether the procedure is carried out during saturation of the tissues with the tracer or clearance of the tracer from the tissues, and the nature of the computational routine employed.

Theory of Tissue Indicator Equilibration Technique

Kety's (1951) principles of inert gas exchange were first applied to the measurement of local cerebral blood flow by Kety and his associates (Landau et al., 1955; Freygang and Sokoloff, 1958). The method is based on the Fick Principle which states that the rate of change of the quantity of a chemically inert tracer in the tissue under study equals the difference between the rates of delivery of the tracer to the tissue in the arterial blood and its removal from the tissue in the venous blood. This relationship can be expressed mathematically by the fundamental equation of the Fick Principle:

$$dQ_i/dt = F(C_a - C_v) \qquad [3]$$

where Q_i is the quantity of tracer in tissue i, F is the rate of blood flow through the tissue, and C_a and C_v are the tracer concentrations in the arterial blood and representative venous blood draining the tissue, respectively.

Dividing Eq. [3] by tissue mass, W:

$$dC_i/dt = F/W\ (C_a - C_v) \qquad [4]$$

where C_i equals the concentration of tracer in tissue i.

For tissues that are homogeneous with respect to tissue perfusion rate and solubility of tracer, Kety (1951) derived the following relationship between the tissue arteriovenous and arterial-tissue concentration differences:

$$(C_a - C_v) = m(C_a - C_i/\lambda) \qquad [5]$$

where λ equals the tissue/blood partition coefficient for the tracer, and m is a constant between 0 and 1 that represents the degree of tracer equilibration between the tissue and its venous drainage. m reflects the net effects of arteriovenous shunts, diffusion limitations, capillary impermeability, and all other factors that might tend to limit the equilibration of the tissue with the blood perfusing it. In the absence of arteriovenous shunts and diffusion limitations, m = 1, and Eq. [5] simplifies to

$$C_v = C_i/\lambda \qquad [6]$$

Substituting Eq. [5] into Eq. [4]:

$$dC_i/dt = \frac{mF}{\lambda W}(\lambda C_a - C_i) \qquad [7]$$

Let $K = mF/\lambda W$, the so-called clearance constant.

Integrating Eq. [7] between 0 time and time T:

$$C_i(T) = C_i(0)e^{-KT} + \lambda K e^{-KT}\int_0^T C_a(t)e^{Kt}\,dt \qquad [8]$$

Equation [8] is the general equation that is applicable for all types of time courses of the arterial curve. In studies carried out during desaturation following a period of saturation, Eq. [8] can be used with zero time set at the onset of desaturation and the second term in the right side of the equation correcting for recirculation of tracer in the arterial blood. This is the procedure generally used with the ^{133}Xe clearance method of measuring regional cerebral blood flow (Lassen and Ingvar, 1972). If C_a drops to zero at zero time after the period of saturation, as is approximately so after an intracarotid injection of radioactive inert gas, then C_a equals 0 and Eq. [8] reduces to

$$C_i(T) = C_i(0)e^{-KT} \qquad [9]$$

and K is determined from the slope of the semi-logarithmic plot of $C_i(T)$ vs. time. Blood flow can then be determined from the K if λ is known. This approach was initiated by Kety (1949) to determine muscle blood flow by clearance of $^{24}NaCl$ after intramuscular injection, but it has been more widely used for measurements of regional cerebral blood flow in man by ^{85}K or ^{133}Xe clearance (Lassen and Ingvar, 1972).

The method is generally applied during saturation when used with autoradiography in animals. In that case zero time is the time at the onset of the administration of the tracer into the circulation, $C_i(0)$ equals zero, and Eq. [8] reduces to

$$C_i(T) = \lambda K e^{-KT} \int_0^T C_a(t) e^{Kt}\, dt \qquad [10]$$

or, alternatively

$$C_i(T) = \lambda K \int_0^T C_a(t) e^{-K(T-t)}\, dt \qquad [11]$$

Equations [10] or [11] are the basis of the autoradiographic techniques for measuring local cerebral blood flow. They state that if the time course of the arterial concentration of the tracer between zero time and time T is measured, the tissue/blood partition coefficient is known, and the local tissue concentration of tracer at time T is measured, then the local clearance constant K can be computed. If there are neither arteriovenous shunts nor diffusion limitations (i.e., $m = 1$), then the local blood flow can be calculated from the clearance constant by the relationship

$$K = mF/\lambda W \qquad [12]$$

If the complete time course of the local tissue tracer concentration can also be measured, as it can with positron emission tomography, then the integrated form of Eq. [10] can be used. Thus,

$$\int_0^T C_i(T)\, dt = \lambda \left[\int_0^T C_a(t) dt - e^{-KT} \int_0^T C_a(t) e^{Kt}\, dt \right] \qquad [13]$$

[^{131}I]Trifluoroiodomethane Technique

The first truly quantitative measurement of local cerebral blood flow by the "tissue indicator equilibration" technique was the [^{131}I]CF_3I ([^{131}I]trifluoroiodomethane) technique (Landau et al., 1955; Freygang and Sokoloff, 1958). The tracer was [^{131}I]CF_3I, a radioactive, chemically stable gas, that diffuses freely through the blood-brain barrier. The species was the cat. The tracer was dissolved in blood or physiological saline and administered by continuous intravenous infusion for one minute during which the time course of the arterial concentration of tracer was measured in a well-type crystal scintillation counter. At one minute after the onset of the infusion the cat was decapitated, and the whole head was frozen as quickly as possible in liquid N_2. Local brain tissue concentrations of the tracer were determined by a unique, quantitative autoradiographic technique. Holes were drilled longitudinally through the extracerebral tissues of the frozen head and filled with 5% gelatin standards containing a range of concentrations of $K^{131}I$, which were calibrated by counting in the same crystal scintillation counter. The head was kept frozen with liquid N_2 and cut into 0.5 cm coronal sections with a band-saw. These sections were then autoradiographed with X-ray film at -75°C. It was essential to keep the head and sections frozen at these low temperatures to avoid loss of the $[^{131}I]CF_3I$, which is volatile at temperatures above -15°C but solid at -75°C.

Local tissue/blood partition coefficients were determined in a separate group of cats. The animals were allowed to breathe a constant concentration of $[^{131}I]CF_3I$ in the inspired air for at least 1 hour until the arterial blood concentration had equilibrated with the inspired air and the cerebral tissues had equilibrated with the arterial blood. Blood samples were then drawn, and the head was decapitated. Blood and local cerebral tissue concentrations of ^{131}I were assayed by crystal scintillation counting and quantitative autoradiography, respectively, as described above, and the local tissue/blood partition coefficients, λ, were determined as the ratios of tissue to blood concentrations at equilibrium. The partition coefficients varied widely among the various structures of the brain with values in white structures generally between 1.0 and 1.5 and values in gray structures between 0.7 and 1.0 (Freygang and Sokoloff, 1958).

From the values of the local partition coefficients, local tissue concentrations of tracer, and the time course of the arterial concentration, the latter corrected for the lag and dead space washout in the arterial sampling catheter (Freygang and Sokoloff, 1958), local clearance constants, K, were calculated by Eq. [10] or [11]. Local rates of blood flow were calculated from the clearance constants by the relationship, $K = mF/\lambda W$, with the assumptions of no arteriovenous shunts in brain, no diffusion limitations with the gaseous tracer, and an m equal to 1.

Local rates of cerebral blood flow. The results obtained with the $[^{131}I]CF_3I$ technique demonstrated a wide distribution of rates of blood flow in the component structures of the conscious cat brain (TABLE 3). Gray matter blood flow was generally 3-4 times that of white matter, but in gray matter there was a wide range with the inferior colliculus exhibiting the highest rate of blood flow in the brain of the cat.

Light thiopental anesthesia depressed blood flow in almost all gray matter structures, particularly in primary sensory cortical regions which in the conscious state had high perfusion rates (TABLE 3). There were no significant changes in blood flow to white matter. The net effect of the anesthesia was to reduce blood flow differentially in the various structures, particularly in the cerebral cortex, and to produce a more uniform distribution of blood flow at a lower level throughout the brain.

Under the conditions of the procedure (i.e., infusion of tracer for 1 min), the distribution of tracer in the various structures of the brain reflects mainly their local perfusion rates. The autoradiograms, which are images of the distribution of the isotope, are, therefore, pictorial representations of the relative rates of blood flow in the component structures of the brain (Fig. 1).

Effects of functional activity on local cerebral blood flow. The $[^{131}I]CF_3I$ technique demonstrated that there is a close relationship between the level of functional activity in discrete structures of the brain and their rates of blood flow. Retinal stimulation by photoflash in the cat produced increases in blood flow selectively in the superior colliculi, lateral geniculate body, and visual cortex (e.g. lateral gyrus) which were clearly visible in the autoradiographic images (Fig. 1) (Sokoloff, 1961). These experiments were, in fact, the first to demonstrate the imaging of local functional activity in the brain.

$[^{14}C]$Iodoantipyrine Technique

The basic principles of "tissue indicator equilibration" techniques were first applied to the measurement of local cerebral blood flow with the chemically inert, radioactive gas, $[^{131}I]$trifluoroiodomethane, but they apply equally well to any freely diffusible, chemically stable tracer. There were a number of technical disadvantages associated with

TABLE 3. Local cerebral blood flow in the cat

Structure	Blood flow: Mean ± S.E.M (ml/g/min) Conscious (10 cats)	Light Thiopental Anesthesia (11 cats)
Superficial Cerebral Structures		
Cortex:		
Sensory-motor	1.38 ± 0.12	0.65 ± 0.07[a]
Auditory	1.30 ± 0.05	0.72 ± 0.07[a]
Visual	1.25 ± 0.06	0.77 ± 0.09[a]
Miscellaneous-association	0.88 ± 0.04	0.67 ± 0.06[a]
Olfactory	0.77 ± 0.06	0.62 ± 0.07
White matter	0.23 ± 0.02	0.26 ± 0.04
Deep Cerebral Structures		
Medial geniculate body	1.22 ± 0.04	0.81 ± 0.09[a]
Lateral geniculate body	1.21 ± 0.08	0.79 ± 0.07[a]
Caudate nucleus	1.10 ± 0.08	0.91 ± 0.11
Thalamus	1.03 ± 0.05	0.71 ± 0.09[a]
Hypothalamus	0.84 ± 0.05	0.55 ± 0.06[a]
Basal ganglia and amygdala	0.75 ± 0.03	0.58 ± 0.05[a]
Hippocampus	0.61 ± 0.03	0.59 ± 0.04
Optic tract	0.27 ± 0.02	0.22 ± 0.08
Midbrain and Pons		
Inferior colliculus	1.80 ± 0.11	1.41 ± 0.14[a]
Superior olive	1.17 ± 0.13	1.56 ± 0.27
Superior colliculus	1.15 ± 0.07	0.82 ± 0.10[a]
Pontine gray	0.88 ± 0.04	0.61 ± 0.03[a]
Reticular formation	0.59 ± 0.05	0.49 ± 0.06
Pontine white	0.24 ± 0.02	0.31 ± 0.04
Cerebellum, Medulla, Spinal Cord		
Cerebellum:		
Nuclei	0.79 ± 0.05	0.56 ± 0.08[a]
Cortex	0.69 ± 0.04	0.57 ± 0.05[a]
White matter	0.24 ± 0.01	0.29 ± 0.06
Medulla:		
Vestibular nuclei	0.91 ± 0.04	0.84 ± 0.10
Cochlear nuclei	0.87 ± 0.07	0.99 ± 0.14
Pyramids	0.26 ± 0.02	0.28 ± 0.03
Spinal cord:		
Gray matter	0.63 ± 0.04	0.53 ± 0.07
White matter	0.14 ± 0.02	0.15 ± 0.06

[a] Significantly different from conscious control values ($p < 0.05$).
From Sokoloff, 1961.

the use of $[^{131}I]CF_3I$. The radioactive gas was not commercially available and because of the 8-day half-life of ^{131}I had to be frequently resynthesized and purified by relatively complex chemical procedures.

Also assay of a volatile tracer in blood and tissues, particularly by quantitative autoradiography, proved to be cumbersome. Also, the tissue/blood partition coefficient, λ, differed widely in the local cerebral tissues and also varied with the hematocrit (Freygang and Sokoloff, 1958). Finally, the autoradiographic resolution that was achieved with the radiations of ^{131}I from 0.5 cm frozen tissue sections was rather limited. A commercially available, nonvolatile ^{14}C-labeled tracer to substitute for [^{131}I]CF_3I was, therefore, sought. Reivich et al. (1969) adapted the method for use with [^{14}C]antipyrine, which at first appeared to satisfy all requirements. It greatly enhanced the convenience of the technique and improved the quality of the images in the autoradiograms. The autoradiographic procedure was modified for use with ^{14}C and applied to dried brain sections, 20 μm in thickness, at room temperature. Because of the long half-life of ^{14}C, permanent ^{14}C-labeled plastic standards could be prepared and calibrated for quantitative densitometry of the autoradiographic images.

Further experience with [^{14}C]antipyrine indicated, however, that it was not completely suitable. Its use led to values for local cerebral blood flow in highly perfused tissues that were considerably below those obtained with radioactive gases (Eklöf et al, 1974). The problems with [^{14}C]antipyrine were shown to arise from limitations in its diffusion through the blood-brain barrier, leading to values for m less than 1.0 and, therefore, values for K that reflected not only blood flow but diffusion limitations as well (Eckman et al., 1975).

The limitations of [^{14}C]antipyrine were overcome by the substitution of 4-iodo-N-[methyl-^{14}C]antipyrine as the tracer (Sakurada et al., 1978). Iodoantipyrine has a higher oil/water partition coefficient and is, therefore, more freely diffusible through the blood-brain barrier. An additional advantage of [^{14}C]iodoantipyrine is that its tissue/blood partition coefficient is uniform (0.8) throughout the central nervous system and does not vary with the hematocrit. The autoradiographic images are, therefore, pictorial representations of the relative rates of blood flow uncontaminated by differences in solubility of the tracer in the tissues or by diffusion barriers. In all respects, the procedure used with [^{14}C]iodoantipyrine is identical to that employed with [^{14}C]antipyrine, but it leads to values for local cerebral blood flow

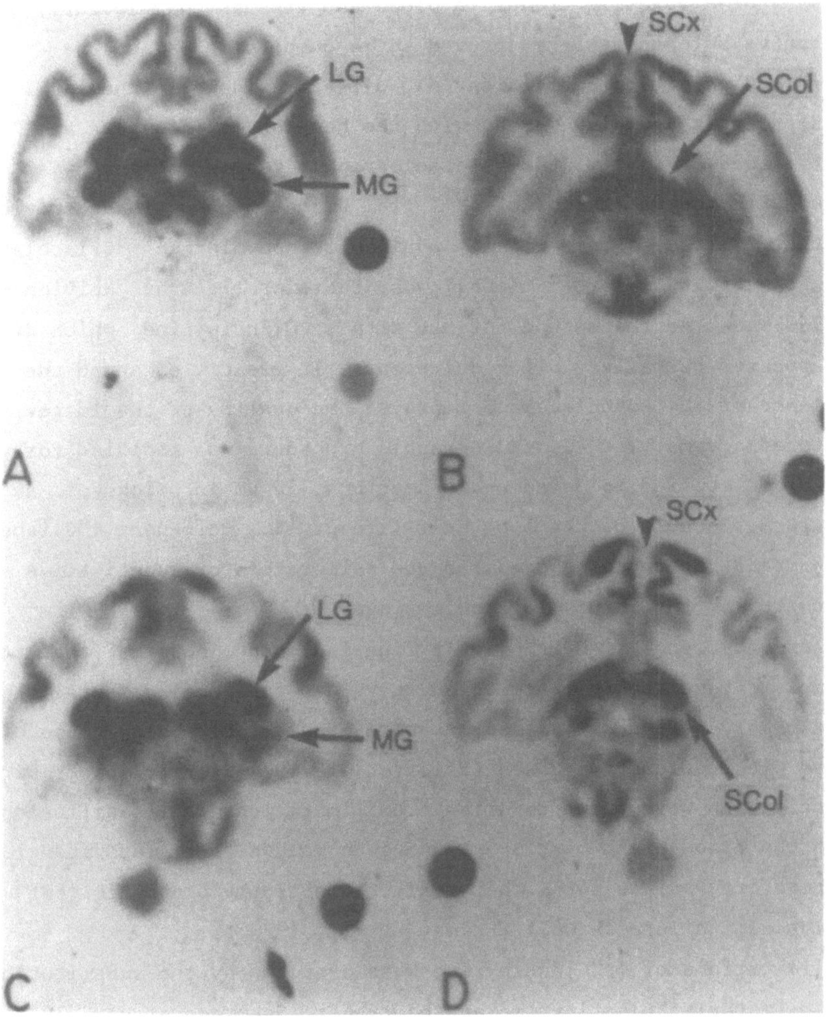

Fig 1. Typical autoradiograms of sections of brains from conscious cats imaging the different concentrations of the radioactive gas, $[^{131}I]CF_3I$, in the various cerebral structures. In general, the differences in optical density reflect differences in blood flow, the greater the density (i.e., darkness), the higher the blood flow (see text). Circular areas of uniform density represent calibrated ^{131}I-containing gelatin solutions used for quantitative densitometric analysis. A and B: Autoradiograms of two sections from blind-folded cat. Note striate cortex (SCx), lateral (LG) and medial (MG) geniculate nuclei, and superior colliculi (SCol). C and D: Autoradiograms of corresponding sections from a cat subjected to photic stimulation of the retina for 5 min. Note marked increases in optical density of striate cortex, the lateral geniculate nuclei, and the superior colliculi relative to that of other structures such as the medial geniculate nuclei. From Sokoloff (1961).

that are essentially the same as those obtained with $[^{131}I]CF_3I$. The $[^{14}C]$iodoantipyrine technique has now become, therefore the standard autoradiographic method for the measurement of local cerebral blood flow.

Indicator Fractionation Techniques

The Fick Principle is also the basis for an alternative approach to the measurement of regional cerebral blood flow, i.e., "indicator fractionation" techniques. From the Fick Principle,

$$dC_i/dt = F/W\ (C_a - C_v) \qquad [14]$$

and

$$F/W = \frac{C_i(T)}{\int_0^T (C_a - C_v)dt} \qquad [15]$$

where F/W = rate of blood flow per unit mass of cerebral tissue.

$C_i(T)$ = concentration of tracer in tissue i at given time T following introduction of tracer into circulation.

C_a and C_v = arterial and cerebral venous tracer concentrations, respectively, at various times t after introduction of the tracer into the circulation.

Multiplying $(C_a\text{-}C_v)$ in the denominator of Eq. [15] by C_A/C_A,

$$F/W = \frac{C_i(T)}{\int_0^T \left[\frac{C_a - C_v}{Ca}\right] C_a\ dt} \qquad [16]$$

But $(C_a\text{-}C_v)/C_a$ equals the net cerebral extraction fraction, E. Thus,

$$F/W = \frac{C_i(T)}{\int_0^T E\ C_a dt} \qquad [17]$$

If E, the net extraction fraction, happens to remain constant with time, then E can be factored out, and

$$F/W = \frac{C_i(T)}{E\int_0^T C_a\, dt} \qquad [18]$$

If E is not only constant with time but is also equal to 1.0, then Eq. [18] reduces to

$$F/W = \frac{Ci(T)}{\int_0^T C_a\, dt} \qquad [19]$$

Equation [19] is the basis of the microsphere techniques in which labeled microspheres that are introduced into the circulation are completely extracted from the blood and trapped in the brain. If, however, E is neither equal to 1.0 nor constant with time, then Eq. [17] must be used, a far more complicated situation. Extensive efforts are being made to find a γ-emitting radiochemical that might be completely extracted and retained by brain tissue so that Eq. [18] or [19] could be used to measure regional cerebral blood flow in man with single photon emission computerized tomography (SPECT); none has yet been found that provides reliable fully quantitative results (Lucignani et al., 1985).

MEASUREMENT OF LOCAL RATES OF CEREBRAL ENERGY METABOLISM

General Principles

Radioisotopes are frequently used to facilitate assay of rates of biochemical reactions. Usually they are used to study chemical reactions *in vitro*. The procedure is to label one of the reactants and to measure the rate of accumulation of a labeled product specific to the reaction of interest. From either assay or prior knowledge of the specific activity of the reactant molecule and the stoichiometry of the reaction, the rate of the overall reaction can be calculated from the rate of radioactive product formation (Fig. 2). Application of these principles of radioassay of biochemical reactions has been applied to measure local rates of energy metabolism in the brain _in vivo_.

$$\text{Rate of Reaction} = \frac{\text{Labeled Product Formed in Interval of Time, 0 to T}}{\begin{bmatrix}\text{"Isotope Effect"} \\ \text{Correction Factor}\end{bmatrix}\begin{bmatrix}\text{Integrated Specific Activity} \\ \text{of Precursor}\end{bmatrix}}$$

Fig. 2. General equation for the determination of the rates of biochemical reactions with radioisotopes (see text).

Quantitative autoradiography measures local concentrations of isotopes in tissues of animals labeled _in vivo_. The judicious selection of an appropriate radiochemical tracer, combined with a rationally designed procedure, makes it possible to use this capability to measure the local rates of a chemical process in tissues of animals _in vivo_. Emission tomography, particularly positron-emission tomography, extends this capability to man and makes it possible to assay rates of biochemical processes in human tissues _in vivo_. Autoradiography or emission tomography do not, however, obviate the need to adhere to established principles of chemical and enzyme kinetics and tracer theory. Generally, all such methods, whether to be used in man with positron-emission tomography or in animals with autoradiography, should be developed by research in animals with autoradiography because it is only in animals that essential measurements needed to validate the model and basic assumptions of the method can be tested and evaluated.

It is a common misconception that the ability to image the distribution of the local concentrations of an isotope in the tissues is by itself sufficient to characterize a physiological or biochemical process in the tissues. It is almost certainly possible to administer radioactive shoe-polish, or some other irrelevant labeled compound, to an animal and observe patterns of distribution of the isotope in the tissues, but it is unlikely that the distribution of the radioactivity would provide any useful information concerning any process of interest in the tissues. The selection of the labeled compound to be used as the tracer and the timing and design of the procedure are of critical importance to identify and assay the process of interest.

The tissue uptake of a tracer circulating in the blood is dependent on at least three processes: 1) blood flow and delivery of the tracer to the tissues; 2) transport of the tracer between the blood and tissues; and 3) metabolism of the tracer compound in the tissue and the fate of the metabolic products. [^{14}C]Antipyrine is a chemically inert diffusible tracer the uptake of which by brain tissue is dependent mainly on blood flow to the tissue and its relative solubility in tissue and blood (Reivich et al., 1969). 3-O-[^{14}C]Methylglucose is distributed to brain tissue according to blood flow, tissue-blood distribution ratio, and hexose carrier-mediated transport through the blood-brain barrier. 2-[^{14}C]Deoxyglucose is distributed to the brain tissue according to blood flow, tissue-blood distribution ratio, carrier mediated blood-brain barrier transport, and hexokinase-catalyzed phosphorylation (Sokoloff et al., 1977). In Fig. 3 are autoradiograms of brain sections obtained 1 and 60 min after each of these labeled compounds was administered as an intravenous pulse; they illustrate the distribution of the label in the cerebral tissues, the darker the density the higher the concentration of ^{14}C. At one min the patterns of distribution of ^{14}C with all three compounds are almost the same, indicating that the influences of blood flow, solubility, transport, and metabolism cannot be discriminated. At 60 minutes after the pulse the autoradiograms obtained with [^{14}C]methylglucose and [^{14}C]antipyrine are essentially homogeneous whereas the [^{14}C]deoxyglucose autoradiogram retains the same pattern as seen 1 min after the pulse. At 60 min the [^{14}C]methylglucose and [^{14}C]antipyrine autoradiograms reflect mainly tissue-blood distribution ratios while that of [^{14}C]deoxyglucose reflects its metabolism. It is clear from these results that not only the chemical species of labeled compound but also the procedure and time are critically important in identifying and defining the relationship between the distribution of the label in the tissues and the nature of the physiological and/or biochemical processes responsible for the distribution.

A chemical reaction is the conversion of one species of molecule to another. The rate of this reaction can be determined by measuring the rate of formation of one or more of the products. The addition of a radioactive tracer to one of the reactants facilitates the measurement of radioactive product formation but does not solve all the problems. The rate of chemical transformation of the labeled species can be measured relatively easily, but this is not the reaction of interest. To derive the rate of the total overall reaction from measurement of the reaction rate of only the labeled species, it is necessary to know the integrated specific activity (i.e., ratio of labeled to total molecules) of the precursor pool. Occasionally the labeled species exhibits a kinetic difference from the natural compound, the so-called isotope effect; this isotope effect can be calibrated and appropriate correction made for it. The general equation for the determination of the rate of a biochemical reaction with radioactive tracers is presented in Fig. 2.

In assays of biochemical reactions in vivo it is impossible to determine the integrated specific activity of the precursor pool directly because that would require measurement of the complete time courses of the concentrations of the labeled and unlabeled precursor molecules in the tissue at the enzyme site. The integrated precursor pool specific

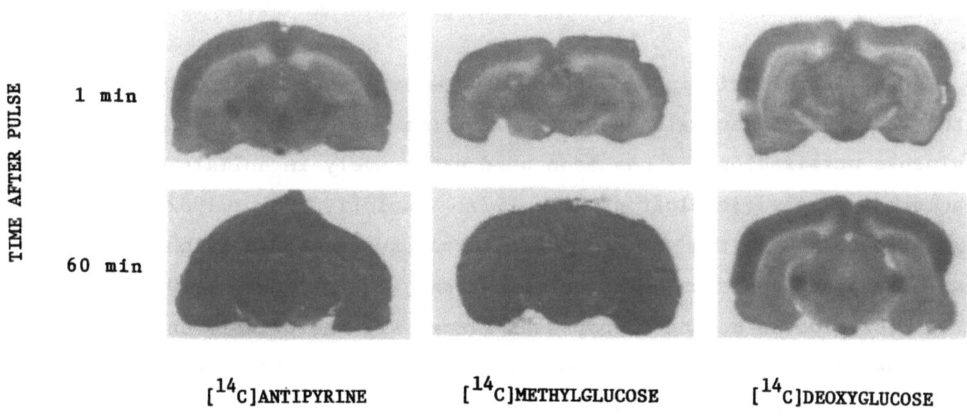

Fig. 3. Autoradiograms of brain sections from rats 1 and 60 min after pulse of tracer doses of [^{14}C]antipyrine, [^{14}C]methylglucose, or [^{14}C]deoxyglucose. Note that at 60 min after the pulse only the [^{14}C]deoxyglucose autoradiogram retains regional heterogeneity, indicating that local metabolism rather than blood flow and transport determines the local tissue concentrations of ^{14}C at long times after a pulse of [^{14}C]deoxyglucose. From Sokoloff et al. (1977).

activity must be determined indirectly from measurements in the blood supplied to the tissue. The time courses of the specific activity in arterial blood or plasma are measured directly, and the integrated specific activity of the precursor pool in the tissue can be calculated from them by correcting for the lag in the equilibration of the precursor pool in the tissue with that in plasma. To apply this correction it is necessary, of course, to know the kinetics of the equilibration process between the precursor pools in the tissue and plasma.

The rate of a chemical reaction is generally determined by measurement of product accumulation. Autoradiography and emission tomography, however, measure only total concentration of radioactive molecules and cannot distinguish between the various chemical species that may be labeled. These may include, in addition to the precursor and specific products of the reaction of interest, other non-specific labeled products of alternative or side reactions. Strategies must, therefore, be developed to ensure that the radioactivity that is measured by the autoradiography or emission tomography is confined exclusively to the precursor and product molecules specific to the chemical reaction under study. The labeled precursor should be so selected that its chemical transformations are limited as much as possible to the specific pathway under study.

These general principles have been more or less successfully applied in the radioactive deoxyglucose method for the measurement of local cerebral glucose utilization. It has been used extensively in animals with autoradiography (Sokoloff et al., 1977; Sokoloff, 1981b, 1982) and adapted to man by the use of $[^{18}F]$fluorodeoxyglucose (Reivich et al.,1979) and positron-emission tomography (Phelps et al., 1979). Because the deoxyglucose method encompasses almost all the principles to be considered in the measurement of biochemical processes _in vivo_, it can serve as a model of how they can be applied.

The Deoxyglucose Method

Theory. The design of the method is based on a kinetic analysis of a model of the biochemical behavior of 2-deoxyglucose (DG) and glucose in brain (Fig. 4A) (Sokoloff et al., 1977; Sokoloff, 1982). Both hexoses are transported bi-directionally between blood and brain by a commom carrier in the blood-brain barrier. In the tissues hexokinase phosphorylates both to their hexose-6-phosphate derivatives (Sols and Crane, 1954). Glucose and 2-deoxyglucose are, therefore, competitive substrates for both blood-brain barrier transport and hexokinase-catalyzed phosphorylation. Unlike glucose-6-phosphate (G-6-P), however, which is rapidly metabolized further eventually to CO_2 and water and to a lesser extent via the pentose phosphate shunt, deoxyglucose-6-P (DG-6-P) cannot be converted to fructose-6-phosphate and is a poor substrate for G-6-P dehydrogenase (Sols and Crane, 1954). There is very little glucose-6-phosphatase (G-6-Pase) activity in brain (Hers, 1957) and even less DG-6-phosphatase (DG-6-Pase) activity (Sokoloff et al., 1977; Fishman and Karnovsky, 1986). DG-6-P can be converted to DG-1-phosphate and then to uridine-diphospho-deoxyglucose (UDP-DG), glycogen, glycolipids and glycoproteins, but these reactions are slow; in mammalian tissues only a tiny fraction of the DG-6-P formed proceeds to these products (Nelson et al., 1984). These compounds are relatively stable products of DG-6-P, and all together represent the ultimate products of phosphorylation of DG. DG-6-P and its derivatives, once formed, are, therefore, trapped in the cerebral tissues, not forever but long enough to allow a fairly long period of measurement before significant loss of product occurs.

If the interval of time following administration of $[^{14}C]$DG is kept short enough (e.g., less than 1 hour) to keep the loss of $[^{14}C]$DG-6-P and its secondary products from the tissues negligible, then the quantity of labeled products accumulated in the tissue at any given time following the injection of $[^{14}C]$DG circulation equals the integral of the rate of $[^{14}C]$DG phosphorylation by hexokinase in that tissue during that interval. This integral is in turn related to the amount of glucose that has been phosphorylated over the same interval, depending on the time courses of the $[^{14}C]$DG and glucose concentrations (e.g., specific activity) in the precursor pools and the Michaelis-Menten kinetic constants for hexokinase with respect to $[^{14}C]$DG and glucose. With cerebral glucose consumption in a steady state, the amount of glucose phosphorylated during the interval of time equals the steady state flux of glucose through the hexokinase-catalyzed step times the duration of the interval, and the net rate of flux of glucose through this step equals the rate of glucose utilization.

These relationships can be mathematically defined and an operational equation derived if the following assumptions are made: 1) steady state for glucose (i.e., constant plasma glucose concentration and constant rate of glucose consumption) during experimental period; 2) homogeneous tissue compartment with respect to blood flow, transport, and concentrations of [^{14}C]DG and glucose; and 3) tracer concentrations of [^{14}C]DG and [^{14}C]DG-6-P (i.e., molecular concentrations essentially equal to zero). The operational equation which defines R_i, the rate of glucose consumption per unit mass in tissue i, in terms of measurable variables is presented in Figure 4B.

The rate constants, K_1^*, k_2^*, k_3^*, are determined in a separate group of animals by a non-linear, least squares best-fit of an equation, which defines the time course of total tissue ^{14}C concentration, C_i^*, in terms of time, history of plasma [^{14}C]DG concentration, and the rate constants, to the experimentally determined time courses of tissue and plasma concentrations of ^{14}C (TABLE 4) (Sokoloff et al., 1977).

The λ, Φ, and Michaelis-Menten constants are grouped together to constitute a single, lumped constant (Fig. 4B). This lumped constant can be shown to equal the asymptotic value of the product of the ratio of the cerebral extraction ratios of [^{14}C]DG and glucose and the ratio of the arterial blood to plasma specific activities when the arterial plasma [^{14}C]DG concentration is maintained constant (Sokoloff et al., 1977). The lumped constant is also determined in a separate group of animals from arterial and cerebral venous blood samples drawn during a programmed intravenous infusion that produces and maintains a constant arterial plasma [^{14}C]DG concentration (Sokoloff et al., 1977). An example of such a determination in a monkey is illustrated in Fig. 5. Under normal conditions the lumped constant appears to be characteristic of the species and does not appear to change significantly in a wide range of physiological conditions (TABLE 5), but it has been found to change in pathophysiological conditions, markedly in severe hypoglycemia (Suda et al., 1981), slightly in hyperglycemia (Schuier et al., 1981), and whenever glucose utilization is limited by glucose supply.

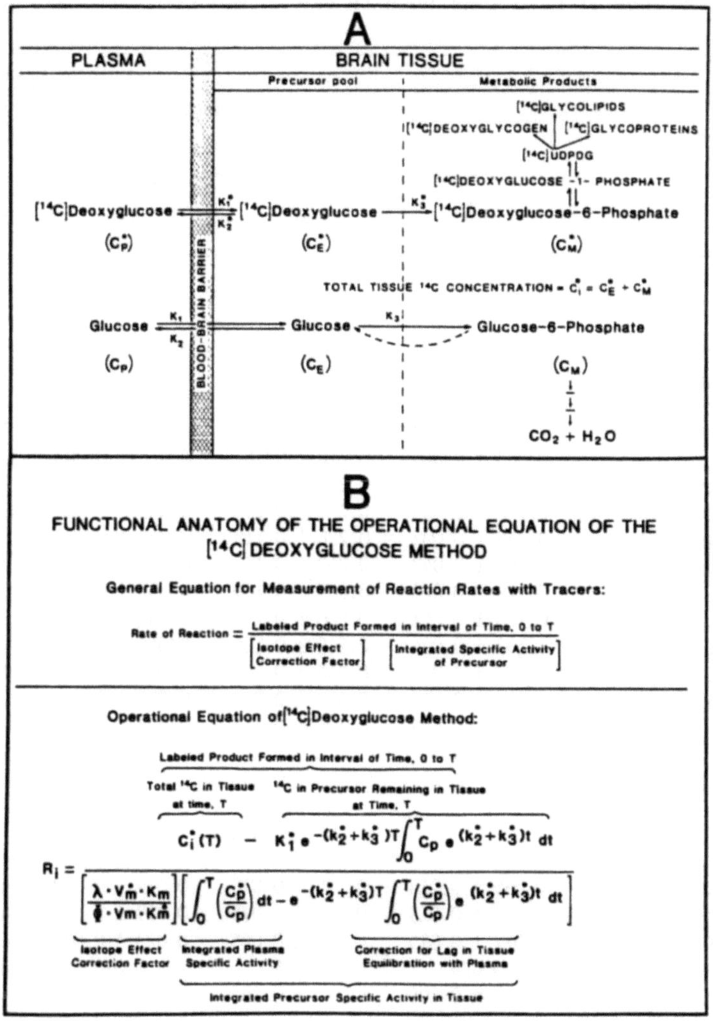

Fig 4. A: Model of DG method. C_i^* = total ^{14}C concentration in tissue; C_P^* and C_P = [^{14}C]DG and glucose concentrations in arterial plasma; C_E^* and C_E are their concentrations in tissue pools that serve as substrates for phosphorylation by hexokinase or transport back to plasma; C_M^* = combined concentrations of [^{14}C]DG-6-P and its derivatives in the tissue. K_1^*, k_2^*, and k_3^* = rate constants for transport of [^{14}C]DG from plasma to tissue and tissue to plasma, and for phosphorylation, respectively; K_1, k_2, and k_3 = equivalent rate constants for glucose. [^{14}C]DG and glucose compete for carrier that transports both between plasma and tissue and also for hexokinase. Dashed arrow represents possibility of G-6-P hydrolysis by G-6-Pase activity, if any. B: Operational equation of DG method. R_i = rate of glucose utilization; T equals duration of experimental period; λ is ratio of distribution space of DG in tissue to that of glucose; Φ equals fraction of G-6-P that continues down glycolytic and pentose phosphate shunt pathways ($\Phi = 1$ in absence of G-6-Pase activity); and K_m^* and V_m^* and K_m and V_m are Michaelis-Menten kinetic constants of hexokinase for DG and glucose, respectively.

TABLE 4. Values of rate constants in the normal conscious albino rat

Structure	Rate Constants (min^{-1})			Half-Life Precursor Pool (min)
	K_1^*	k_2^*	k_3^*	$\log_e 2/(k_2^*+k_3^*)$
		Gray Matter		
Visual cortex	0.189 ± 0.048	0.279 ± 0.176	0.063 ± 0.040	2.03
Auditory cortex	0.226 ± 0.068	0.241 ± 0.198	0.067 ± 0.057	2.25
Parietal cortex	0.194 ± 0.051	0.257 ± 0.175	0.062 ± 0.045	2.17
Sensory-motor cortex	0.193 ± 0.037	0.208 ± 0.112	0.049 ± 0.035	2.70
Thalamus	0.188 ± 0.045	0.218 ± 0.144	0.053 ± 0.043	2.56
Medial geniculate body	0.219 ± 0.055	0.259 ± 0.164	0.055 ± 0.040	2.21
Lateral geniculate body	0.172 ± 0.038	0.220 ± 0.134	0.055 ± 0.040	2.52
Hypothalamus	0.158 ± 0.032	0.226 ± 0.119	0.043 ± 0.032	2.58
Hippocampus	0.169 ± 0.043	0.260 ± 0.166	0.056 ± 0.040	2.19
Amygdala	0.149 ± 0.028	0.235 ± 0.109	0.032 ± 0.026	2.60
Caudate-putamen	0.176 ± 0.041	0.200 ± 0.140	0.061 ± 0.050	2.66
Superior colliculus	0.198 ± 0.054	0.240 ± 0.166	0.046 ± 0.042	2.42
Pontine gray matter	0.170 ± 0.040	0.246 ± 0.142	0.037 ± 0.033	2.45
Cerebellar cortex	0.225 ± 0.066	0.392 ± 0.229	0.059 ± 0.031	1.54
Cerebellar nucleus	0.207 ± 0.042	0.194 ± 0.111	0.038 ± 0.035	2.99
Mean	0.189	0.245	0.052	2.39
± S.E.M.	± 0.012	± 0.040	± 0.010	± 0.40
		White Matter		
Corpus callosum	0.085 ± 0.015	0.135 ± 0.075	0.019 ± 0.033	4.50
Genu of corpus callosum	0.075 ± 0.013	0.131 ± 0.075	0.019 ± 0.034	4.62
Internal Capsule	0.077 ± 0.015	0.134 ± 0.085	0.023 ± 0.039	4.41
Mean	0.079	0.133	0.020	4.51
± S.E.M.	± 0.008	± 0.046	± 0.020	± 0.90

From Sokoloff et al. (1977)

Despite its complex appearance, the operational equation in Figure 4B is really nothing more than a re-statement of the standard relationship by which rates of enzyme-catalyzed reactions are determined from measurements made with radioactive tracers (Fig. 2). The numerator of the equation represents the amount of radioactive product formed in a given interval of time· it is equal to C_i^*, the sum of the concentrations of [^{14}C]DG, [^{14}C]DG-6-P, and the further products of the latter in the tissue at time, T, measured by quantitative autoradiography, less a term that represents the free unmetabolized [^{14}C]DG still remaining in the tissue. The denominator represents the integrated specific activity of the precursor pool times a factor, the lumped constant, which corresponds to a correction factor for an isotope effect. The term with the exponential factor in the denominator takes into account the lag in the equilibration of the tissue precursor pool with that of the plasma.

By the use of labeled 2-deoxyglucose as a probe, a single biochemical reaction, the first step in the pathway of glucose metabolism, has been isolated and essentially purified in the midst of the morass of chemical reactions going on *in vivo* (Fig. 6). This reaction is the hexokinase-catalyzed phosphorylation of glucose. The total amount of radioactive product formed and the integrated specific activity of the precursor at the enzyme site can be experimentally determined. From these data and the use of the "lumped constant" to account for the difference in the kinetic behavior of deoxyglucose and glucose, the net rate of phosphorylation of glucose by hexokinase can be calculated by the operational equation (Fig. 4B). In a steady state the net rate through any one step in a pathway equals the net rate through any other step and the rate of the overall pathway. The DG method, therefore, always measures *in vivo* the local net rates of glucose phosphorylation, and in a steady state these are the net rates of the entire glycolytic pathway (Fig. 6).

Procedure. The deoxyglucose method is based on the mathematical analysis of a kinetic model of the biochemical behavior of [^{14}C]DG and glucose in brain (Fig. 4A). Models, however, are mainly products of the human mind; they almost always represent idealized over-simplified situations that cannot possibly account for all the known, let alone unknown, properties of the complex biological systems existing in the brain *in vivo*. The procedure of the method was, therefore, designed to minimize, as much as possible, effects of imperfections in the model and to make the method as model-independent and measurement-dependent as possible. To do so, however, it was necessary to make some compromises, particularly with respect to time resolution, but these compromises were compensated by improved accuracy, precision, and reliability.

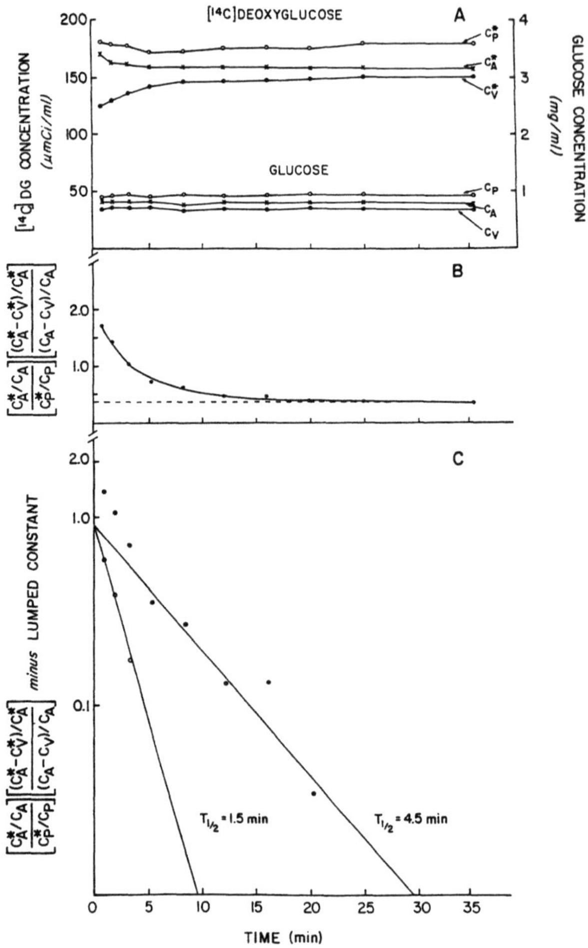

Fig. 5. Example of determination of lumped constant in monkey. A: Time courses of [^{14}C]DG and glucose concentrations in arterial blood, C_A^* and C_A; arterial plasma, C_P^* and C_P; and cerebral venous blood, C_V^* and C_V; respectively, during programmed infusion that maintains constant arterial [^{14}C]DG concentration. **B**: Arithmetic plot of function (i.e., formula in ordinate) of the variables measured in (A) vs. time. This function declines exponentially with a rate constant equal to $(k_2^* + k_3^*)$ until it reaches an asymptotic value equal to the lumped constant, 0.35 in this experiment (dashed line). **C**: Semi-logarithmic plot of curve in (B) less the lumped constant, i.e., its asymptotic value. ●, Actual values. This curve is analyzed by curve-peeling technique into 2 components to yield 2 straight lines representing the two compartments. o, Points for fast compartment, obtained by subtracting values for slow compartment from solid circles. The rate constants for these 2 components represent the values for $(k_2^* + k_3^*)$ for fast and slow compartments that are assumed to represent gray and white matter. The values for $(k_2^* + k_3^*)$ in this experiment were 0.462 (T1/2 = 1.5 min) and 0.154 (T1/2 = 4.5 min) in gray and white matter, respectively. From Kennedy et al. (1978).

TABLE 5. Values of lumped constant in various species

Animal	No. of Animals	Mean ± S.D.	S.E.M.
Albino rat:			
Conscious	15	0.464 ± 0.099a	± 0.026
Anesthetized	9	0.512 ± 0.118a	± 0.039
Conscious (5% CO2)	2	0.463 ± 0.122a	± 0.086
Combined	26	0.481 ± 0.119	± 0.023
Rhesus monkey (Conscious)	7	0.344 ± 0.095	± 0.036
Cat (Anesthetized	6	0.411 ± 0.013	± 0.005
Dog (beagle puppy) (Conscious)	7	0.558 ± 0.082	± 0.031
Sheep:			
Fetus	5	0.416 ± 0.031	± 0.014
Newborn	4	0.382 ± 0.024	± 0.012
Combined	9	0.400 ± 0.033	± 0.011
Humans			
Conscious	6	0.568 ± 0.105	± 0.043

a No statistically significant difference between normal conscious and anesthetized rats ($0.3 < p < 0.4$) and conscious rats breathing 5% CO_2 ($p > 0.9$).

Values obtained as follows: Rat, Sokoloff et al.(1977); monkey, Kennedy et al. (1978); cat, M. Miyaoka, J. Magnes, C.Kennedy, M. Shinohara, and L. Sokoloff (unpublished data); dog, Duffy et al. (1982); sheep, Abrams et al. (1984); humans, Reivich et al. (1985).

Theoretical and practical considerations in design of procedure. The operational equation method (Fig. 4B) specifies the variables to be measured and the parameters to be known in order to determine R_i, the rate of glucose consumption. Variables measured in each experiment are: 1) the entire history of the arterial plasma [^{14}C]DG concentration, C_p^*, from zero time to time T; 2) the steady state arterial plasma glucose concentration, C_p; and 3) the concentration of ^{14}C in the tissue at the time of killing, $C_i^*(T)$. The rate constants, K_1^*, k_2^*, and k_3^*, and the lumped constant, $\lambda V_m^* k_m / \Phi V_m k_m^*$, are not measured in each experiment; their values are those determined separately in other groups of animals as described above and presented in TABLES 4 and 5. It is in the role of the rate constants and the lumped constant in the operational equation that the sensitivities to imperfections in the model reside.

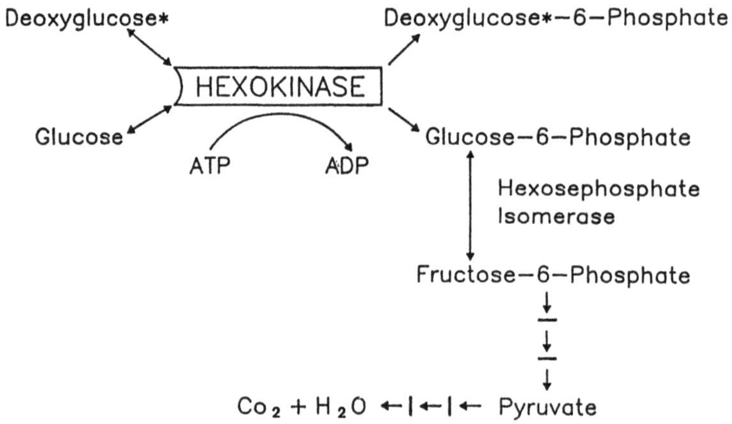

Fig. 6. Schema illustrating fundamental principles of radioactive DG method for measurement of local cerebral glucose utilization. Glucose utilization commences with hexokinase-catalyzed phosphorylation by ATP, but the product of this reaction, glucose-6-P, is not retained in the tissues. Instead, it is metabolized further to products such as CO_2 and H_2O that rapidly leave the tissue. Deoxyglucose, an analogue and competitive substrate with glucose in the hexokinase reaction, leads to a product, DG-6-P, that does accumulate in the tissue quantitatively for a reasonable length of time. By labeling the DG, it is possible to determine the rate of labeled DG-6-P formation. From knowledge of the time courses of the relative concentrations of labeled DG and glucose in the tissue at the enzyme site (determined indirectly from measurements of the time courses in the plasma) and the relative Michaelis-Menten constants of hexokinase for the two substrates, it is possible to calculate the amount of glucose phosphorylated during the same period in which the amount of measured DG-6-P was produced. In a steady state of glucose metabolism the net rate of glucose phosphorylation is equal to the net rate of the glycolytic pathway.

Rate constants: The operational equation is generally applicable with all types of arterial plasma [^{14}C]DG time courses, but its configuration suggests that a declining curve approaching zero by the time of killing is the choice to minimize potential errors associated with the rate constants. The rate constants, K_1^*, k_2^*, and k_3^*, are not measured in the same animals in which glucose consumption is being measured. Their values in normal conscious animals (TABLE 4) vary from structure to structure, particularly between gray and white matter, and they may not apply to all physiological, pharmacological, and pathological states. In fact, K_1^* and k_2^* are influenced by both blood flow and transport of [^{14}C]DG across the blood-brain barrier; also, because of the competition for the transport carrier, the glucose concentrations in the plasma and tissue affect the transport of [^{14}C]DG and, therefore, K_1^* and k_2^*. k_3^* is related to phosphorylation of [^{14}C]DG and certainly changes when glucose utilization is altered. The rate constants can be expected, therefore, to vary from condition to condition. The procedure was designed to minimize errors arising from fluctuations in the values of the rate constants.

Quantitative autoradiography measures only total ^{14}C concentration in tissue and does not distinguish between [^{14}C]DG-6-P and [^{14}C]DG. It is [^{14}C]DG-6-P concentration, however, that must be known to determine glucose consumption (Fig. 2). [^{14}C]DG-6-P concentration is calculated in the numerator of the operational equation by subtracting from $C_i^*(T)$, the total tissue ^{14}C content, the [^{14}C]DG concentration present in the tissue, which is estimated by the second term in the numerator containing the exponential factor, $e^{-(k_2^*+k_3^*)T}$ (Fig. 4B). The denominator of the operational equation also contains a term with the same exponential factor; this term serves to correct for the lag in the tissue behind the plasma and allows the integrated precursor specific activity in the tissue at the enzyme site to be calculated from the measured plasma specific activities (Fig. 4B). It is only in these two terms that the rate constants appear in the operational equation, and both of these exponential terms have the property of approaching zero with increasing time if C_p^* is also allowed to approach zero.

One solution to the problem of uncertainties in the exact values of the rate constants is to redetermine the rate constants for each structure in each condition to be studied as has been done for ischemia (Hawkins et al., 1981) and hyperglycemia (Orzi et al., 1988). Alternatively, [^{14}C]DG can be administered as a single intravenous pulse at zero time and sufficient time allowed for the free [^{14}C]DG to clear from the plasma and tissues and for the two exponential terms containing the rate constants to fall to sufficiently low values to have relatively little influence on the final result. Eventually these terms would reach zero, and the rate constants would become negligible and disappear from the equation. To prolong the experimental period until these terms reach zero is not practical, however, because loss of [^{14}C]DG-6-P from the tissues, though small, is finite and becomes more significant with increasing time after the pulse (Sokoloff, 1982; Nelson et al., 1986). A reasonable period is 45 minutes. On the basis of the values of ($k_2^* + k_3^*$) in TABLE 4, the exponential factors have then declined through at least 6-10 half-lives, and plasma and tissue concentrations of [^{14}C]DG have fallen to very low levels. The exponential terms in the numerator and denominator then also become very small, particularly in the denominator, and effects of even large inaccuracies in the values of the rate constants on the final estimates of local glucose utilization are markedly attenuated (Fig. 7).

The virtue of the use of a pulse followed by a relatively long experimental period on the accuracy of the determination of the integrated precursor specific activity in the tissue from measurements in the plasma is graphically illustrated in Fig. 8. From typical arterial plasma [^{14}C]DG and glucose concentration curves obtained in a normal conscious rat, the portion of the denominator of the operational equation underlined by the heavy bar was computed with a wide range of values for ($k_2^* + k_3^*$) as a function of time. The values for ($k_2^* + k_3^*$) are presented in the Figure as their equivalent half-lives (calculated by dividing their values into $\text{Log}_e 2$). The values of ($k_2^* + k_3^*$) were varied from infinite (i.e., T 1/2 = 0 min) to 0.14 per min (i.e., T 1/2 = 5 min) and more than cover the range of values occurring under physiological conditions. The portion of the equation underlined and computed represents the integral of the precursor pool specific activity in the tissue. The curves represent the time course of this function, one each for every value of ($k_2^* + k_3^*$) examined. It can be seen that these curves differ markedly at early times but rise and converge with increasing time until at 45 minutes the differences between them over the entire range of $(k_2^* + k_3^*)$ equal only a small fraction of the value of the integral. These curves demonstrate that at short times enormous errors can occur if the values of the rate constants are not precisely known, but only negligible errors occur at 45 minutes, even over a wide range of rate constants of several fold. In fact, it was precisely for this reason that [^{14}C]deoxyglucose rather than [^{14}C]glucose was selected as the tracer for glucose metabolism. The relationships are similar for glucose. Because the products of [^{14}C]glucose metabolism are so rapidly lost from the tissues, it is necessary with [^{14}C]glucose to limit the experimental period to short times when enormous errors in the estimates of the precursor pool specific activity can occur if the rate constants are not precisely known for each individual structure, an impossible requirement. A common misconception is that [^{14}C]DG must be used only with long experimental periods. Quite the contrary, [^{14}C]DG can be used with experimental periods as short as those with [^{14}C]glucose if the same degree of inaccuracy is accepted. The advantage of [^{14}C]DG is that, because of its stable products, it permits the prolongation of the experimental period to times when inaccuracies in rate constants have relatively little effect on the final result and the method becomes largely model-independent.

In pathological conditions, such as severe ischemia or hyperglycemia, the rate constants may fall far below the range examined in Fig. 8. Also, in conditions of markedly depressed cerebral glucose utilization and/or slow clearance of the [^{14}C]DG from the plasma, the second term in the numerator which contains the rate constants may not be much smaller than $C_i^*(T)$, the first term in the numerator. Inaccuracies in the rate constants, particularly K_1^*, may then produce considerable errors in the numerator. In such abnormal conditions the rate constants should be redetermined for the particular condition under study as, for example, has been done for hyperglycemia (Orzi et al., 1988). Another alternative is to prolong the experimental period to allow the exponential terms to diminish further. To correct for the loss of [^{14}C]DG-6-P that might then occur, additional rate constants for the loss, e.g., a k_4^* (Phelps et al., 1979; Huang et al., 1980; Sokoloff, 1982) or possibly even a k_4^* and k_5^* (Schmidt et al., 1989) can be incorporated into the operational equation.

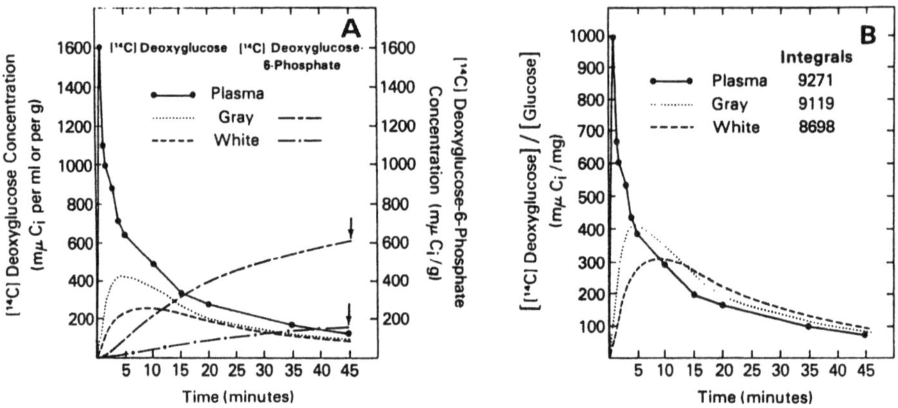

Fig. 7. Graphical representation of significant variables in equation (Fig. 4B) used to calculate cerebral glucose utilization. **A**: Time courses of $[^{14}C]$DG concentrations in arterial plasma and in average gray and white matter and $[^{14}C]$DG-6-P concentrations in average gray and white tissues following intravenous pulse of 50 μCi of $[^{14}C]$DG. Plasma [14C]DG concentrations were measured. Tissue $[^{14}C]$DG concentrations were calculated from the plasma curve and the mean values of K_1^*, k_2^*, and k_3^* for gray and white matter in TABLE 4 according to the second term in the numerator of the operational equation. The concentrations of $[^{14}C]$DG-6-P in the tissues were calculated from the mean values of k_3^* and the integral of the $[^{14}C]$DG concentrations in the tissues. The arrows point to the concentrations of $[^{14}]$DG and $[^{14}C]$DG-6-P in the tissues at the time of killing; quantitative autoradiography measures only total ^{14}C content (i.e., sum of $[^{14}]$DG and $[^{14}C]$DG-6-P contents), which is equal to C_i^* (T), the first term in the numerator of the operational equation. Note that at the time of killing, the total ^{14}C content represents mainly $[^{14}C]$DG-6-P concentration, especially in gray matter. **B**: Time courses of ratios of $[^{14}C]$DG to glucose concentrations (i.e. specific activities) in plasma and average gray and white matter. The curve for plasma was determined by division of the plasma curve in (A) by the measured plasma glucose concentrations. The curves for the tissues were calculated by the derivative of the function in the right set of brackets in the denominator of the operational equation (Fig. 4B). The integrals in (B) are the integrated tissue specific activities and are equal to the areas under the corresponding curves. These integrals equal all of the denominator of the operational equation for each of the tissue types, except for the lumped constant. Note that by the time of killing, the integrals of the tissue specific activities approach equality with each other and that of the plasma curve, indicating that the exponential term in the denominator has become negligible. From Sokoloff et al. (1977).

INFLUENCE OF RATE CONSTANTS ON INTEGRATED POOL SPECIFIC ACTIVITY

OPERATIONAL EQUATION:

$$R_i = \frac{C_i(T) - K_i^* e^{-(k_2^* + k_3^*)T} \int_0^T C_p^* e^{(k_2^* + k_3^*)t}\, dt}{\left(\frac{\lambda V_m^* K_m}{\Phi V_m K_m^*}\right) \underline{\left[\int_0^T (C_p^*/C_p)\, dt - e^{-(k_2^* + k_3^*)T} \int_0^T (C_p^*/C_p)\, e^{(k_2^* + k_3^*)t}\, dt\right]}}$$

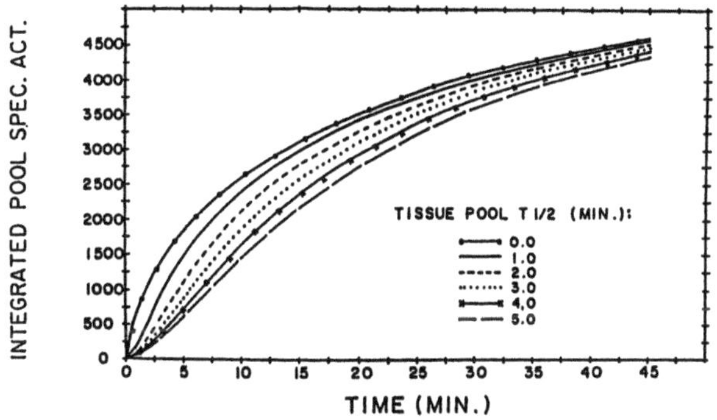

Fig. 8. Influence of time and rate constants, $(k_2^* + k_3^*)$, on integrated precursor pool specific activity in tissue calculated from time course of arterial plasma specific activity measured in normal conscious rat given standard tracer dose of [^{14}C]DG at zero time. The underlined portion of the operational equation, corresponding to the integrated tissue precursor pool specific activity, was computed as a function of time with the range of values of $(k_2^* + k_3^*)$ indicated by their equivalent half-lives. The half-lives were calculated by $T_{1/2} = 0.693/(k_2^* + k_3^*)$. From Sokoloff (1982).

Glucose-6-Phosphatase: The key advantage of the radioactive deoxyglucose method is that, contrary to labeled glucose, it allows the experimental period after a pulse of the tracer to be sufficiently prolonged to minimize errors arising from inaccuracies in the difficult-to-determine model-dependent rate constants without significant loss of product. This is possible because DG-6-P is not metabolized further down the glycolytic pathway, and G-6-Pase activity is very low in brain (Hers, 1957). Although low, some activity is present, but experimental evidence has demonstrated that it has negligible influence on the DG method if the experimental period is limited to the prescribed 45 min. In fact, it was the results of experimental efforts to detect the effects of G-6-Pase activity that established the 45 min limit (Sokoloff et al, 1977; Sokoloff, 1982). After 45 min the effects of G-6-Pase activity begin to appear and increase in magnitude with increasing time.

G-6-Pase activity would affect the results by hydrolyzing [^{14}C]DG-6-P and causing loss of product, and its effect would be cumulative with increasing time. If this activity were significant, it would cause the rates of glucose utilization calculated by the original operational equation (Fig. 4B), which assumes no loss of product, to be too low and to become progressively lower with increasing time. On the other hand, as long as there is no significant loss of product, the calculated rate of glucose utilization would remain constant. In rats studied over 20, 30, and 45 minute periods, the local rates of glucose utilization in essentially all structures remained constant with time (Sokoloff et al., 1977; Sokoloff, 1982; Nelson et al., 1987; Schmidt et al., 1989). Also, the average glucose utilization of all the structures in the brain weighted for their relative sizes agree very well and are certainly no lower than those obtained with the Kety-Schmidt method (Kety and Schmidt, 1948), the recognized standard method for measuring average energy metabolism of the brain as a whole (Siesjö, 1978) (TABLE 6). There is, therefore, substantive evidence that loss of product is too minimal to have detectable effects on the results obtained with the DG method within the 45 min experimental period. After 45 min, however, effects of G-6-Pase activity, and/or other causes of loss of label from product molecules, begin to appear and become progressively greater with increasing time (Sokoloff et al., 1982; Nelson et al., 1987; Schmidt et al., 1989). The time course of the effects is compatible with the intracellular distributions of the [^{14}C]DG-6-P and G-6-Pase. [^{14}C]DG-6-P is formed in the cytosol, but the G-6-Pase is localized on the inner surface of the membranes of the cisterns of the endoplasmic reticulum. The [^{14}C]DG-6-P must first be formed in the cytosol and then transported across the endoplasmic reticular membrane by a specific carrier before G-6-Pase can hydrolyze it. The kinetics of this process, i.e., a lag period with undetectable effects of G-6-Pase activity followed by progressively increasing activity, are those to be expected from the separate compartmentation of substrate and enzyme and a rate-limiting transport of the substrate across the membrane to the site of the enzyme. This compartmentation allows a period of grace before G-6-Pase can act on DG-6-P, and it is a prolonged period of grace because the stereospecific translocase that transports G-6-P and DG-6-P across the endoplasmic reticular membrane in tissues rich in G-6-Pase activity (e.g., liver and kidney) is absent in brain (Fishman and Karnovsky, 1986).

If the experimental period is extended beyond 45 minutes, as is sometimes needed with double-label autoradiographic techniques and with slow PET scanners, then correction must be made for the effects of product loss. It is a common misconception that phosphatase activity would invalidate the DG method. It would not. In the original version the model and operational equation did not take product loss into account because the evidence described above indicated that there was no reason to do so, as long as the experimental period was limited to 45 min, which was the maximum period contemplated. It is a relatively simple matter, however, to modify slightly the model and the derived operational equation to include a k_4^*, a rate constant for product loss (Phelps et al., 1979; Huang et al., 1980; Sokoloff, 1982), which can be used in studies with prolonged experimental periods beyond 45 minutes. Schmidt et al.(1989) have recently introduced a more comprehensive model, including a k_4^* and k_5^*, to take into account the intracellular compartmentation of the DG-6-P and the G-6-Pase.

Despite the evidence above, and other evidence as well, that G-6-Pase activity has little, if any, influence on the results obtained with the DG method if the experimental period is limited to the prescribed 45 min, there have been several allegations that G-6-Pase activity is a major source of error at all times (Hawkins and Miller, 1978; Sacks et al., 1983; Huang and Veech, 1982, 1985). The bases of these claims have been intensively and comprehensively examined, and all have been demonstrated to be without foundation. The first (Hawkins and Miller, 1978) was shown to be based on miscalculations with an invalid value of the lumped constant (Sokoloff et al., 1982; Nelson et al., 1986; Nelson et al., 1987). The second (Sacks et al., 1983) was shown to be based on misinterpretations of data that really had no bearing on the issue of G-6-Pase activity (Nelson et al., 1986). The others (Huang and Veech, 1982, 1985) have been found to be due to methodological deficiencies leading to radiochemical impurities and incomplete recoveries of critical labeled products that were assayed (Nelson et al., 1985; Dienel et al., 1988). The issue of the effects of G-6-Pase activity on the DG method has recently been reviewed (Hawkins and Miller, 1987; Nelson et al., 1987).

Lumped Constant: The lumped constant is composed of 6 separate constants (Fig. 4B). One of these, Φ, is a measure of the steady-state hydrolysis of G-6-P to free glucose and phosphate. Because in normal brain tissue there is little such phosphohydrolase activity (Hers, 1957), Φ is normally approximately equal to unity. The other components are arranged in three ratios: λ, which is the ratio of the distribution spaces in the tissue for free DG and glucose; V_m^*/V_m; and K_m/K_m^*. Although each individual constant may vary from structure to structure and condition to condition, it is likely that the ratios tend to remain stable under normal physiological conditions. For reasons described in detail previously (Sokoloff et al., 1977; Sokoloff, 1982) and verified experimentally (Gjedde and Diemer, 1983), the lumped constant is the same throughout the brain and more or less characteristic of the species of animal under normal physiological conditions. Empirical experience thus far indicates that it is generally so, except in some pathophysiological states. The greatest experience has been accumulated in the albino rat. In this species the lumped constant for the brain as a whole has been determined under a variety of conditions. In the normal conscious rat local cerebral glucose utilization, determined by the [^{14}C]DG method with the single value of the lumped constant for the brain as a whole, correlates almost perfectly ($r = 0.96$) with local cerebral blood flow measured by the [^{14}C]iodoantipyrine method, an entirely independent method (Sokoloff, 1978). It is generally recognized that local blood flow is adjusted to local metabolic rate, but if the single value of the lumped constant did not apply to the individual structures studied, then errors in local glucose utilization would occur that might be expected to obscure the correlation. Also, the lumped constant has been directly determined in the albino rat in the normal conscious state, under barbiturate anesthesia, and during the inhalation of 5% CO_2; no significant differences were observed (TABLE 5). The lumped constant does vary with the species of animal because of species differences in the kinetic properties of the hexokinase and possibly also of the blood-brain barrier transporter (TABLE 5), but the values for local rates of glucose utilization determined with these lumped constants in the various species are close to what would be expected from measurement of cerebral energy metabolism in these species by other methods.

Although the lumped constant changes only negligibly under physiological conditions, it certainly does change in pathophysiological states. For example, in severe hypoglycemia there is a progressive rise in the lumped constant with falling plasma glucose concentration (Suda et al., 1981), and in severe hyperglycemia there is a small decrease (Schuier et al., 1981). In severe seizure activity, when the rate of glucose consumption is so increased that it exceeds the rate of resupply, the lumped constant also rises (Diemer and Gjedde, 1983). Indeed, theoretically the lumped constant should change whenever there is an alteration in the balance between glucose supply to the tissue and the tissue's rate of glucose utilization. It is only when the rate of glucose utilization becomes limited by the supply, that a major change in the lumped constant occurs. Also, tissue damage may disrupt normal intracellular compartmentation, and the ratio of the distribution spaces for $[^{14}C]$DG and glucose in damaged tissue may differ from that in normal tissue. In pathological states lysosomal acid-hydrolases may be released that might hydrolyze G-6-P and thus alter the value of Φ. It is advisable, therefore, to redetermine the lumped constant whenever pathological or pathophysiological states are studied.

Experimental Protocol. The experimental procedure was designed with full consideration of the theoretical and practical issues just discussed. The animals are prepared for the experiment by insertion of polyethylene catheters into any artery and vein. The catheterizations are carried under anesthesia; anesthetic agents without long-lasting after-effects should be used. Light halothane anesthesia with or without supplementation with nitrous oxide is quite satisfactory. At least 2 hours are allowed for recovery from the surgery and anesthesia before initiation of the experimental procedure. At zero time a pulse of 125 μCi (contained in no more than 2.5 μmoles to adhere to tracer conditions) of $[^{14}C]$DG per kg of body weight is injected intravenously. Arterial sampling is initiated with the onset of the pulse, and timed samples of arterial blood are collected consecutively as rapidly as possible during the early period so as not to miss the peak of the arterial curve. Arterial sampling is continued at less frequent intervals later in the experimental period but at sufficient frequency to define fully the time course of the arterial curve. The arterial blood samples are immediately centrifuged to separate the plasma, which is stored on ice until assayed for [^{14}C]DG concentrations by liquid scintillation counting and glucose concentrations by standard enzymatic methods. At approximately 45 minutes the animal is decapitated, and the brain is removed and frozen in Freon XII or isopentane maintained between -50° and -75°C with liquid N_2. When fully frozen, the brain is stored at -70°C until sectioned and autoradiographed. The experimental period may be limited to 30 minutes. This is theoretically permissible and may sometimes be necessary for reasons of experimental expediency, but greater errors due to possible inaccuracies in the rate constants may result.

The ^{14}C concentrations in localized regions of the brain are measured by a modification of the quantitative autoradiographic technique previously described (Reivich et al., 1969). The frozen brain is coated with chilled embedding medium (Lipshaw Manufacturing Co., Detroit MI) and fixed to object-holders appropriate to the microtome to be used. Brain sections, precisely 20 μm in thickness, are cut in a cryostat maintained at -20°C to -22°C. The brain sections are picked up on glass cover-slips, dried on a hot plate at 60°C for at least 5 minutes, and placed sequentially in an X-ray cassette. A set of [^{14}C]methylmethacrylate standards (Amersham Corp., Arlington heights, IL), which include a blank and a series of progressively increasing ^{14}C concentrations, are also placed in the cassette. These standards must previously have been calibrated for their autoradiographic equivalence to the ^{14}C concentrations in brain sections, 20 μm in thickness, prepared as described above. The method of calibration has been previously described (Reivich et al., 1969).

Autoradiograms are prepared from these sections directly in the X-ray cassette with Kodak Medical X-ray Film, Type SB-5 (Eastman Kodak Co., Rochester, NY). The exposure time is generally 5-6 days with the doses used as described above, and the exposed films are developed according to the instructions supplied with the film. SB-5 X-ray film is rapid but coarse grained. For finer grained autoradiograms and, therefore, better defined images with higher resolution, it is possible to use mammographic films, such as DuPont LoDose or Kodak OM-1 films, or fine grain panchromatic film, such as Kodak Plus-X, but the exposure times are then considerably longer. The autoradiograms provide a pictorial representation of the relative ^{14}C concentrations in the various cerebral structures and the plastic standards (Fig. 9). A calibration curve of the relationship between optical density and tissue ^{14}C concentration for each film is obtained by densitometric measurements of the portions of the film representing the various standards. The local tissue concentrations are then determined from the calibration curve and the optical densities of the film in the regions representing the cerebral structures of interest. Local cerebral glucose utilization is calculated from the local tissue concentrations of ^{14}C and the plasma [^{14}C]DG and glucose concentrations according to the operational equation (Fig. 4B). Image-processing systems can be used to carry out the quantitative densitometry automatically and even to convert the autoradiograms into quantitative color-coded maps of the distribution of local rates of glucose utilization in the brain (Goochee et al., 1980).

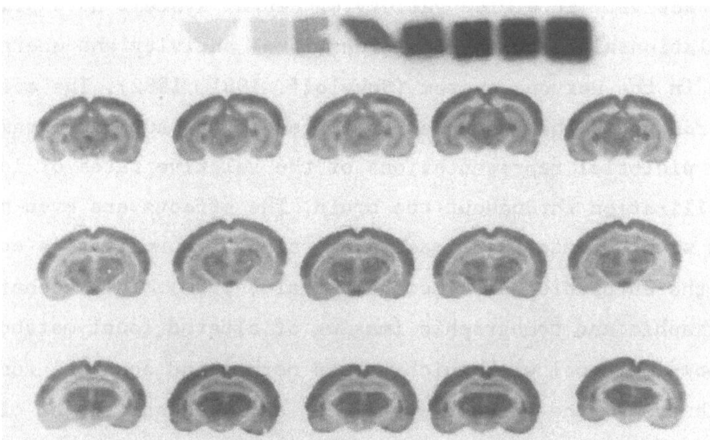

Fig. 9. [^{14}C]Deoxyglucose autoradiograms of coronal sections of conscious rat brain and of [^{14}C]methylmethacrylate standards used to quantify ^{14}C concentrations in tissues by quantitative densitometry.

Local Rates of Cerebral Glucose Utilization. Local rates of cerebral glucose utilization in normal conscious adult rats vary widely throughout the brain (TABLE 6). The values in white structures tend to group closely together and are always considerably below those of gray structures. The values in gray matter are more diverse and range from approximately 2 to 5 times those in white matter. The individual values in gray matter vary from approximately 50 to 200 μmoles/100g/min. The highest values are in structures involved in auditory functions with the inferior colliculus clearly the most metabolically active structure in the brain.

The rates of local cerebral glucose utilization in the conscious monkey exhibit similar heterogeneity, but they are generally one-third to one-half of the values in corresponding structures of the rat brain (TABLE 6). The differences in rates in the rat and monkey brain are consistent with the different cellular packing densities in the brains of these species.

In both cases, the weighted average rates of glucose utilization in the brain as a whole are in good agreement with those obtained by measurements of average cerebral energy metabolism in these species with the Kety-Schmidt method (TABLE 6).

Metabolic Mapping of Local Functional Activity. Numerous studies with the DG method involving experimentally induced alterations of local functional activity in a wide variety of neural systems have demonstrated a close relationship between local functional activity and energy metabolism in the nervous system (Sokoloff, 1981, 1982). The effects are often so pronounced that they can be imaged in the autoradiograms, which are really pictorial representations of the relative rates of local glucose utilization throughout the brain. The effects are even more pronounced with computerized image-processing or quantitative color-coding of the autoradiograms (Goochee et al., 1980). This technique of autoradiographic and tomographic imaging of altered local metabolism offers a powerful tool with which to map normal and abnormal functional neural pathways in the nervous system. An early demonstration of its potential was its dramatic imaging of the extent and distribution of the optical dominance columns in the striate cortex of primates with binocular visual systems (Kennedy et al., 1976) (Fig. 9).

The [^{18}F]Fluorodeoxyglucose Technique

Because the [^{14}C]DG method requires measurement of local concentrations of radioactivity in brain, its use in man as originally designed was precluded by the impossibility of using autoradiography in man or of detecting β-radiation of ^{14}C outside the body. The development of computerized emission tomography, however, made it possible to measure local tissue concentrations of γ-emitting isotopes *in vivo* in man.

TABLE 6. Representative values for local cerebral glucose utilization in normal conscious albino rat and monkey

Structure	Albino Rat (10)[a]	Monkey (7)[b]
Gray Matter		
Visual Cortex	107 ± 6	59 ± 3
Auditory Cortex	162 ± 5	79 ± 4
Parietal Cortex	112 ± 5	48 ± 4
Sensory-Motor Cortex	120 ± 5	44 ± 3
Thalamus: Lateral Nucleus	116 ± 5	54 ± 2
Thalamus: Ventral Nucleus	109 ± 5	43 ± 2
Medial Geniculate Body	131 ± 5	65 ± 3
Lateral Geniculate Body	96 ± 5	39 ± 1
Hypothalamus	54 ± 2	25 ± 1
Mamillary Body	121 ± 5	57 ± 3
Hippocampus	79 ± 3	39 ± 2
Amygdala	52 ± 2	25 ± 2
Caudate-Putamen	110 ± 4	52 ± 3
Nucleus Accumbens	82 ± 3	36 ± 2
Globus Pallidus	58 ± 2	26 ± 2
Substantia Nigra	58 ± 3	29 ± 2
Vestibular Nucleus	128 ± 5	66 ± 3
Cochlear Nucleus	113 ± 7	51 ± 3
Superior Olivary Nucleus	133 ± 7	63 ± 4
Inferior Colliculus	197 ± 10	103 ± 6
Superior Colliculus	95 ± 5	55 ± 4
Pontine Gray Matter	62 ± 3	28 ± 1
Cerebellar Cortex	57 ± 2	31 ± 2
Cerebellar Nucleus	100 ± 4	45 ± 2
White Matter		
Corpus Callosum	40 ± 2	11 ± 1
Internal Capsule	33 ± 2	13 ± 1
Cerebellar White Matter	37 ± 2	12 ± 1
Weighted Average for Whole Brain		
	68 ± 3	36 ± 1

Note: The values are Means ± S.E.M. (in μmoles/100g/min) of values determined in number of animals indicated in parentheses.

[a] From Sokoloff et al. (1977)   [b] From Kennedy et al. (1978)

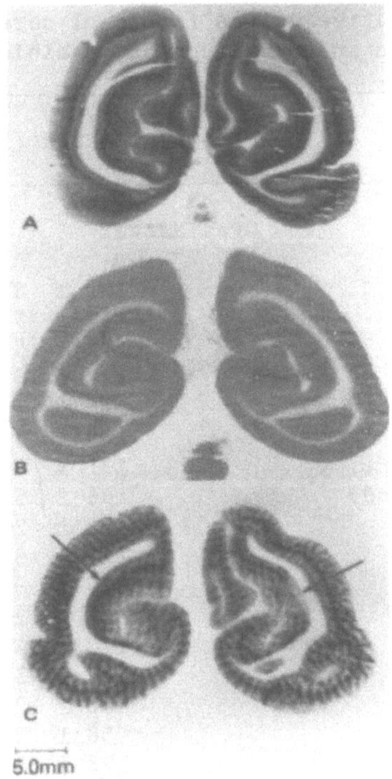

Fig. 10. Autoradiograms of coronal brain sections from Rhesus monkeys at level of striate cortex. **A:** Animal with normal binocular vision. Note laminar distribution of the density; the dark band corresponds to Layer IV. **B:** Animal with bilateral visual occlusion. Note the almost uniform and reduced relative density, especially the virtual disappearance of the dark band corresponding to Layer IV. **C:** Animal with right eye occluded. The half-brain on the left side of the photograph represents the left hemisphere contralateral to the occluded eye. Note alternate dark and light striations, each approximately 0.3-0.4 mm in width that represent the ocular dominance columns. These columns are most apparent in the dark band (Layer IV) but extend through the entire thickness of the cortex. The arrows point to regions of bilateral asymmetry where the ocular dominance columns are absent. These are presumably areas with normally only monocular input. The one on the left, contralateral to occluded eye, has a continuous dark lamina corresponding to Layer IV which is completely absent on the side ipsilateral to the occluded eye. These regions are believed to be the loci of the cortical representations of the blind spots of the visual fields. From Kennedy et al. (1976).

A γ-emitting analogue of deoxyglucose, 2-[^{18}F]fluoro-2-deoxy-D-glucose, was, therefore, synthesized and found to retain the necessary biochemical properties of 2-DG. It was first used to measure local cerebral glucose utilization in man with a single photon scanner (Reivich et al., 1979). Because ^{18}F is a positron-emitter, the method was adapted further for use with positron emission tomography (PET), which provides better spatial resolution and more accurate estimates of local concentrations of isotope in the tissues (Phelps et al., 1979). Recently, [^{11}C]deoxyglucose, a positron-emitting version of 2-deoxyglucose, has been synthesized, and its lumped and rate constants for the human brain have been determined (Reivich et al., 1985). The 20-minute physical half-life of [^{11}C]DG offers considerable advantages because it allows repeated studies in the same subject within a relatively short time-span.

REFERENCES

Abrams R, Ito M, Frisinger JE, Patlak, CS, Pettigrew KD, Kennedy C (1984) Local cerebral glucose utilization in fetal and neonatal sheep. **Am J Physiol** 246:R608

Diemer NH, Gjedde A (1983) Autoradiographic determination of brain glucose content and visualization of the regional lumped constant. **J Cereb Blood Flow Metab** 3(Suppl 1):S79-S80

Dienel GA, Nelson T, Cruz NF, Jay T, Crane, AM, Sokoloff L (1988) Over-estimation of glucose-6-phosphatase activity in brain _in vivo_. Apparent difference in rates of [2-^{3}H]glucose and [U-^{14}C]glucose utilization is due to contamination of precursor pool with ^{14}C-labeled products and incomplete recovery of ^{14}C-labeled metabolites. **J. Biol Chem** 263:19697-19708

Duffy TE, Cavazzuti M, Cruz NF, Sokoloff L (1982) Local cerebral glucose metabolism in newborn dogs: effects of hypoxia and halothane anesthesia. **Ann Neurol** 11:233-246

Eckman WW, Phair RD, Fenstermacher JD, Patlak CS, Kennedy C, Sokoloff L (1975) Permeability limitation in estimation of local brain blood flow with [^{14}C]antipyrine. **Am J Physiol** 229:215-221

Eklöf B, Lassen NA, Nilsson L, Norberg K, Siesjö BK, Torlöf P (1974) Regional cerebral blood flow in the rat measured by the tissue sampling technique: a critical evaluation using four indicators, C^{14}-antipyrine, C^{14}-ethanol, H^{3}-water, and xenon^{133}. **Acta Physiol Scand** 91:1-10

Fishman RS, Karnovsky ML (1986) Apparent absence of a translocase in the cerebral glucose-6-phosphatase system. **J Neurochem** 46:371

Freygang WH, Jr, Sokoloff L (1958) Quantitative measurement of regional circulation in the central nervous system by the use of radioactive inert gas. **Adv Biol Med Physics** 6:263-279

Gjedde A and Diemer NH (1983) Autoradiographic determination of regional brain glucose content. **J Cereb Blood Flow Metab** 4:303-310

Goochee C, Rasband W, Sokoloff L (1980) Computerized densitometry and color-coding of [^{14}C]deoxyglucose autoradiographs. **Ann Neurol** 7:359-370

Hawkins RA, Miller AL (1978) Loss of radioactive 2-deoxy-D-glucose-6-phosphate from brains of conscious rats: implications for quantitative autoradiographic determination of regional glucose utilization. **Neuroscience** 3:251-258

Hawkins RA, Miller AL (1987) Deoxyglucose-6-phosphate stability _in vivo_ and the deoxyglucose method. **J Neurochem** 49:1941-1960

Hawkins R, Phelps M, Huang SC, Kuhl D (1981) Effect of ischemia upon quantification of local cerebral metabolic rates for glucose with 2-(F-18)fluoro-deoxyglucose (FDG). **J Cereb Blood Flow Metab** 1 (Suppl 1):S9-S10

Hers HG, (1957) **Le Metablisme du Fructose**, Bruxelles, Editions Arscia, p 102

Huang SC, Phelps ME, Hoffman EJ, Sideris K, Selin CJ, Kuhl DE (1980) Noninvasive determination of local cerebral metabolic rate of glucose in man. **Am J Physiol** 238:E69-E82

Huang M-T, Veech RL (1982) The quantitative determination of the in vivo dephosphorylation of glucose 6-phosphate in rat brain. **J Biol Chem** 257:11358-11363

Huang M-T, Veech RL (1985) Metabolic fluxes between 2-[^{14}C]deoxyglucose and 2-[^{14}C]deoxyglucose-6-phosphate in brain in vivo. **J Neurochem** 44:567-573

Kennedy C, Des Rosiers MH, Sakurada O, Shinohara M, Reivich M, Jehle JW, Sokoloff L (1976) Metabolic mapping of the primary visual system of the monkey by means of the autoradiographic [^{14}C]deoxyglucose technique. **Proc Natl Acad Sci USA** 73:4230-4234

Kennedy C, Sakurada O, Shinohara M, Jehle J, Sokoloff L (1978) Local cerebral glucose utilization in the normal conscious Macaque monkey. **Ann Neurol** 4:293-301

Kety SS (1949) Measurement of regional circulation by the local clearance of radioactive sodium. **Amer. Heart J** 38:321-328

Kety SS (1951) The theory and applications of the exchange of inert gas at the lungs and tissues. **Pharmacol Rev** 3:1-41

Kety SS (1957) The general metabolism of the brain in vivo. In: Richter D (ed) **The Metabolism of the Nervous System**. Pergamon Press, London, p 221-237

Kety SS, Schmidt CF (1948) The nitrous oxide method for the quantitative determination of cerebral blood flow in man: theory, procedure, and normal values. **J Clin Invest** 27:476-483

Krebs HA, Williamson DH, Bates MW, Page MA, Hawkins RA (1971) The role of ketone bodies in caloric homeostasis. **Adv Enzyme Regul** 9:387-409

Landau WH, Freygang WH, Rowland LP, Sokoloff L, Kety SS (1955) The local circulation of the living brain: values in the unanesthetized and anesthetized cat. **Trans Am Neurol Assoc** 80:125-129

Lassen NA, Ingvar DH (1972) Radioisotopic assessment of regional cerebral blood flow. **Prog Nucl Med** 1:376-490

Lucignani G, Nehlig A, Blasberg R, Patlak CS, Anderson L, Fieschi, C, Fazio F, and Sokoloff L (1985) Metabolic and kinetic considerations in the use of [^{125}I]HIPDM for quantitative measurement of regional blood flow. **J Cereb Blood Flow Metab** 5:86-96

Nelson T, Kaufman EE, Sokoloff L. (1984) 2-Deoxyglucose incorporation into rat brain glycogen during measurement of local cerebral glucose utilization by the 2-deoxyglucose method. **J Neurochem** 43:949-956

Nelson T, Lucignani G, Atlas S, Crane AM, Dienel GA, Sokoloff L (1985) Reexamination of glucose-6-phosphatase activity in the brain in vivo: no evidence for a futile cycle. **Science** 229:60-62

Nelson T, Lucignani G, Goochee J, Crane AM, Sokoloff L (1986) Invalidity of criticisms of the deoxyglucose method based on alleged glucose-6-phosphatase activity in brain. **J Neurochem** 46: 905-919

Nelson T, Dienel GA, Mori K, Cruz NF, Sokoloff L (1987) Deoxyglucose-6-phosphate stability in vivo and the deoxyglucose method: response to comments of Hawkins and Miller. **J Neurochem** 49:1949-1960

Orzi F, Lucignani G, Dow-Edwards D, Namba H, Nehlig A, Patlak CS, Pettigrew K, Schuier, F, And Sokoloff L (1988) Local cerebral glucose utilization in controlled graded levels of hyperglycemia in the conscious rat. **J Cereb Blood Flow Metab** 8:346-356

Owen OE, Morgan AP, Kemp HG, Sullivan JM, Herrera MG, Cahill GF (1967) Brain metabolism during fasting. **J Clin Invest** 46:1589-1595

Phelps ME, Huang SC, Hoffman EJ, Selin C, Sokoloff L, Kuhl DE (1979) Tomographic measurement of local cerebral glucose metabolic rate in humans with (F-18)2-fluoro-2-deoxy-D-glucose: validation of method. **Ann Neurol** 6:371-388

Reivich M, Alavi A, Wolf A, Fowler J, Russell J, Arnett C, MacGregor RR, Shiue CY, Atkins H, Anand A, Dann R, Greenberg JH, (1985) Glucose metabolic rate kinetic model parameter determination in humans: the lumped constants and rate constants for [^{18}F]fluorodeoxyglucose and [^{11}C]deoxyglucose. **J Cereb Blood Flow Metab** 5:179-192

Reivich M, Jehle J, Sokoloff L, Kety SS (1969) Measurement of regional cerebral blood flow with antipyrine-^{14}C in awake cats. **J Appl Physiol** 27:296-300

Reivich M, Kuhl D, Wolf A, Greenberg J, Phelps M, Ido T, Cassella V, Fowler J, Hoffman E, Alavi A, Som P, Sokoloff L (1979) The [^{18}F]fluoro-deoxyglucose method for the measurement of local cerebral glucose utilization in man. **Circ Res** 44:127-137

Sacks W, Sacks S, Fleischer A (1983) A comparison of the cerebral uptake and metabolism of labeled glucose and deoxyglucose in vivo in rats. **Neurochem Res** 8:661-685

Sakurada O, Kennedy C, Jehle J, Brown, JD, Carbin, GL, Sokoloff L (1978) Measurement of local cerebral blood flow with [^{14}C]iodoantipyrine. **Am J Physiol** 234:H59-H66

Schuier F, Orzi F, Suda S, Kennedy C, Sokoloff L (1981) The lumped constant for the [^{14}C]deoxyglucose method in hyperglycemic rats. **J Cereb Blood Flow Metab** 1:S63

Schmidt K, Lucignani G, Mori K, Jay T, Palombo, E., Nelson T, Pettigrew, K, Holden JE, and Sokoloff L (1989) Refinement of the kinetic model of the 2-[^{14}C]deoxyglucose method to incorporate effects of intracellular compartmentation in brain. **J Cereb Blood Flow Metab** In press

Siesjö B (1978) **Brain Energy Metabolism**. John Wiley & Sons, New York, p 107-109

Sokoloff L (1960) The metabolism of the cerebral nervous system in vivo. In: Field J, Magoun HW, Hall VE (eds) **Handbook of Physiology: Neurophysiology, Vol III**. American Physiological Society, Washington, D.C., p 1843-1864

Sokoloff L (1978) Local cerebral energy metabolism: its relationships to local functional activity and blood flow. In: Purves MJ and Elliott K (eds) **Cerebral Vascular Muscle and Its Control**. Elsevier/North-Holland, Amsterdam, p 171-197

Sokoloff L (1981a) Circulation and energy metabolism of the brain. In: Siegel GJ, Albers RW, Katzman R, Agranoff BW (eds) **Basic Neurochemistry, 3rd ed**. Little Brown, Boston, p 471-495

Sokoloff L (1981b) Localization of functional activity in the central nervous system by measurement of glucose utilization with radioactive deoxyglucose. **J Cereb Blood Flow Metab** 1:7-36

Sokoloff L (1982) The radioactive deoxyglucose method: theory, procedure, and applications for the measurement of local glucose utilization in the central nervous system. In: Agranoff BW, Aprison MH (eds) **Advances in Neurochemistry, Vol 4**. Plenum Publishing Corp., New York, p 1-82

Sokoloff L, Reivich M, Kennedy C, Des Rosiers MH, Patlak CS, Pettigrew KD, Sakurada O, Shinohara M (1977) The [^{14}C]deoxyglucose method for the measurement of local cerebral glucose utilization: theory, procedure, and normal values in the conscious and anesthetized albino rat. **J Neurochem** 28:897-916

Sols A, Crane RK (1954) Substrate specificity of brain hexokinase. **J Biol Chem** 210:581-595

Suda S, Shinohara M, Miyaoka M, Kennedy C, Sokoloff L (1981) Local cerebral glucose utilization in hypoglycemia. **J Cereb Blood Flow Metab** 1:S62

DISCUSSION

Chaired by Cunningham V. and Fazio F.

B.K. Siesjo:

What about the use of tritiated deoxyglucose for autoradiography?

L. Sokoloff:

In principle there should be no difference between 3H and 14C. The problem, however, is that with 3H there is significant differential self-absorption of radiation within the various structures of the brain. White matter contains approximately 70% water, and gray matter contains about 80% water. The content of solid material is, therefore, approximately 50% greater in white matter than in gray matter. This difference in the contents of solid material is responsible for the differential self-absorption that interferes with the quantitative autoradiography of dried brain sections. Because of the greater solid residue in white matter after drying, the self-absorption is greater in white matter than in gray matter. The difference in solid residues in gray and white matter is probably due mainly to differences in their lipid contents. Some workers have extracted the lipids from the dried sections with lipid solvents; this does reduce differential self-absorption but also washes out some of the radioactivity. We are currently trying to develop a procedure by which we could extract the lipids with negligible loss of radioactivity or else take the losses into account in the calibration of the autoradiographic standards. Alternatively, we could carry out the autoradiography with whole tissue sections in the frozen state rather than with dried tissue sections, but 3H-autoradiography with frozen sections at low temperatures is very susceptible to artifacts due to water condensation.

R. Orkand:

We know that the action potentials in an excitatory neuron look the same as in an inhibitory neuron. Can inhibitory or excitatory activity be distinguished on the basis of metabolic studies.

L. Sokoloff:

This question is a common one which we once could not answer definitively. It seemed that increased metabolism occurred with both excitation and inhibition, and the speculation was that the processes of inhibition consumed as much energy as those of excitation. We now know that the component of energy metabolism in a given structure that changes with altered functional activity in that structure is related to the electrical activity in the terminals of the afferent pathways to that structure. Since the action potentials are the same whether an excitatory or inhibitory neurotransmitter is released at the terminals, increased metabolic activity in the structure in question cannot distinguish which type of neurotransmitter is released. To do so, it is necessary to look at the next synapses in the efferent projection zones of the structure; increased metabolic activity in the terminal fields would indicate that excitation had occurred at the previous synapse, and decreased activity would suggest inhibition.

F. Fazio:

Do you think that there is always coupling between glucose and oxygen consumption in the brain?
In other organs, such as the heart, uncoupling can occur under certain conditions.

L. Sokoloff:

Almost all organs other than brain can use substrates other glucose for oxidative metabolism. For example, skeletal and heart muscle can utilize pyruvate, fatty acids, etc., in addition to, or in place of, glucose. Because of the blood-brain barrier the brain is normally limited almost entirely to glucose as the substrate for its energy metabolism; in ketotic states it can use ketone bodies, at least in part, in place of glucose. It is highly unlikely that in the presence of adequate oxygen levels normal brain can increase its glucose utilization without also increasing its oxygen consumption. the same intracellular biochemical changes that activate glycolysis at the phosphofructokinase step (e.g. rises in ADP/ATP ratio and inorganic phosphate concentration) are also those that stimulate electron transport and turn on state-3 mitochondrial respiration. There may be temporal differences in the time courses of the changes in glucose utilization and oxygen consumption that may give rise to an apparent uncoupling between the two, but these would be transient until a new steady state has been reached. I do not believe that in the presence of a normal oxygen supply steady state oxygen consumption and glucose utilization are ever uncoupled.

B.K. Siesjo:

I agree on the improbability that there would be any sustained uncoupling between glucose and O2 utilization.
It has been shown that during anoxia the rate of glycolysis may increase 6-7 times while O2 consumption can never increase more than 3 times. This means that initially during anoxia there is a fast rate accumulation of lactate. Theoretically, this may represent an advantage in the sense of sparing ATP.
However, the problem is the existence in the brain of the BBB, which prevents the elimination of lactate causing its

accumulation in the central nervous system. For this reason I believe that the brain needs to work according to a coupled O_2-glucose mechanism.

USE OF POSITRON EMISSION TOMOGRAPHY TO STUDY PATTERNS OF BRAIN METABOLISM IN RELATION TO AGE AND DISEASE: A CORRELATION MATRIX APPROACH

Stanley I. Rapoport and Barry Horwitz
Laboratory of Neurosciences
National Institute on Aging
National Institutes of Health
Bldg. 10, Rm. 6C-103
Bethesda, Md. 20892 USA

ABSTRACT

Region-by-region comparisons of mean $rCMR_{glc}$ data have proven uninformative for understanding behavior and cognition in patients with Down syndrome and childhood onset autism, and in the healthy elderly. Such comparisons ignore an important dimension of brain metabolism, namely patterns of regional cerebral metabolic rates. This paper shows that a correlation matrix approach can be used to determine these patterns, and to relate them to underlying functional neuroanatomy in health and disease.

INTRODUCTION

Since Roy and Sherrington (1890) observed that regional cerebral blood flow (rCBF) is coupled locally to neuronal activity via products of cerebral metabolism, physiologists have measured rCBF, as well as regional cerebral metabolic rates for glucose ($rCMR_{glc}$) and for oxygen, to examine cerebral functional activity. Quantitative autoradiography in animals allows measurements at levels of resolution of 100 microns or less, whereas positron emission tomography (PET) provides measurements in humans with spatial resolutions approaching 5 mm.

To measure $rCMR_{glc}$ in humans, the positron emitting tracer ^{18}F-2-deoxy-D-glucose (^{18}FDG) is injected intravenously, and plasma radioactivity and glucose concentration are determined periodically until regional brain radioactivity is determined at 45 min with PET. The rate of local accumulation of phosphorylated ^{18}FDG is used to calculate $rCMR_{glc}$ in individual brain regions by the equation of Sokoloff et al (1977), modified to take into account dephosphorylation of ^{18}FDG-6-phosphate (Huang and Phelps, 1980). In animals, ^{14}C-2-deoxy-D-glucose, rather than ^{18}FDG, is employed as the intravenous tracer, and brain radioactivity is measured using quantitative autoradiography (Sokoloff et al, 1977).

Region-by-Region Analysis of Mean $rCMR_{glc}$ Data in Alzheimer's

NATO ASI Series, Vol. H 33
Regulatory Mechanisms of Neuron to Vessel Communication in the Brain. Edited by F. Battaini et al.

Disease, Down Syndrome, Childhood-onset Autism and Healthy Aging

Our laboratory has used PET to examine $rCMR_{glc}$ in different clinical groups. Most measurements were obtained with a low resolution ECAT II tomograph (ORTEC, Life Sciences, Oak Ridge, TN; with in-plane and axial resolutions of 17 mm), but some recent data have been obtained with a PC 1024 tomograph (Scanditronix, Uppsala, Sweden; with an in-plane resolution of 6 mm and an axial resolution of 10 mm). Studies were conducted on subjects in a "resting" state -- awake, not taking medication, eyes covered and ears plugged to minimize effects of extraneous sensory inputs or group differences in visual and auditory acuities. In these subjects, somatosensory afferent input were dominant and cognitive activity was quite variable.

Table 1. Relative Metabolic Rates in Alzheimer's Disease

Metabolic Ratio	Control(30)[a]	Mild AD(12)	Mod AD(15)	Severe AD(8)
Parietal/Sensorimotor	0.93±0.05	0.83±0.09[b]	0.85±0.09[b]	0.72±0.08[b]
Frontal/Sensorimotor	0.97±0.07	0.94±0.06	0.92±0.11	0.87±0.18[b]
Lat. Temporal/Occipital	0.85±0.08	0.76±0.11[b]	0.77±0.11[b]	0.66±0.09[b]

[a]Number of subjects in parenthesis. Data from Haxby et al (1986).
[b]Mean±S.D. differs from control by Bonferroni t test ($p < 0.05$).

In patients with a diagnosis of Alzheimer's disease, region-by-region analyses of mean $rCMR_{glc}$ values, or of mean $rCMR_{glc}/CMR_{glc}$ ratios (where division by the mean metabolic rate for glucose, CMR_{glc}, reduces the coefficient of variation of and normalizes the data), have identified regional brain dysfunction in relation to dementia severity. For example, $rCMR_{glc}$ was found to be reduced in parietal and lateral temporal association areas, relative to primary sensory and motor areas, early and throughout the course of the disease (Haxby et al, 1986) (Table 1). Early in vivo dysfunction is consistent with selective cortical degeneration, indicated by the selective distribution of neurofibrillary tangles with paired helical filaments in association but not primary sensory and motor neocortices of the post-mortem Alzheimer brain (see below) (Brun, 1983; Pearson et al, 1985; Rogers and Morrison, 1985), and with the regional specificity of neocortically mediated cognitive deficits (Haxby et al, 1985).

Although useful for the understanding of Alzheimer's disease, region-by-region comparisons of mean $rCMR_{glc}$ values (and of mean

$rCMR_{glc}/CMR_{glc}$ values) between healthy elderly and healthy young men (Duara et al, 1984), and between adult Down subjects (aged 20 to 35 yr) and age-matched controls (Schapiro et al, unpublished results), showed no statistically significant differences. Furthermore, although mean $rCMR_{glc}$ was generally 15-20% higher in adults with childhood-onset autism than in age-matched controls (Rumsey et al, 1985; Horwitz et al, 1988), there were no specific regional differences indicative of selective vulnerability. Despite these observations, however, we know that the healthy elderly demonstrate reduced performance on various aspects of "fluid" intelligence, including visuospatial function; that young adult Down subjects are retarded and have selective language and attentional deficits; and that autistics show abnormal directed attention and language abilities (see below).

Region-by-region comparisons of mean $rCMR_{glc}$ data may be uninformative with respect to subtle group differences in behavior or cognition because they ignore an important dimension of PET data, namely the correlated patterns of individual metabolic rates for the brain as a whole. To examine such patterns, and to relate them to underlying functional neuroanatomy in health and disease, new methods are needed. In this paper, we describe one such method, the correlation matrix method, and examine results obtained with it and discuss its properties. This method is based on considering the brain as a constellation of networks of brain regions which mediate cognition and behavior.

Network Hypothesis

Network models have been enunciated by Luria (1973) and Mesulam (1981), among others. Luria (1973) proposed that human mental activities are mediated by "internal" structures -- networks of interconnected brain regions. According to him, "simple" mental activities include visual perception, auditory perception, somatosensory perception and motor action. More "complex" activities include planned movement, directed attention, memory (nonspecific and modality-specific), speech (receptive and expressive), and logical thinking (constructive and discursive). The network hypothesis implies that lesions confined to a single cortical region can cause multiple functional deficits if the region contributes to several networks. Conversely, the same function may be impaired as a consequence of different lesions within the network mediating that function (Luria, 1973; Mesulam, 1981).

Network components of "simple" mental activities, such as auditory perception, have been identified in animals using quantitative autoradiography. For example, auditory deprivation or excitation in rats selectively modifies glucose and lipid metabolism in the following regions -- cochlear nucleus, superior olivary complex, lateral lemnisci, inferior colliculus, medial geniculate body and primary auditory cortex -- components of an ascending auditory system (Sharp et al, 1981; Woolf et al, 1983; Tone et al, 1988). On the other hand, the internal structures of the "complex" mental activities identified by Luria (1973) are not well defined. As these activities characterize higher primates and, in the case of language, only humans (Rapoport, 1988a), they can be elucidated by examining patterns of regional cerebral blood flow or metabolism using PET, given the coupling of these local metabolic measures with local functional activity (Roy and Sherrington, 1890).

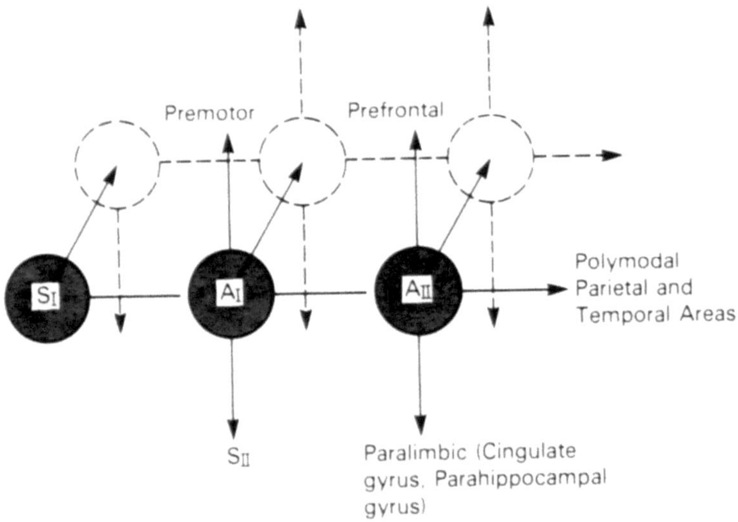

Figure 1. Serial arrangement of sensory (somatosensory, visual or auditory) central processing. Primary sensory sensory area (S_1) send connections to first order association area (A_1), which in turn send connections to second-order association regions (A_{11}), as well as to secondary sensory areas (S_{11}). Polymodal association areas receive inputs from each of second-order association regions. Connections with paralimbic structures place processes in memory and give them emotional content. Regions in each hemisphere are connected to homotopic regions in the contralateral hemisphere via the corpus callosum. Adapted from Pandya and Seltzer (1982).

Figures 1 and 2 illustrate two intricate networks that have been proposed to mediate complex sensory processing and directed attention, respectively. Figure 1 identifies regions involved in sequential processing of auditory, visual or somatosensory information that has been received by a specific primary sensory cortical region (S_1) (Pandya and Seltzer, 1982). This primary area sends fibers to an adjacent first-order association area (A_1), which in turn sends fibers to modality-specific second-order association areas (A_{11}) and to second primary sensory areas (S_{11}). First-order association regions also project to the premotor region of the frontal cortex, and second-order association regions project to the prefrontal association cortex, to the superior temporal sulcus (a polymodal association area involved with language), and to paralimbic areas. The sequences of sensory processing in each hemisphere are integrated via long fiber tracts within the corpus callosum, which connect homotopic left-right regions, except for the striate and somatosensory hand areas (Myers, 1965; Pandya and Seltzer, 1982).

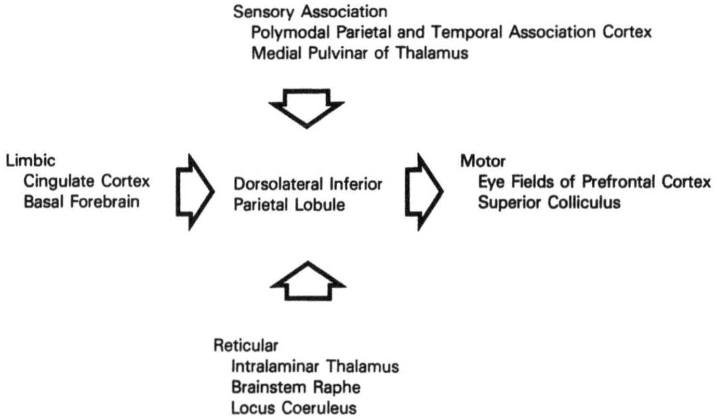

Figure 2. Regions thought to participate in directed attention in higher primates. Sensory input is via polymodal association cortex (see Figure 1) and from the medial pulvinar, a thalamic association nucleus. After Mesulam (1981).

Figure 2 identifies components of a neural network proposed to subserve directed attention (Mesulam, 1981). Its core is the posterior parietal cortex, a polymodal association area which receives inputs from the cingulate cortex and brain stem, from second-order sensory association areas, and from the nonspecific pulvinar and lateroposterior nuclei of the thalamus; and which provides input to the frontal eye fields in the prefrontal association cortex.

Matrix Analysis of Metabolic Patterns with Networks

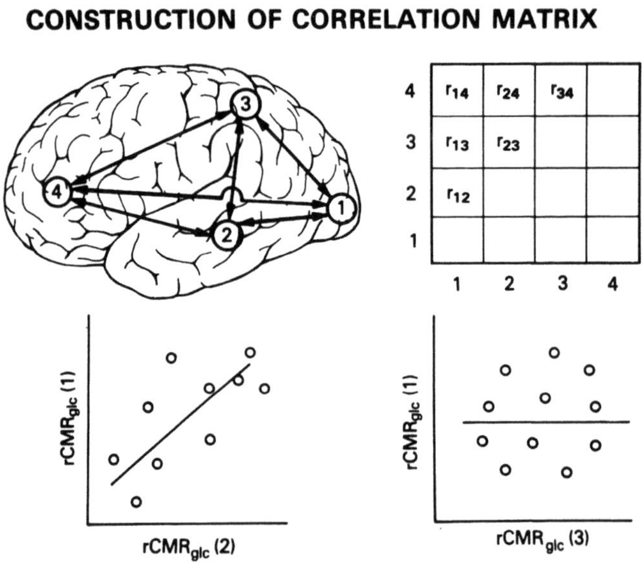

Figure 3. Drawings illustrating fundamental hypothesis of correlation method. The correlation matrix is shown in the upper right. r_{ij} is the correlation coefficient for the ij'th region pair. Note that because r_{ii} = 1 and $r_{ij} = r_{ji}$; only the correlations above the diagonal are indicated. For the specific cases represented by the plots in the lower half of the figure, r_{12} would have a large statistically significant value, whereas r_{13} would have a value close to zero. From Horwitz (1987).

In our laboratory, Horwitz et al (1984, 1986a, b, 1987) elaborated a correlation matrix method to examine patterns of cerebral metabolism or blood flow. Its fundamental hypothesis, illustrated in Figure 3, is that if two brain regions, such as 1 and 2, are functionally coupled, so that the functional activity of one depends on the activity of the second, a plot of $rCMR_{glc}$ in region 1 against $rCMR_{glc}$ in region 2 (each data point represents a value taken from a single subject) will have a statistically significant correlation coefficient. The size of the correlation coefficient represents the strength of functional coupling between the two regions. However, if two regions such as 1 and 3 are not functionally coupled, the correlation between their $rCMR_{glc}$ values will be small and statistically insignificant. This correlation approach was applied to subject groups -- healthy aged, Down syndrome, autistics -- for whom region-by-region comparisons of $rCMR_{glc}$ data had proved uninformative (see above).

Healthy aging. For the healthy men whose $rCMR_{glc}$ and $rCMR_{glc}/CMR_{glc}$ values were shown to be not significantly correlated with age (see above) (Duara et al, 1984), correlation coefficients were calculated between pairs of $rCMR_{glc}$ values after partialing out global CMR_{glc} to eliminate its common influence on individual rates (Horwitz et al., 1986a). Criteria were established for statistical significance and "reliability," using "bootstrap" and "jackknife" methods to minimize effects of outliers. A total of 59 regional metabolic rates from ECAT II-derived data were correlated pairwise (giving 1711 partial correlation coefficients). The significant ($p < 0.01$), positive coefficients from each group are arranged in Figure 4 within lobar submatrices or boxes (left and right parietal, frontal, temporal and occipital cortices, and left and right noncortical submatrices (including the caudate nucleus, thalamus, lenticular nucleus, insula and cerebellum).

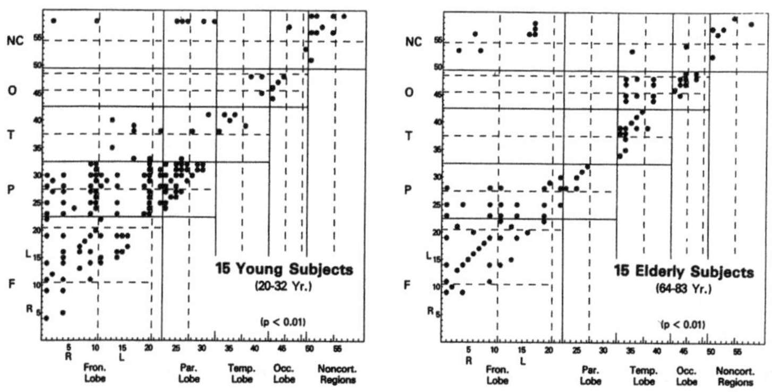

Matrices of positive partial correlation coefficients between pairs of $rCMR_{glc}$ values for young and old healthy men. Significant ($p < 0.01$), reliable correlations are illustrated. Regions are arranged according to whether they fall in the left (L) or right (R) frontal (F), parietal (P), temporal (T), occipital (O) or noncortical (NC) domains. ECAT II data. See Horwitz et al (1984) for identities of each numbered region.

The matrices of the old and young men have two common characteristics: (a) there are many positive, significant correlations between metabolic rates in homologous right and left regions, which appear along the diagonal in the appropriate right-left lobar submatrices, (2) many correlations occur between frontal and parietal regions on the one hand, and between occipital and temporal regions on the other, but the frontal-parietal and occipital-temporal domains interact minimally. The two matrices differ in one important way, however. The matrix for the old men has fewer fewer significant correlations between frontal and parietal regions, and within the parietal lobe, than does the matrix for the younger subjects (Table 2). The regions in the parietal-frontal domain with fewer correlations include associational (prefrontal and premotor; posterior parietal), as removal of sensorimotor areas from the matrix comparison does not affect the significance of the age-differences between the matrices.

Table 2. Differences in Numbers of Significant, Reliable Partial Positive Correlations between Matrices from Healthy Old and Young Men

Correlated Regions	Number of Correlations			χ^2	Statistical Significance
	Young	Elderly	Total Possible		
Frontal-parietal	36	13	220	12.2	$p < 0.001$
Minus Sensorimotor	9	0	108	9.5	$p < 0.05$
Parietal-parietal	36	13	45	10.0	$p < 0.002$

As illustrated in Figure 1, prefrontal and second-order association cortices are connected reciprocally by long fiber tracts that integrate their combined activities. The observed differences between the correlation matrices from the old and young men imply that this integration is reduced or less "coherent" in the healthy elderly, despite the fact that individual $rCMR_{glc}$ values are not significantly reduced (see above). Age-reductions in integrated activity of parietal and frontal association regions may underlie known age-differences in visuospatial function and other aspects of "fluid" intelligence (see above), which depend on integrity of these regions (Horn and Cattell, 1967; Benton, 1985; Haxby et al, 1985).

Childhood-onset autism. Childhood-onset autism is a developmental disorder characterized by deviant social behavior and an inability to shift attention appropriately, indicative of abnormal directed attention mechanisms (Damasio and Maurer, 1978; Rumsey et al, 1985; Horwitz et al, 1988). Microscopic neuropathology is indeterminate in autistics, and quantitative computer assisted tomography (CT) demonstrates no abnormality in any brain structure (Creasey et al, 1986).

As noted above, although mean resting $rCMR_{glc}$ generally is 15-20% higher in autistic subjects than in controls, there is no specific regional metabolic difference between the two groups (Rumsey et al, 1985). Nevertheless, a correlation-matrix analysis of $rCMR_{glc}/CMR_{glc}$ data demonstrated selectively impaired interactions among frontal/parietal association regions and between these regions and subcortical nuclei (basal ganglia, the thalamus) (Figure 5) (Horwitz et al, 1988). There were fewer ($p < 0.05$) large ($r > 0.5$), positive correlation coefficients in the following lobar boxes of the autistic subjects than of the controls: frontal-frontal, frontal-parietal, frontal-subcortical, parietal-parietal, and parietal-subcortical. (The subcortical submatrix included the thalamus, caudate nucleus, lenticular nucleus and insula.) These differences are consistent with disturbed functional relations between regions within the neural network said to subserve directed attention (Figure 2).

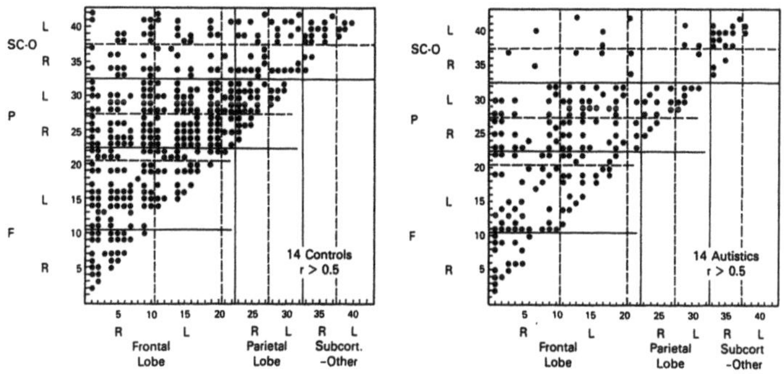

Figure 5. Matrices of positive correlations coefficients, using $rCMR_{glc}/CMR_{glc}$ values, from adult subjects with childhood-onset autism and from controls. ECAT II data from Horwitz et al (1988).

<u>Down syndrome (trisomy 21)</u>. As with autism, the neuropathological basis for mental retardation in Down syndrome is not certain. Deficits in the areas of language (Hartley, 1982; Haxby, in press) are particularly prominent in Down subjects who, in contrast to autistic subjects, show reduced inhibition and are more distractible. Some authors have reported that brains of Down subjects are abnormally small, particularly the frontal and temporal lobes (Benda, 1969), but quantitative CT of Down adults has not demonstrated any anatomic difference from controls that could not be ascribed to short stature (Schapiro et al, 1987). On the other hand, trisomy 21 neurons have fundamental abnormalities in the action potential and potassium currents that probably underlie retardation (Orozco et al, 1987; Nieminen et al, in press).

$rCMR_{glc}$ values (obtained with the Scanditronix tomograph) do not differ significantly between healthy adult Down subjects (age 20-35 yr) and controls (Schapiro et al, unpublished observations). [Our earlier report (Schwartz et al, 1982) that 4 young adult Down subjects had excessive cerebral metabolic rates probably reflected the use of a standard operator-drawn ellipse with the ECAT II tomograph, and not a transmission scan for attenuation correction, causing a calculation error due to the small head size of the Down subjects (Schapiro et al, unpublished observations).] Nevertheless, as compared with the correlation matrix of age-matched controls, the matrix of the young adult Down subjects (derived from ECAT II $rCMR_{glc}/CMR_{glc}$ data, Figure 6) had significantly more correlations with lower values between region pairs within the frontal lobe (74%) and the parietal lobe (87%), and between frontal and parietal regions (83%) (Horwitz et al, submitted). Specifically, the left inferior frontal gyrus, which includes Broca's area (region 17 of Figure 6) had 20 significantly lower correlations out of 58 possible, whereas the right thalamus (region 51) had 18 of 58, and the left thalamus (region 56) had 15 of the 58 which were significantly reduced in magnitude in the Down matrix. These differences are consistent with abnormal language and directed attention (cf. Figure 2) in Down syndrome.

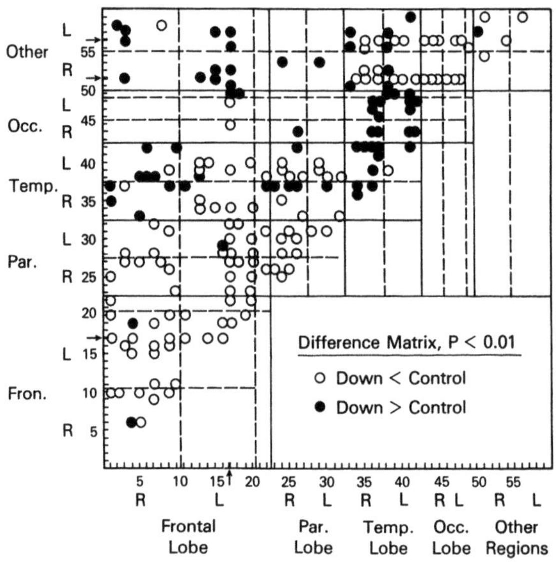

Figure 6. Matrix of significant differences ("difference matrix") between correlation coefficients between matrices from young adult Down subjects and controls. Arrows identify left inferior frontal gyrus (region 17, Broca's area) and right and left thalami (regions 51 and 56, respectively). ECAT II data from Horwitz et al (submitted).

Considerations of Correlation Matrix Approach

There are several issues related to the use and interpretation of the correlation matrix analysis of resting $rCMR_{glc}$ data. One is technical. Recently, Ford (1986) argued that partial correlation coefficients may not always reflect the underlying intrasubject correlation pattern. He used a theoretical example in which the brain was arranged in two distinct clusters of strongly interacting regions, resulting in a bimodal distribution of correlation values. However, in all clinical groups that we have studied so far, as well as in awake rats, correlation coefficients have been found to have unimodal frequency distributions (Soncrant et al, 1986a; Horwitz and Rapoport, 1988), in contrast to Ford's (1988) theoretical model.

It also is important to realize that $rCMR_{glc}$ may not reflect underlying disease in a brain allowed to compensate to a chronic insult. For example, Taniguchi (1980) demonstrated that unilateral auditory deprivation in immature mice initially reduces $rCMR_{glc}$ by 60% in the contralateral inferior colliculus. However, by 40 days post-deprivation, $rCMR_{glc}$ has returned to baseline, even though the inferior colliculus is histologically abnormal at this time (Webster, 1983), has abnormal lipid metabolism (Tone et al, 1988), and does not process contralateral auditory input. $rCMR_{glc}$ in the human cerebral cortex also can recover with time after an ipsilateral thalamic stroke (Baron et al, 1986), despite maintained cortical deafferentation. In chronic pathological conditions, therefore, the correlation matrix may prove more informative than region-by-region analysis of $rCMR_{glc}$ data.

Finally, we derived correlation-matrices from PET data, obtained on resting subjects with reduced visual and auditory inputs, whereas we tried to relate the matrix-results to group differences in performance. During PET, for example, Down subjects were not asked to use language skills, elderly subjects were not asked to perform a visuospatial task, and autistics were not given tests requiring directed attention.

To see whether extrapolations from resting data to performance are valid, performance could be initiated during PET, and metabolism might be examined in lesioned brains. With regard to the former approach, our laboratory is examining rCBF responses in subjects during visuospatial as compared with facial-recognition tasks (Haxby et al., 1988), so as to identify visual association pathways which may be altered in the elderly and in Alzheimer's disease, as far as the latter approach is concerned. We also have applied correlation analyses to $rCMR_{glc}$ data from Alzheimer patients, and from awake rats which had undergone sectioning of the corpus callosum (see below).

Alzheimer's disease. As noted above, post-mortem neuropathology and cognitive and metabolic measurements during life show that Alzheimer's disease causes neurodegeneration of the association as compared with primary sensory and motor neocortices, and of subcortical regions with which they are connected -- -- posterior hippocampus, entorhinal cortex, lateral basal nucleus of amygdala, nucleus basalis of Meynert (Rapoport et al, 1986; Rapoport, 1988a, b). In the Alzheimer brain, neurofibrillary tangles with paired helical filaments are found in large pyramidal neurons within layers II-III and V of the association cortex, but rarely in primary sensory or motor regions (Pearson et al, 1985; Rogers and Morrison, 1985).

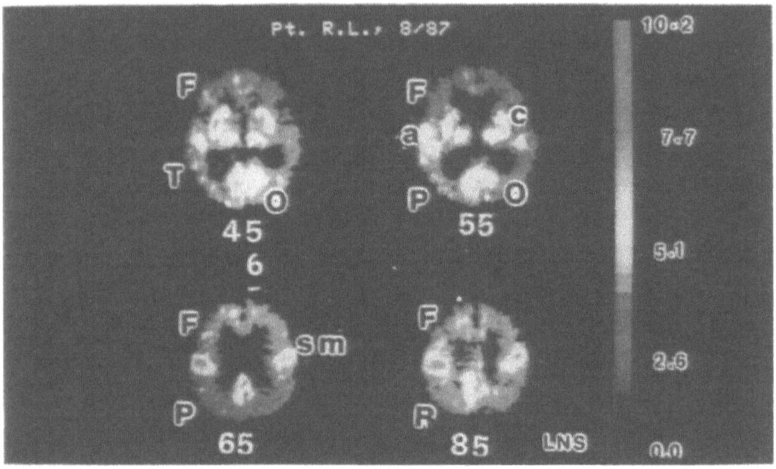

Figure 7. PET scans from severely demented patient with Alzheimer disease. $rCMR_{glc}$, derived with a Scanditronix tomograph, in units of $mg.100g^{-1}.min^{-1}$. High metabolic rates are retained in the primary sensorimotor cortex (s), occipital cortex (o), auditory cortex (a), caudate and lenticular nuclei (c), but are reduced elsewhere in the brain. P, parietal lobe; F, frontal cortex; T, temporal cortex, O, occipital cortex. Laboratory of Neurosciences, National Institute on Aging.

As shown in Table 1, mean $rCMR_{glc}$ is reduced in parietal and lateral temporal association areas, relative to primary sensory and motor areas, early and throughout the course of Alzheimer's disease (Haxby et al, 1985, 1986). Indeed, PET scans late in the disease course show that metabolic integrity is retained only in the primary sensory and motor cortices and in the caudate and lenticular nuclei (Figure 7). Furthermore, the regional metabolic decrements in Alzheimer's disease correspond to excessive right-left asymmetries (difference between right and left homotopic $rCMR_{glc}$ values, divided by mean) in the frontal, parietal and temporal association cortices, but not in the primary somatosensory or visual cortices (Figure 8) (Haxby et al, 1985). These abnormal asymmetries, in turn, are correlated with defects in cognitive functions that are mediated by the right or left neocortex, such that, in moderately-demented Alzheimer patients, a lower left- than right-sided $rCMR_{glc}$ corresponds to worse language than visuospatial ability in a given individual, whereas a lower right- than left-sided $rCMR_{glc}$ corresponds to relatively worse visuospatial ability (Haxby et al, 1985, 1986).

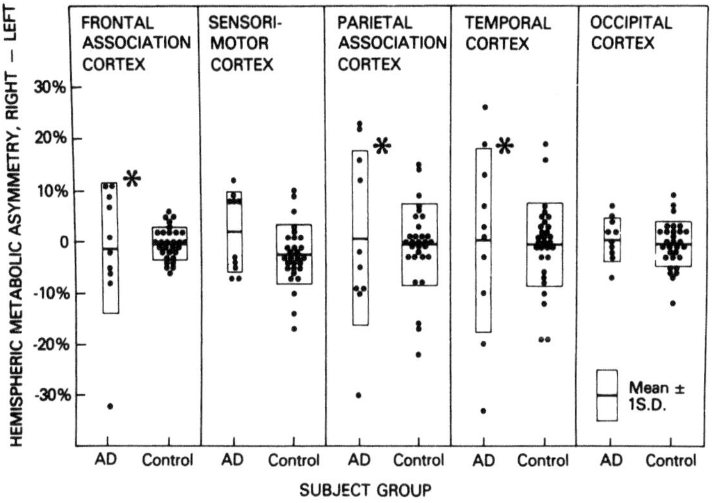

Figure 8. Metabolic asymmetry in each of 5 cortical regions in mildly-moderated demented DAT patients and controls. Values are right minus left $rCMR_{glc}$ values, divided by the mean, in percent. *, coefficient of variation differs from control asymmetry. Data from Haxby et al (1985).

Table 3. Differences Between Numbers of Significant Positive Partial Correlations between Mildly-Moderately Demented Alzheimer Patients and Controls

	TOTAL POSSIBLE	CONTROL	ALZHEIMER
Frontal-Parietal	220	32	12*
Homologous (left-right)	28	22	14*

*Significantly different from control by χ^2 ($p < 0.05$)

Data from Horwitz et al (1987)

The known lesions of large pyramidal, cortical association neurons in Alzheimer's disease explain the observed reduction of intrahemispheric and well as of interhemispheric significant correlations in the matrix calculated from resting $rCMR_{glc}$ data (partialing out CMR_{glc}) from mildly-to-moderately demented Alzheimer patients (Horwitz et al, 1987; Rapoport et al., 1986) (Figure 9). As summarized by Table 3, there were significantly fewer significant correlation coefficients between frontal-parietal regions and between left and right homolateral regions in the Alzheimer than in the control matrix. The latter reduction is evidenced as a depletion of significant correlations in the diagonals of left-right lobar boxes of the correlation matrix (Figure 9).

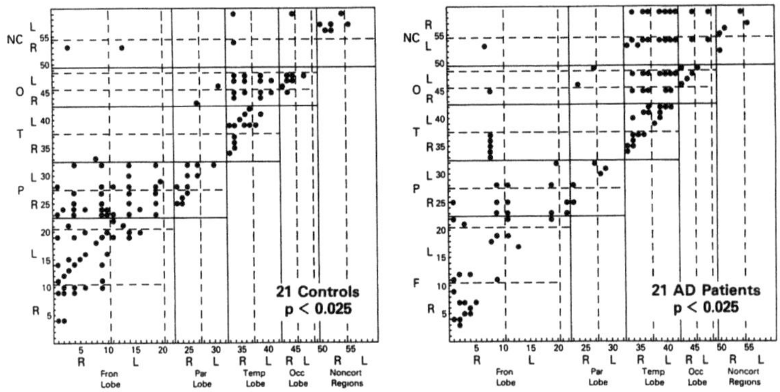

Figure 9. Matrices of positive partial correlation coefficients between pairs of $rCMR_{glc}$ values for control and Alzheimer subjects. Significant correlations at $p < 0.025$ are illustrated. See Figure 4 and Horwitz et al (1987) for explanations.

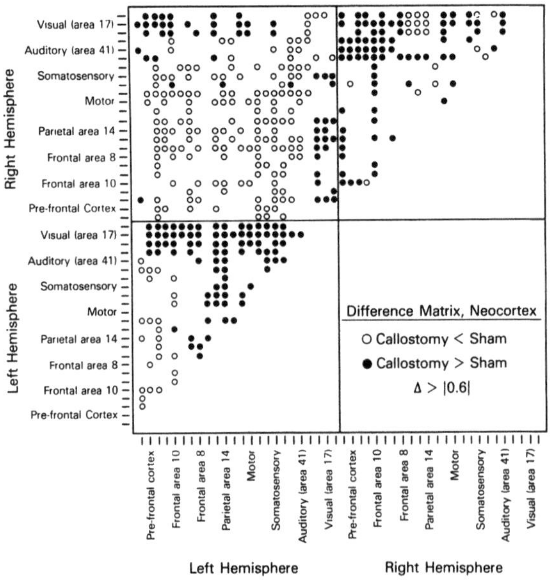

Figure 10. Difference matrix derived from partial correlation matrixes of $rCMR_{glc}$ data from callosotomized and control rats. Only neocortical regions are shown. Differences shown differed by $\Delta > 0.6$. From Soncrant et al (1986b and unpublished).

Corpus callosotomy. Unlike the Alzheimer matrix, the matrix of rats whose corpus callosum had been severed demonstrates reduced values of interhemispheric correlations, but higher than normal values of intrahemispheric correlations. Partial correlation coefficients were derived from $rCMR_{glc}$ values obtained from awake rats that had been callostomized 10 days earlier, and were compared via a difference matrix with parietal correlations obtained from measurements in sham-operated animals (Figure 10) (Soncrant et al, 1986b, in preparation). As summarized in Table 4, callostomy reduced the values of interhemispheric cortex-cortex correlations in two thirds of cases, but increased intrahemispheric cortex-cortex correlations in three-fourths of the cases. The results suggest that the corpus callosum enhances functional interactions between cortical regions in opposite hemispheres, but reduces functional interactions between cortical regions in the same hemisphere.

Table 4. Comparison of Partial Correlations Between Callosotomized and Sham-Operated Rats

	PARTIAL CORRELATION COEFFICIENTS AFTER CALLOSOTOMY		
Region Pairs	Number Increased	Number Decreased	χ^2
CORTEX-CORTEX			
Interhemispheric	195	379	58*
Intrahemispheric	425	126	161*

*$p < 0.001$. Data from Soncrant et al (in preparation).

CONCLUSIONS

The correlation-matrix approach to PET-derived data offers an insight into patterns of brain function which is not provided by region-to-region comparisons of mean values of $rCMR_{glc}$ or of $rCMR_{glc}/CMR_{glc}$. It is based on the assumption that the brain is composed of networks of neurons whose integrated activities subserve cognition and behavior, and which account for the correlation patterns that are identified by matrix analysis. When applied to different subject groups, matrix analysis suggests that functional interactions between ipsilateral parietal and frontal association areas are reduced in the elderly, consistent with reduced "fluid" intelligence and visuospatial ability; that cortical-cortical and noncortical-cortical interrelations are altered in autism, consistent with altered directed attention; and that Broca's area is functionally disconnected in Down syndrome, as are thalamic regions, consistent with language problems and increased distractibility.

In rats whose corpus callosum has been lesioned, the matrix approach indicates that many interhemispheric cortical correlations depend on callosal integrity, whereas intrahemispheric correlations are increased by corpus callosotomy. In light of these observations, reductions of both interhemispheric and intrahemispheric correlations within the Alzheimer brain are consistent with a primary defect of pyramidal association neurons which contribute to both sets of correlations.

Up to now, correlation matrices have been derived from resting metabolic rates, but have been interpreted in terms of defective performance that was not specifically evoked during PET. In the future, hypotheses concerning network abnormalities, derived from resting matrix data, will have to be tested in subjects performing tasks proposed to be mediated by those networks.

REFERENCES

Baron JC, D'Anotana R, Serdaru M, Pantano P, Bousser MG, Samson Y (1986) Hypometabolisme cortical après lésion thalamique chez l'homme; étude par la tomographie à positons. Rev Neurol (Paris) 142:465-474.

Benda CE (1969) Down syndrome. Grune and Stratton, New York, 139.

Benton A (1985) Visuoperceptual, visuospatial and visuocontructive disorders. In: Heilman KM, Valenstein E (eds) Clinical neuropsychology, 2nd edn. Oxford University Press, Oxford, p 151.

Brun A (1983) An overview of light and electron microscopic changes. In: Reisberg B (ed) Alzheimer's disease: the standard reference. Free Press, New York, p 37.

Creasey H, Rumsey JM, Schwartz M, Duara R, Rapoport JL, Rapoport SI (1986) Brain morphometry in autistic men as measured by volumetric computed tomography. Arch Neurol 43:669-672.

Damasio AR, Maurer RG (1978) A neurological model for childhood autism. Arch Neurol 35:777-786.

Duara R, Grady CL, Haxby JV, Ingvar D, Sokoloff L, Margolin RA, Manning RG, Cutler NR, Rapoport SI (1984) Human brain glucose utilization and cognitive function in relation to age. Ann Neurol 16:702-713.

Ford I (1986) Confounded correlations: statistical limitations in the analysis of interregional relationships of cerebral metabolic activity. J Cereb Blood Flow Metab 6:385-388.

Hartley XY (1982) Receptive language processing of Down's syndrome in children. J Ment Defic Res 26:263-269.

Haxby JV (in press) Neuropsychological evaluation of adults with Down syndrome: patterns of selective impairment in nondemented old adults. J Ment Defic Res

Haxby JV, Duara R, Grady CL, Cutler NR, Rapoport SI (1985) Relations between neuropsychological and cerebral metabolic asymmetries in early Alzheimer's disease. J Cerebr Blood Flow Metab 5:193-200.

Haxby JV, Grady CL, Duara R, Schlageter N, Berg G, Rapoport SI (1986) Neocortical metabolic abnormalities precede nonmemory cognitive defects in early Alzheimer's-type dementia. Arch Neurol 43:882-885.

Haxby JV, Grady CL, Horwitz B, Schapiro MB, Carson RE, Ungerleider LG, Mishkin M, Herscovitch P, Friedland RP, Rapoport SI (1988) Mapping two visual pathways in man with regional cerebral blood flow(RCBF) as measured by positron emission tomography (PET) and H215O. Abstr Soc Neurosci 14:750.

Horn JL, Cattell RB (1967) Age differences in fluid and crystallized intelligence. Acta Psychol (Amst) 26:107-129.

Horwitz G (1987) Brain metabolism and blood flow during aging. In: Elligson RJ, Murray NMF, Halliday AM (eds) The London Symposium. EEG (Suppl 39): 396-402.

Horwitz B, Duara R, Rapoport SI (1984) Intercorrelations of glucose metabolic rates between brain regions: application to healthy males in a state of reduced sensory input. J Cerebr Blood Flow Metab 4:484-499.

Horwitz B, Duara R, Rapoport SI (1986a) Age differences in intercorrelations between regional cerebral metabolic rates for glucose. Ann Neurol 19:60-67.

Horwitz B, Grady CL, Schlageter NL, Duara R, Rapoport SI (1987) Intercorrelations of regional cerebral glucose metabolic rates in Alzheimer's disease. Brain Res 407:294-306.

Horwitz B, Rapoport SI (1988) Partial correlation coefficients approximate the real intrasubject correlation pattern in the analysis of interregional relations of cerebral metabolic activity. J Nucl Med 29:392-399.

Horwitz B, Rumsey J, Grady C, Rapoport SI (1988) The cerebral metabolic landscape in autism: intercorrelations of regional glucose utilization. Arch Neurol 45:749-755.

Horwitz B, Schapiro MB, Grady CL, Rapoport SI (submitted) Cerebral metabolic pattern in young adult Down syndrome: altered intercorrelations between regional cerebral metabolic rates of glucose utilization.

Horwitz B, Soncrant TT, Rapoport SI (1986b) New approach to the examination of data derived with PET. In Battistin L, Gerstenbrand F (eds) PET and NMR: New perspectives in neuroimaging and clinical neurochemistry Alan R. Liss, New York, p 151.

Huang S-C, Phelps ME, Hoffman EJ, Sideris K, Selin CJ, Kuhl DE (1980) Noninvasive determination of local cerebral metabolic rate of glucose in man. Am J Physiol 238:E69-E82.

Luria AR (1973) The working brain. Basic Books, New York.

Mesulam MM (1981) A cortical network for directed attention and unilateral neglect. Ann Neurol 10:309-325.

Myers RE (1965) Phylogenetic studies of commissural connexions. In: Ettlinger EG (ed) Functions of the Corpus callosum. Ciba Foundation Study Group, No. 20. London, J & A Churchill, p 138.

Nieminen K, Suarez-Isla BA, Rapoport SI (in press) Electrical properties of cultured dorsal root ganglion neurons from normal and trisomy 21 human fetal tissue. Dev Brain Res

Orozco CB, Smith SA, Epstein CJ, Rapoport SI (1987) Electrophysiological properties of cultured dorsal root ganglion and spinal cord neurons of normal and trisomy 16 fetal mice. Dev Brain Res 32:111-122.

Pandya DN, Seltzer B (1982) Association areas of the cerebral cortex. Trends Neurosci 5:386-390.

Pearson RC, Esiri MM, Hiorns RW, Wilcock GK, Powell TPS (1985) Anatomical correlates of the distribution of the pathological changes in the neocortex in Alzheimer's disease. Proc Natl Acad Sci 82:4531-4534.

Rapoport SI (1988a) Brain evolution and Alzheimer's disease. Rev Neurol 144:79-90.

Rapoport SI (1988b) Alzheimer's disease: phylogenetic vulnerability of associative neocortex and its connections. In: Molecular neuropathology of Aging, Banbury Report No. 27. Cold Spring Harbor Laboratory, Cold Spring Harbor, NY, p 37.

Rapoport SI, Horwitz B, Haxby JV, Grady CL (1986) Alzheimer's disease: metabolic uncoupling of associative brain regions. Can J Neurol Sci 13:540-545.

Rogers J, Morrison JH (1985) Quantitative morphology and regional and laminar distributions of senile plaques in Alzheimer's disease. J Neurosci 5:2801-2808.

Roy CS, Sherrington CS (1890): On the regulation of the blood supply to the brain. J Physiol (Lond) 11:85-108.

Rumsey JM, Duara R, Grady CL, Rapoport JL, Margolin RA, Rapoport SI, Cutler NR (1985) Brain metabolism in autism: resting cerebral glucose utilization rates as measured with positron emission tomography. Arch Gen Psych 42:448-455.

Schapiro MB, Creasey H, Schwartz M, Haxby JV, White B, Moore A, Rapoport SI (1987) Quantitative CT analysis of brain morphometry in adult Down's syndrome at different ages. Neurol 37:1424-1427.

Schwartz M, Duara R, Haxby J, Grady C, White BJ, Kessler RM, Kay AD, Cutler NR, Rapoport SI (1983) Down's syndrome in adults: brain metabolism. Science (Wash) 221:781-783.

Sharp FR, Ryan AF, Goodwin P, Woolf NK (1981) Increasing intensities of wide band noise increase [14C]2-deoxyglucose uptake in gerbil central auditory structures. Brain Res 230:87-96.

Sokoloff L, Reivich M, Kennedy C, Des Rosiers MH, Patlak CS, Pettigrew KD, Sakurada O, Shinohara M (1977) The [14C]deoxyglucose method for the measurement of local cerebral glucose utilization: theory, procedure, and normal values in the conscious and anesthetized albino rat. J Neurochem 28:897-916.

Soncrant TT, Horwitz B, Holloway HW, Rapoport SI (1986a) The pattern of functional coupling of brain regions in the awake rat. Brain Res 369:1-11.

Soncrant TT, Horwitz B, Sato S, Holloway HW, Rapoport SI (1986b) Left-right regional functional interactions are disrupted by corpus callosotomy in the rat. Abstr Soc Neurosci 12:177.

Soncrant TT, Sato S, Holloway HW, Horwitz B, Rapoport SI (in preparation) Corpus callosotomy reduces interhemispheric and increases intrahemispheric functional interactions in the rat cerebral cortex.

Taniguchi I (1980): Changes in metabolic activity in the inferior colliculus following removal of auditory input. Biomed Res 1:510-516.

Tone O, Miller JC, Bell JM, Rapoport SI (1988) Regional cerebral palmitate incorporation after unilateral auditory deprivation in immature and adult Fischer-344 rats. Exp Neurol 100:491-505.

Webster DB (1983) A critical period during postnatal auditory development of mice. Int J Pediatr Otorhinolaryngol 6:107-118.

Woolf NK, Sharp FR, Davidson TM, Ryan AF (1983) Cochlear and middle ear effects on metabolism in the central auditory pathway during silence: A 2-deoxyglucose study. Brain Res 274:119-127.

SUBJECT INDEX

NATO ASI Series H

Vol. 1: **Biology and Molecular Biology of Plant-Pathogen Interactions.** Edited by J.A. Bailey. 415 pages. 1986.

Vol. 2: **Glial-Neuronal Communication in Development and Regeneration.** Edited by H.H. Althaus and W. Seifert. 865 pages. 1987.

Vol. 3: **Nicotinic Acetylcholine Receptor: Structure and Function.** Edited by A. Maelicke. 489 pages. 1986.

Vol. 4: **Recognition in Microbe-Plant Symbiotic and Pathogenic Interactions.** Edited by B. Lugtenberg. 449 pages. 1986.

Vol. 5: **Mesenchymal-Epithelial Interactions in Neural Development.** Edited by J.R. Wolff, J. Sievers, and M. Berry. 428 pages. 1987.

Vol. 6: **Molecular Mechanisms of Desensitization to Signal Molecules.** Edited by T.M. Konijn, P.J.M. Van Haastert, H. Van der Starre, H. Van der Wel, and M.D. Houslay. 336 pages. 1987.

Vol. 7: **Gangliosides and Modulation of Neuronal Functions.** Edited by H. Rahmann. 647 pages. 1987.

Vol. 8: **Molecular and Cellular Aspects of Erythropoietin and Erythropoiesis.** Edited by I.N. Rich. 460 pages. 1987.

Vol. 9: **Modification of Cell to Cell Signals During Normal and Pathological Aging.** Edited by S. Govoni and F. Battaini. 297 pages. 1987.

Vol. 10: **Plant Hormone Receptors.** Edited by D. Klämbt. 319 pages. 1987.

Vol. 11: **Host-Parasite Cellular and Molecular Interactions in Protozoal Infections.** Edited by K.-P. Chang and D. Snary. 425 pages. 1987.

Vol. 12: **The Cell Surface in Signal Transduction.** Edited by E. Wagner, H. Greppin, and B. Millet. 243 pages. 1987.

Vol. 13: **Toxicology of Pesticides: Experimental, Clinical and Regulatory Perspectives.** Edited by L.G. Costa, C.L. Galli, and S.D. Murphy. 320 pages. 1987.

Vol. 14: **Genetics of Translation. New Approaches.** Edited by M.F. Tuite, M. Picard, and M. Bolotin-Fukuhara. 524 pages. 1988.

Vol. 15: **Photosensitisation. Molecular, Cellular and Medical Aspects.** Edited by G. Moreno, R.H. Pottier, and T.G. Truscott. 521 pages. 1988.

Vol. 16: **Membrane Biogenesis.** Edited by J.A.F. Op den Kamp. 477 pages. 1988.

Vol. 17: **Cell to Cell Signals in Plant, Animal and Microbial Symbiosis.** Edited by S. Scannerini, D. Smith, P. Bonfante-Fasolo, and V. Gianinazzi-Pearson. 414 pages. 1988.

Vol. 18: **Plant Cell Biotechnology.** Edited by M.S.S. Pais, F. Mavituna, and J.M. Novais. 500 pages. 1988.

Vol. 19: **Modulation of Synaptic Transmission and Plasticity in Nervous Systems.** Edited by G. Hertting and H.-C. Spatz. 457 pages. 1988.

Vol. 20: **Amino Acid Availability and Brain Function in Health and Disease.** Edited by G. Huether. 487 pages. 1988.

NATO ASI Series H

Vol. 21: **Cellular and Molecular Basis of Synaptic Transmission.**
Edited by H. Zimmermann. 547 pages. 1988.

Vol. 22: **Neural Development and Regeneration. Cellular and Molecular Aspects.**
Edited by A. Gorio, J.R. Perez-Polo, J. de Vellis, and B. Haber. 711 pages. 1988.

Vol. 23: **The Semiotics of Cellular Communication in the Immune System.**
Edited by E.E. Sercarz, F. Celada, N.A. Mitchison, and T. Tada. 326 pages. 1988.

Vol. 24: **Bacteria, Complement and the Phagocytic Cell.**
Edited by F.C. Cabello und C. Pruzzo. 372 pages. 1988.

Vol. 25: **Nicotinic Acetylcholine Receptors in the Nervous System.**
Edited by F. Clementi, C. Gotti, and E. Sher. 424 pages. 1988.

Vol. 26: **Cell to Cell Signals in Mammalian Development.**
Edited by S.W. de Laat, J.G. Bluemink, and C.L. Mummery. 322 pages. 1989.

Vol. 27: **Phytotoxins and Plant Pathogenesis.**
Edited by A. Graniti, R.D. Durbin, and A. Ballio. 508 pages. 1989.

Vol. 28: **Vascular Wilt Diseases of Plants. Basic Studies and Control.**
Edited by E.C. Tjamos and C.H. Beckman. 590 pages. 1989.

Vol. 29: **Receptors, Membrane Transport and Signal Transduction.**
Edited by A.E. Evangelopoulos, J.P. Changeux, L. Packer, T.G. Sotiroudis, and K.W.A. Wirtz. 387 pages. 1989.

Vol. 30: **Effects of Mineral Dusts on Cells.**
Edited by B.T. Mossman and R.O. Bégin. 470 pages. 1989.

Vol. 31: **Neurobiology of the Inner Retina.**
Edited by R. Weiler and N.N. Osborne. 529 pages. 1989.

Vol. 32: **Molecular Biology of Neuroreceptors and Ion Channels.**
Edited by A. Maelicke. 675 pages. 1989.

Vol. 33: **Regulatory Mechanisms of Neuron to Vessel Communication in Brain.**
Edited by F. Battaini, S. Govoni, M.S. Magnoni, and M. Trabucchi. 416 pages. 1989.

MIX
Papier aus verantwortungsvollen Quellen
Paper from responsible sources
FSC® C105338

If you have any concerns about our products,
you can contact us on
ProductSafety@springernature.com

In case Publisher is established outside the EU,
the EU authorized representative is:
Springer Nature Customer Service Center GmbH
Europaplatz 3, 69115 Heidelberg, Germany

Printed by Libri Plureos GmbH
in Hamburg, Germany